# AN OUTLINE FOR THE STUDY OF CALCULUS

# Volume III

# AN OUTLINE FOR THE STUDY OF CALCULUS Volume III

by

John H. Minnick
DE ANZA COLLEGE

Edited by

Louis Leithold

HARPER & ROW, PUBLISHERS
New York Hagerstown San Francisco London

Sponsoring Editor: George J. Telecki
Designer: Rita Naughton
Production Supervisor: Will C. Jomarrón
Compositor: T. McNabney Composition Service
Printer: The Murray Printing Company
Binder: The Murray Printing Company
Art Studio: J & R Technical Services Inc.

AN OUTLINE FOR THE STUDY OF CALCULUS Volume III



**Library of Congress Cataloging in Publication Data**

Minnick, John Harper, Date –
An outline for the study of calculus.

"The definitions, theorems, and exercises are taken from The calculus with analytic geometry, third edition, by Louis Leithold."

1. Calculus--Outlines, syllabi, etc. I. Title.
QA303.M685 515'.02'02 76-8190
ISBN 0-06-044546-7 (v. III)

# Contents

# Preface

Each section of the outline includes all of the most important definitions and theorems that are usually found in a course in calculus and analytic geometry. Often these are followed by a discussion that elaborates the concepts and presents a summary of problem solving techniques. A selection of exercises with complete and detailed solutions, including all graphs, is given for each section. At the end of each chapter there is a set of review exercises, also with complete solutions. In the Appendix there is a test for each chapter with a time limit indicated, followed by solutions for the test.

For those exercises that are more easily solved by using a computer, general flow charts that show how to apply the computer are given. Each flow chart is followed by a sample program, written in BASIC, that illustrates the solution of a particular exercise. The computer solutions are found in Chapters 7, 16, and 21.

The outline may be used for self study or to supplement any standard three semester course in calculus. Volume I contains Chapters 1-8, Volume II contains Chapters 9-16, and Volume III contains Chapters 17-21. The definitions, theorems, and exercises are taken from *The Calculus with Analytic Geometry, third edition,* by Louis Leithold. The chapter and section numbers and the exercise numbers agree with those used in Leithold. However, the chapter tests found in the Appendix are compiled from test questions that I have used with my own students at De Anza College.

J.H.M

# 17

# Vectors in the plane and parametric equations

**17.1 VECTORS IN THE PLANE** A *vector in the plane* is an ordered pair of real numbers $\langle x, y\rangle$. The numbers $x$ and $y$ are called the *components* of the vector $\langle x, y\rangle$.

**17.1.4 Definition** The *sum* of two vectors $\mathbf{A} = \langle a_1, a_2\rangle$ and $\mathbf{B} = \langle b_1, b_2\rangle$ is the vector $\mathbf{A} + \mathbf{B}$, defined by

$$\mathbf{A} + \mathbf{B} = \langle a_1 + b_1, a_2 + b_2\rangle$$

**17.1.5 Definition** If $\mathbf{A} = \langle a_1, a_2\rangle$, then the vector $\langle -a_1, -a_2\rangle$ is defined to be the *negative* of $\mathbf{A}$, denoted by $-\mathbf{A}$.

**17.1.6 Definition** The *difference* of the two vectors $\mathbf{A}$ and $\mathbf{B}$, denoted by $\mathbf{A} - \mathbf{B}$, is the vector obtained by adding $\mathbf{A}$ to the negative of $\mathbf{B}$; that is,

$$\mathbf{A} - \mathbf{B} = \mathbf{A} + (-\mathbf{B})$$

**17.1.7 Definition** If $c$ is a scalar and $\mathbf{A}$ is the vector $\langle a_1, a_2\rangle$, then the *product* of $c$ and $\mathbf{A}$, denoted by $c\mathbf{A}$, is a vector and is given by

$$c\mathbf{A} = c\langle a_1, a_2\rangle = \langle ca_1, ca_2\rangle$$

If $\mathbf{A} = \langle a_1, a_2\rangle$, then the directed line segment $\overrightarrow{OA}$, where $O$ is the origin and $A$ is the point $(a_1, a_2)$, is called the *position representation* of the vector $\mathbf{A}$. If $P = (p_1, p_2)$ and $Q = (q_1, q_2)$, then the directed line segment $\overrightarrow{PQ}$ is a *representation* of the vector $\langle q_1 - p_1, q_2 - p_2\rangle$. Thus, by Definition 17.1.6, it follows that the vector represented by the directed line segment $\overrightarrow{PQ}$ is the difference of the vectors whose position representations are $\overrightarrow{OQ}$ and $\overrightarrow{OP}$. In symbols,

$$\begin{aligned} \mathbf{V}(\overrightarrow{PQ}) &= \mathbf{V}(\overrightarrow{OQ}) - \mathbf{V}(\overrightarrow{OP}) \\ &= \langle q_1 - p_1, q_2 - p_2\rangle \end{aligned}$$

**17.1.2 Definition** The *magnitude* of a vector is the length of any of its representations, and the *direction* of a nonzero vector is the direction of any of its representations.

The magnitude of the vector **A** is denoted by $|\mathbf{A}|$.

**17.1.3 Theorem** If **A** is the vector $\langle a_1, a_2\rangle$, then $|\mathbf{A}| = \sqrt{a_1^2 + a_2^2}$

If $\mathbf{A} = \langle a_1, a_2\rangle$, and $|\mathbf{A}| \neq 0$, the *direction* of **A** is the radian measure of the angle $\theta$ with $0 \leqslant \theta < 2\pi$, such that

$$\cos\theta = \frac{a_1}{|\mathbf{A}|} \quad \text{and} \quad \sin\theta = \frac{a_2}{|\mathbf{A}|}$$

Thus, if $a_1 = 0$, we have

$$\tan\theta = \frac{a_2}{a_1}$$

## *Exercises 17.1*

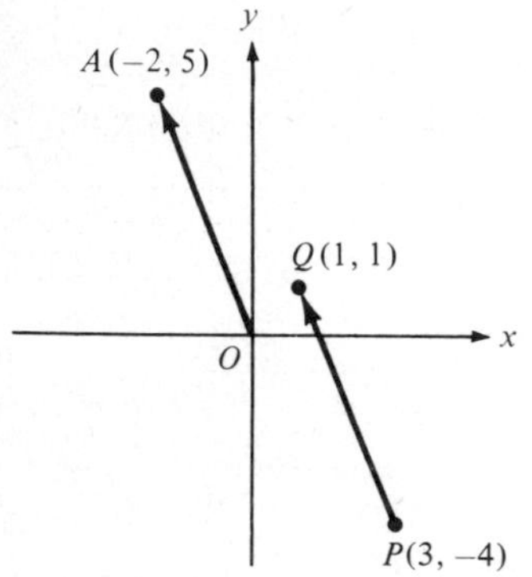

Figure 17.1.2

**2.** Draw the position representation of the given vector **A** and also the particular representation through the given point $P$; find the magnitude of **A**.

$$\mathbf{A} = \langle -2, 5\rangle \qquad P = (3, -4)$$

SOLUTION: Let $A$ be the point $(-2, 5)$. Then the directed line segment $\overrightarrow{OA}$ is the position representation of the vector **A**, as shown in Fig. 17.1.2. Let $\overrightarrow{PQ}$ be a representation of the vector **A**. If $Q = (x, y)$, then

$$x - 3 = -2 \quad \text{and} \quad y + 4 = 5$$
$$x = 1 \qquad\qquad y = 1$$

Thus, $Q = (1, 1)$. The representation $\overrightarrow{PQ}$ is also shown in Fig. 17.1.2. By Theorem 17.1.3, the magnitude of the vector **A** is given by

$$|\mathbf{A}| = \sqrt{(-2)^2 + 5^2}$$
$$= \sqrt{29}$$

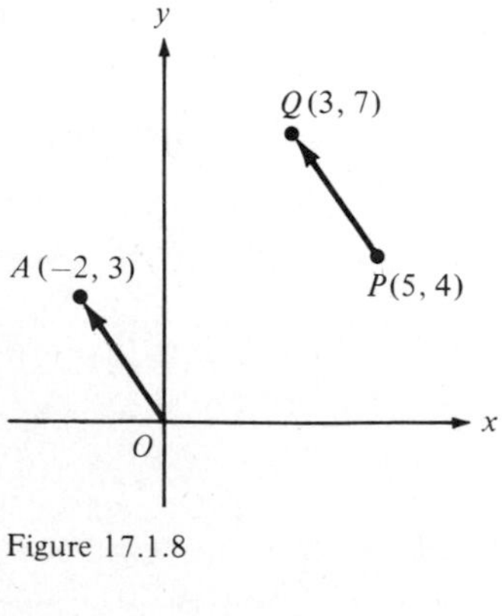

Figure 17.1.8

**8.** Find the vector **A** having $\overrightarrow{PQ}$ as a representation. Draw $\overrightarrow{PQ}$ and the position representation of **A**.

$$P = (5, 4) \qquad Q = (3, 7)$$

SOLUTION: Let $\mathbf{A} = \langle a_1, a_2\rangle$. Then since $\overrightarrow{PQ}$ is a representation of **A**, we have

$$a_1 = 3 - 5 \quad \text{and} \quad a_2 = 7 - 4$$
$$= -2 \qquad\qquad = 3$$

Thus, $\mathbf{A} = \langle -2, 3\rangle$. The representation $\overrightarrow{PQ}$ and the position representation $\overrightarrow{OA}$ of the vector **A** are shown in Fig. 17.1.8.

**14.** Find the point $S$ so that $\overrightarrow{PQ}$ and $\overrightarrow{RS}$ are each representations of the same vector.

$$P = (-1, 4) \qquad Q = (2, -3) \qquad R = (-5, -2)$$

SOLUTION: Let $S = (x, y)$. Then

$$x - (-5) = 2 - (-1) \quad \text{and} \quad y - (-2) = -3 - 4$$
$$x = -2 \qquad\qquad y = -9$$

Thus, $S = (-2, -9)$.

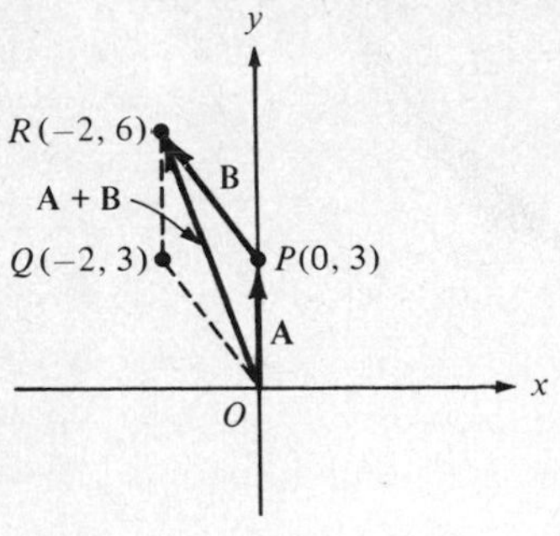

Figure 17.1.18

**18.** Find the sum of the given pair of vectors and illustrate geometrically

$\langle 0, 3\rangle \quad \langle -2, 3\rangle$

SOLUTION: Let $\mathbf{A} = \langle 0, 3\rangle$ and $\mathbf{B} = \langle -2, 3\rangle$. By Definition 17.1.4

$$\begin{aligned}\mathbf{A}+\mathbf{B} &= \langle 0-2, 3+3\rangle \\ &= \langle -2, 6\rangle\end{aligned}$$

In Fig. 17.1.18 quadrilateral $OPRQ$ is a parallelogram. $\overrightarrow{OA}$ is the position representation of the vector **A**. The directed line segment $\overrightarrow{OQ}$ is the position representation of the vector **B**; thus, $\overrightarrow{PR}$ is also a representation of the vector **B**. $\overrightarrow{OR}$, the diagonal of the parallelogram, is the position representation of the vector **A + B**.

**24.** Subtract the second vector from the first and illustrate geometrically.

$\langle 0, 5\rangle \quad \langle 2, 8\rangle$

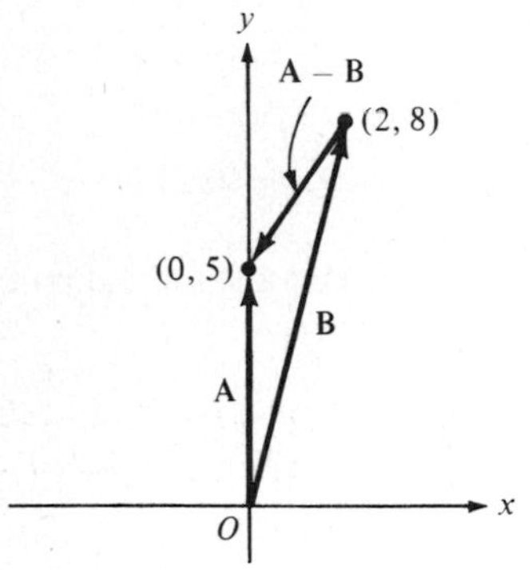

Figure 17.1.24

SOLUTION: Let $\mathbf{A} = \langle 0, 5\rangle$ and $\mathbf{B} = \langle 2, 8\rangle$. By Definition 17.1.6,

$$\begin{aligned}\mathbf{A}-\mathbf{B} &= \langle 0-2, 5-8\rangle \\ &= \langle -2, -3\rangle\end{aligned}$$

In Fig. 17.1.24 we show representations of the vectors **A**, **B**, and their difference **A − B**.

**34.** Let $\mathbf{A} = \langle 2, 4\rangle$ and $\mathbf{B} = \langle 4, -3\rangle$. Find $2\mathbf{A} + 3\mathbf{B}$.

SOLUTION: By Definition 17.1.7

$$\begin{aligned}2\mathbf{A} &= 2\langle 2, 4\rangle \quad \text{and} \quad & 3\mathbf{B} &= 3\langle 4, -3\rangle \\ &= \langle 4, 8\rangle & &= \langle 12, -9\rangle\end{aligned}$$

Thus,

$$\begin{aligned}2\mathbf{A}+3\mathbf{B} &= \langle 4, 8\rangle + \langle 12, -9\rangle \\ &= \langle 4+12, 8-9\rangle \\ &= \langle 16, -1\rangle\end{aligned}$$

**36.** Given $\mathbf{A} = \langle 3, 2\rangle$, $\mathbf{C} = \langle 8, 8\rangle$, and $\mathbf{A} + \mathbf{B} = \mathbf{C}$, find $|\mathbf{B}|$.

SOLUTION: Because $\mathbf{A} + \mathbf{B} = \mathbf{C}$, then

$$\begin{aligned}\mathbf{B} &= \mathbf{C} - \mathbf{A} \\ &= \langle 8, 8\rangle - \langle 3, 2\rangle \\ &= \langle 5, 6\rangle\end{aligned}$$

Thus,

$$\begin{aligned}|\mathbf{B}| &= \sqrt{5^2+6^2} \\ &= \sqrt{61}\end{aligned}$$

## 17.2 PROPERTIES OF VECTOR ADDITION AND SCALAR MULTIPLICATION

**17.2.1 Theorem** If **A**, **B**, and **C** are any vectors in $V_2$, and $c$ and $d$ are any scalars, then vector addition and scalar multiplication satisfy the following properties:

(i) $\mathbf{A}+\mathbf{B} = \mathbf{B}+\mathbf{A}$ (commutative law)
(ii) $\mathbf{A}+(\mathbf{B}+\mathbf{C}) = (\mathbf{A}+\mathbf{B})+\mathbf{C}$ (associative law)
(iii) There is a vector $\mathbf{0}$ in $V_2$ for which
$\mathbf{A}+\mathbf{0} = \mathbf{A}$ (existence of additive identity)

(iv) There is a vector $-\mathbf{A}$ in $V_2$ such that
$\mathbf{A} + (-\mathbf{A}) = \mathbf{0}$ (existence of negative)
(v) $(cd)\mathbf{A} = c(d\mathbf{A})$ (associative law)
(vi) $c(\mathbf{A} + \mathbf{B}) = c\mathbf{A} + c\mathbf{B}$ (distributive law)
(vii) $(c + d)\mathbf{A} = c\mathbf{A} + d\mathbf{A}$ (distributive law)
(viii) $1(\mathbf{A}) = \mathbf{A}$ (existence of scalar multiplicative identity)

**17.2.2 Definition** A *real vector space* $V$ is a set of elements, called *vectors,* together with a set of real numbers, called *scalars,* with two operations called *vector addition* and *scalar multiplication* such that for every pair of vectors $\mathbf{A}$ and $\mathbf{B}$ in $V$ and for every scalar $c$, a vector $\mathbf{A} + \mathbf{B}$ and a vector $c\mathbf{A}$ are defined so that properties (i)-(viii) of Theorem 17.2.1 are satisfied.

Any vector that has magnitude one is called a *unit* vector. We define the unit vectors $\mathbf{i}$ and $\mathbf{j}$ which form a basis for the vector space $V_2$ as follows.

$$\mathbf{i} = \langle 1, 0 \rangle \quad \text{and} \quad \mathbf{j} = \langle 0, 1 \rangle$$

Thus, for any vector $\langle a_1, a_2 \rangle$ in $V_2$, we have

$$\langle a_1, a_2 \rangle = a_1\mathbf{i} + a_2\mathbf{j}$$

**17.2.3 Theorem** If the nonzero vector $\mathbf{A} = a_1\mathbf{i} + a_2\mathbf{j}$, then the unit vector $\mathbf{U}$ having the same direction as $\mathbf{A}$ is given by

$$\mathbf{U} = \frac{a_1}{|\mathbf{A}|}\mathbf{i} + \frac{a_2}{|\mathbf{A}|}\mathbf{j}$$

## *Exercises 17.2*

In Exercises 1-8, let $\mathbf{A} = 2\mathbf{i} + 3\mathbf{j}$ and $\mathbf{B} = 4\mathbf{i} - \mathbf{j}$.

**4.** Find $|\mathbf{A}||\mathbf{B}|$.

SOLUTION: Because $\mathbf{A} = 2\mathbf{i} + 3\mathbf{j} = \langle 2, 3 \rangle$ then

$$|\mathbf{A}| = \sqrt{2^2 + 3^2} = \sqrt{13}$$

Similarly, $\mathbf{B} = 4\mathbf{i} - \mathbf{j} = \langle 4, -1 \rangle$. Thus

$$|\mathbf{B}| = \sqrt{4^2 + (-1)^2} = \sqrt{17}$$

Therefore,

$$|\mathbf{A}||\mathbf{B}| = \sqrt{13}\sqrt{17} = \sqrt{221}$$

**8.** Find $|3\mathbf{A}| - |2\mathbf{B}|$.

SOLUTION: Because

$$3\mathbf{A} = 3(2\mathbf{i} + 3\mathbf{j}) = 6\mathbf{i} + 9\mathbf{j}$$

and

$$2\mathbf{B} = 2(4\mathbf{i} - \mathbf{j}) = 8\mathbf{i} - 2\mathbf{j}$$

we have

$$\begin{aligned} |3\mathbf{A}| - |2\mathbf{B}| &= \sqrt{6^2 + 9^2} - \sqrt{8^2 + (-2)^2} \\ &= 3\sqrt{13} - 2\sqrt{17} \end{aligned}$$

**10.** Given $\mathbf{A} = -8\mathbf{i} + 7\mathbf{j}$, $\mathbf{B} = 6\mathbf{i} - 9\mathbf{j}$, and $\mathbf{C} = -\mathbf{i} - \mathbf{j}$, find $|2\mathbf{A} - 3\mathbf{B} - \mathbf{C}|$.

SOLUTION:

$$\begin{aligned} 2\mathbf{A} - 3\mathbf{B} - \mathbf{C} &= 2(-8\mathbf{i} + 7\mathbf{j}) - 3(6\mathbf{i} - 9\mathbf{j}) - (-\mathbf{i} - \mathbf{j}) \\ &= (-16\mathbf{i} + 14\mathbf{j}) + (-18\mathbf{i} + 27\mathbf{j}) + (\mathbf{i} + \mathbf{j}) \\ &= (-16 - 18 + 1)\mathbf{i} + (14 + 27 + 1)\mathbf{j} \\ &= -33\mathbf{i} + 42\mathbf{j} \end{aligned}$$

Thus,

$$\begin{aligned} |2\mathbf{A} - 3\mathbf{B} - \mathbf{C}| &= \sqrt{(-33)^2 + 42^2} \\ &= 3\sqrt{317} \end{aligned}$$

**12.** **(a)** Write the vector $3\mathbf{i} - 3\mathbf{j}$ in the form $r(\cos\theta\ \mathbf{i} + \sin\theta\ \mathbf{j})$, where $r$ is the magnitude of the vector and $\theta$ is the radian measure of the angle giving the direction of the vector.
**(b)** Find a unit vector having the same direction.

SOLUTION:

**(a)** Let $\mathbf{A} = 3\mathbf{i} - 3\mathbf{j}$. Then

$$\begin{aligned} r &= |\mathbf{A}| \\ &= \sqrt{3^2 + (-3)^2} \\ &= 3\sqrt{2} \end{aligned}$$

If

$$\mathbf{A} = r(\cos\theta\ \mathbf{i} + \sin\theta\ \mathbf{j}) \tag{1}$$

then

$$\cos\theta = \frac{a_1}{|\mathbf{A}|} = \frac{3}{3\sqrt{2}} = \frac{1}{2}\sqrt{2} \quad \text{and} \quad \sin\theta = \frac{a_2}{|\mathbf{A}|} = \frac{-3}{3\sqrt{2}} = -\frac{1}{2}\sqrt{2}$$

Substituting the values of $r$, $\cos\theta$, and $\sin\theta$ in (1), we obtain

$$3\mathbf{i} - 3\mathbf{j} = 3\sqrt{2}\left(\frac{1}{2}\sqrt{2}\,\mathbf{i} - \frac{1}{2}\sqrt{2}\,\mathbf{j}\right) \tag{2}$$

**(b)** Using Theorem 17.2.3, the required unit vector is

$$\begin{aligned} \mathbf{U} &= \frac{3}{3\sqrt{2}}\mathbf{i} - \frac{3}{3\sqrt{2}}\mathbf{j} \\ &= \frac{1}{2}\sqrt{2}\,\mathbf{i} - \frac{1}{2}\sqrt{2}\,\mathbf{j} \end{aligned}$$

We note that the unit vector $\mathbf{U}$ is the vector $\cos\theta\ \mathbf{i} + \sin\theta\ \mathbf{j}$ that is in the parentheses in Eq. (2)

**18.** Prove Theorem 17.2.1(vii).

SOLUTION: Let $\mathbf{A} = \langle a_1, a_2 \rangle$. By the distributive law for real numbers,

$$(c + d)a_1 = ca_1 + da_1 \quad \text{and} \quad (c + d)a_2 = ca_2 + da_2$$

Thus,

$$(c + d)\mathbf{A} = (c + d)\langle a_1, a_2 \rangle$$

$$
\begin{aligned}
&= \langle (c+d)a_1, (c+d)a_2 \rangle \\
&= \langle ca_1 + da_1, ca_2 + da_2 \rangle \\
&= \langle ca_1, ca_2 \rangle + \langle da_1, da_2 \rangle \\
&= c\langle a_1, a_2 \rangle + d\langle a_1, a_2 \rangle \\
&= c\mathbf{A} + d\mathbf{A}
\end{aligned}
$$

## 17.3 DOT PRODUCT

**17.3.1 Definition** If $\mathbf{A} = \langle a_1, a_2 \rangle$ and $\mathbf{B} = \langle b_1, b_2 \rangle$ are two vectors in $V_2$, then the *dot product* of $\mathbf{A}$ and $\mathbf{B}$, denoted by $\mathbf{A} \cdot \mathbf{B}$, is given by

$$\mathbf{A} \cdot \mathbf{B} = \langle a_1, a_2 \rangle \cdot \langle b_1, b_2 \rangle = a_1 b_1 + a_2 b_2$$

Expressed in terms of the unit vectors $\mathbf{i}$ and $\mathbf{j}$, the definition of the dot product of two vectors is as follows,

$$(a_1\mathbf{i} + a_2\mathbf{j}) \cdot (b_1\mathbf{i} + b_2\mathbf{j}) = a_1 b_1 + a_2 b_2$$

We note that the dot product of two vectors is a scalar. We use the dot product to find the work done by a force $\mathbf{F}$ that causes a displacement $\mathbf{D}$ of an object. We have

$$W = \mathbf{F} \cdot \mathbf{D}$$

**17.3.4 Definition** If $\overrightarrow{OP}$ is the position representation of $\mathbf{A}$ and $\overrightarrow{OQ}$ is the position representation of $\mathbf{B}$, then the *angle between the vectors* $\mathbf{A}$ and $\mathbf{B}$ is defined to be the angle of positive measure between $\overrightarrow{OP}$ and $\overrightarrow{OQ}$ interior to the triangle $POQ$. If $\mathbf{A} = c\mathbf{B}$, where $c$ is a scalar, then if $c > 0$, the angle between the vectors has radian measure 0; if $c < 0$, the angle between the vectors has radian measure $\pi$.

**17.3.5 Theorem** If $\alpha$ is the radian measure of the angle between the two nonzero vectors $\mathbf{A}$ and $\mathbf{B}$, then

$$\mathbf{A} \cdot \mathbf{B} = |\mathbf{A}||\mathbf{B}| \cos \alpha$$

Thus,

$$\mathbf{A} \cdot \mathbf{B} > 0 \text{ if } 0 \leqslant \alpha < \frac{1}{2}\pi$$

$$\mathbf{A} \cdot \mathbf{B} = 0 \text{ if } \alpha = \frac{1}{2}\pi$$

$$\mathbf{A} \cdot \mathbf{B} < 0 \text{ if } \frac{1}{2}\pi < \alpha \leqslant \pi$$

**17.3.6 Definition** Two vectors are said to be *parallel* if and only if one of the vectors is a scalar multiple of the other.

**17.3.7 Definition** Two vectors $\mathbf{A}$ and $\mathbf{B}$ are said to be *orthogonal* (*perpendicular*) if and only if $\mathbf{A} \cdot \mathbf{B} = 0$.

The *scalar projection* of the vector $\mathbf{A}$ onto the nonzero vector $\mathbf{B}$, represented by $\mathrm{A}_{\mathrm{B}}$, is given by

$$\mathrm{A}_{\mathrm{B}} = \frac{\mathbf{A} \cdot \mathbf{B}}{|\mathbf{B}|}$$

The *vector projection* of the vector $\mathbf{A}$ onto the nonzero vector $\mathbf{B}$, represented by $\mathbf{A}_{\mathbf{B}}$, is given by

$$\mathbf{A}_{\mathbf{B}} = \frac{\mathbf{A} \cdot \mathbf{B}}{|\mathbf{B}|^2}\mathbf{B}$$

If $\mathbf{U}$ is a unit vector, then $A_{\mathbf{U}} = \mathbf{A} \cdot \mathbf{U}$, and the scalar projection of $\mathbf{A}$ onto $\mathbf{U}$ is called the *component* of the vector $\mathbf{A}$ in the direction of $\mathbf{U}$. As a special case, for the vector $a_1\mathbf{i} + a_2\mathbf{j}$ the numbers $a_1$ and $a_2$ are the components of the vector $\mathbf{A}$ in the directions of $\mathbf{i}$ and $\mathbf{j}$, respectively.

**17.3.2 Theorem** If $\mathbf{A}, \mathbf{B}$, and $\mathbf{C}$ are any vectors in $V_2$, then

(i) $\mathbf{A} \cdot \mathbf{B} = \mathbf{B} \cdot \mathbf{A}$ (commutative law)
(ii) $\mathbf{A} \cdot (\mathbf{B} + \mathbf{C}) = \mathbf{A} \cdot \mathbf{B} + \mathbf{A} \cdot \mathbf{C}$ (distributive law)

**17.3.3 Theorem** If $\mathbf{A}$ and $\mathbf{B}$ are any vectors in $V_2$ and $c$ is any scalar, then

(i) $c(\mathbf{A} \cdot \mathbf{B}) = (c\mathbf{A}) \cdot \mathbf{B}$
(ii) $\mathbf{0} \cdot \mathbf{A} = 0$
(iii) $\mathbf{A} \cdot \mathbf{A} = |\mathbf{A}|^2$

## *Exercises 17.3*

4. Find $\mathbf{A} \cdot \mathbf{B}$ with $\mathbf{A} = -2\mathbf{i}$ and $\mathbf{B} = -\mathbf{i} + \mathbf{j}$.

SOLUTION:

$$\begin{aligned} \mathbf{A} \cdot \mathbf{B} &= (-2\mathbf{i}) \cdot (-\mathbf{i} + \mathbf{j}) \\ &= \langle -2, 0\rangle \cdot \langle -1, 1\rangle \\ &= (-2)(-1) + 0 \cdot 1 \\ &= 2 \end{aligned}$$

8. Prove Theorem 17.3.3(i).

SOLUTION: Let $\mathbf{A} = \langle a_1, a_2\rangle$ and $\mathbf{B} = \langle b_1, b_2\rangle$. By the associative property of real numbers,

$$c(a_1b_1) = (ca_1)b_1 \quad \text{and} \quad c(a_2b_2) = (ca_2)b_2$$

Thus,

$$\begin{aligned} c(\mathbf{A} \cdot \mathbf{B}) &= c(\langle a_1, a_2\rangle \cdot \langle b_1, b_2\rangle) \\ &= c(a_1b_1 + a_2b_2) \\ &= c(a_1b_1) + c(a_2b_2) \\ &= (ca_1)b_1 + (ca_2)b_2 \\ &= \langle ca_1, ca_2\rangle \cdot \langle b_1, b_2\rangle \\ &= c\langle a_1, a_2\rangle \cdot \langle b_1, b_2\rangle \\ &= (c\mathbf{A}) \cdot \mathbf{B} \end{aligned}$$

12. If $\alpha$ is the radian measure of the angle between $\mathbf{A}$ and $\mathbf{B}$, find $\cos \alpha$.

$\mathbf{A} = \langle -2, -3\rangle \qquad \mathbf{B} = \langle 3, 2\rangle$

SOLUTION: By Theorem 17.3.5

$$\mathbf{A} \cdot \mathbf{B} = |\mathbf{A}||\mathbf{B}| \cos \alpha$$

Thus,

$$\cos \alpha = \frac{\mathbf{A} \cdot \mathbf{B}}{|\mathbf{A}||\mathbf{B}|} \tag{1}$$

We have

$$\begin{aligned} \mathbf{A} \cdot \mathbf{B} &= \langle -2, -3\rangle \cdot \langle 3, 2\rangle \\ &= (-2)(3) + (-3)(2) \\ &= -12 \end{aligned}$$

Furthermore,

$$|\mathbf{A}| = \sqrt{(-2)^2 + (-3)^2} \quad \text{and} \quad |\mathbf{B}| = \sqrt{3^2 + 2^2}$$
$$= \sqrt{13} \qquad\qquad = \sqrt{13}$$

Thus,

$$\cos \alpha = \frac{-12}{\sqrt{13}\sqrt{13}}$$
$$= -\frac{12}{13}$$

**16.** Given $\mathbf{A} = k\mathbf{i} - 2\mathbf{j}$ and $\mathbf{B} = k\mathbf{i} + 6\mathbf{j}$, where $k$ is a scalar, find $k$ so that $\mathbf{A}$ and $\mathbf{B}$ are orthogonal.

SOLUTION: We use Definition 17.3.7. We have

$$\begin{aligned} \mathbf{A} \cdot \mathbf{B} &= (k\mathbf{i} - 2\mathbf{j}) \cdot (k\mathbf{i} + 6\mathbf{j}) \\ &= k \cdot k + (-2)(6) \\ &= k^2 - 12 \end{aligned}$$

Because $\mathbf{A}$ and $\mathbf{B}$ are orthogonal if and only if $\mathbf{A} \cdot \mathbf{B} = 0$, then $\mathbf{A}$ and $\mathbf{B}$ are orthogonal if and only if

$$k^2 = 12$$
$$k = \pm 2\sqrt{3}$$

**20.** Find two unit vectors each having a representation whose initial point is $(2, 4)$ and which are tangents to the parabola $y = x^2$ there.

SOLUTION: Because $D_x y = 2x$, the slope of the line tangent to the curve $y = x^2$ at the point $(2, 4)$ is 4. Let $\mathbf{U} = \langle u_1, u_2 \rangle$ be one of the required unit vectors, and let $\theta$ be the radian measure of the direction angle of $\mathbf{U}$. Because the slope of any representation of $\mathbf{U}$ is 4, we have

$$\tan \theta = 4$$

Thus

$$\frac{u_2}{u_1} = 4$$

$$u_2 = 4u_1 \tag{1}$$

Moreover, because $\mathbf{U}$ is a unit vector, then

$$\sqrt{{u_1}^2 + {u_2}^2} = 1 \tag{2}$$

Solving (1) and (2) simultaneously, we have

$$u_1 = \pm \frac{1}{17}\sqrt{17}$$

$$u_2 = \pm \frac{4}{17}\sqrt{17}$$

Thus, the two required unit vectors are

$$\left\langle \frac{1}{17}\sqrt{17}, \frac{4}{17}\sqrt{17} \right\rangle \quad \text{and} \quad -\left\langle \frac{1}{17}\sqrt{17}, \frac{4}{17}\sqrt{17} \right\rangle$$

**24.** Find the vector projection of $\mathbf{B}$ onto $\mathbf{A}$ for the vectors

$$\mathbf{A} = -8\mathbf{i} + 4\mathbf{j} \quad \text{and} \quad \mathbf{B} = 7\mathbf{i} - 6\mathbf{j}$$

SOLUTION: The vector projection of **B** onto **A**, represented by $\mathbf{B_A}$, is given by

$$\mathbf{B_A} = \frac{\mathbf{B} \cdot \mathbf{A}}{|\mathbf{A}|^2}\mathbf{A} \tag{1}$$

We have

$$\begin{aligned}\mathbf{B} \cdot \mathbf{A} &= (7\mathbf{i} - 6\mathbf{j}) \cdot (-8\mathbf{i} + 4\mathbf{j}) \\ &= 7(-8) + (-6)(4) \\ &= -80\end{aligned}$$

and

$$\begin{aligned}|\mathbf{A}|^2 &= (-8)^2 + 4^2 \\ &= 80\end{aligned}$$

Substituting in (1), we obtain

$$\begin{aligned}\mathbf{B_A} &= \frac{-80}{80}(-8\mathbf{i} + 4\mathbf{j}) \\ &= 8\mathbf{i} - 4\mathbf{j}\end{aligned}$$

**26.** For the vectors $\mathbf{A} = 5\mathbf{i} - 6\mathbf{j}$ and $\mathbf{B} = 7\mathbf{i} + \mathbf{j}$, find the component of the vector **B** in the direction of vector **A**.

SOLUTION: The component of the vector **B** in the direction of vector **A** is $B_\mathbf{A}$, the scalar projection of **B** onto **A**. Because

$$B_\mathbf{A} = \frac{\mathbf{B} \cdot \mathbf{A}}{|\mathbf{A}|}$$

we have

$$\begin{aligned}B_\mathbf{A} &= \frac{(5\mathbf{i} - 6\mathbf{j}) \cdot (7\mathbf{i} + \mathbf{j})}{\sqrt{5^2 + (-6)^2}} \\ &= \frac{5 \cdot 7 + (-6)(1)}{\sqrt{61}} \\ &= \frac{29}{\sqrt{61}}\end{aligned}$$

**28.** Two forces represented by the vectors $\mathbf{F}_1$ and $\mathbf{F}_2$ act on a particle and cause it to move along a straight line from the point $(2, 5)$ to the point $(7, 3)$. If $\mathbf{F}_1 = 3\mathbf{i} - \mathbf{j}$ and $\mathbf{F}_2 = -4\mathbf{i} + 5\mathbf{j}$, the magnitudes of the forces are measured in pounds, and distance is measured in feet, find the work done by the two forces acting together.

SOLUTION: Let $\mathbf{D} = a_1\mathbf{i} + b_1\mathbf{j}$ be the displacement vector for the force. Because the motion is from the point $(2, 5)$ to the point $(7, 3)$, we have

$$\begin{aligned}a_1 &= 7 - 2 \quad \text{and} \quad a_2 = 3 - 5 \\ &= 5 \qquad\qquad\quad\; = -2\end{aligned}$$

Thus, $\mathbf{D} = 5\mathbf{i} - 2\mathbf{j}$. The number of foot-pounds in the work done by the two forces $\mathbf{F}_1$ and $\mathbf{F}_2$ is given by

$$\begin{aligned}W &= \mathbf{F}_1 \cdot \mathbf{D} + \mathbf{F}_2 \cdot \mathbf{D} \\ &= (3\mathbf{i} - \mathbf{j}) \cdot (5\mathbf{i} - 2\mathbf{j}) + (-4\mathbf{i} + 5\mathbf{j}) \cdot (5\mathbf{i} - 2\mathbf{j}) \\ &= (3)(5) + (-1)(-2) + (-4)(5) + (5)(-2) \\ &= -13\end{aligned}$$

Thus, the work done is $-13$ foot-pounds.

32. Prove by vector analysis that the line segment joining the midpoints of the nonparallel sides of a trapezoid is parallel to the parallel sides and that its length is one-half the sum of the lengths of the parallel sides.

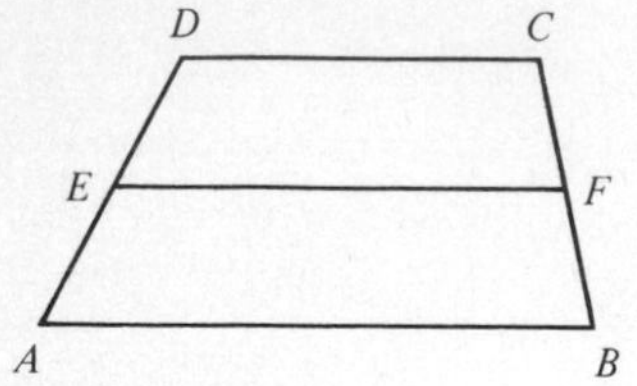

Figure 17.3.32

SOLUTION: Let $ABCD$ be a trapezoid with parallel sides $AB$ and $DC$. Let $E$ be the midpoint of side $AD$, and let $F$ be the midpoint of side $BC$. Refer to Fig. 17.3.32. We let $\mathbf{V}(\overrightarrow{AB})$ be the vector having the directed line segment $\overrightarrow{AB}$ as a representation, as in Example 4 of the text. We use similar notation for other vectors with representations shown in the figure. Because $E$ is the midpoint of side $AD$, we have

$$\mathbf{V}(\overrightarrow{AE}) = \mathbf{V}(\overrightarrow{ED}) \tag{1}$$

Similarly, because $F$ is the midpoint of side $BC$, then

$$\mathbf{V}(\overrightarrow{BF}) = \mathbf{V}(\overrightarrow{FC}) \tag{2}$$

First, we show that

$$\mathbf{V}(\overrightarrow{EF}) = \frac{1}{2}(\mathbf{V}(\overrightarrow{AB}) + \mathbf{V}(\overrightarrow{DC}))$$

We have

$$\begin{aligned}\mathbf{V}(\overrightarrow{AB}) &= \mathbf{V}(\overrightarrow{AE}) + \mathbf{V}(\overrightarrow{EF}) + \mathbf{V}(\overrightarrow{FB})\\ &= \mathbf{V}(\overrightarrow{AE}) + \mathbf{V}(\overrightarrow{EF}) - \mathbf{V}(\overrightarrow{BF})\end{aligned} \tag{3}$$

and

$$\begin{aligned}\mathbf{V}(\overrightarrow{DC}) &= \mathbf{V}(\overrightarrow{DE}) + \mathbf{V}(\overrightarrow{EF}) + \mathbf{V}(\overrightarrow{FC})\\ &= -\mathbf{V}(\overrightarrow{ED}) + \mathbf{V}(\overrightarrow{EF}) + \mathbf{V}(\overrightarrow{FC})\end{aligned} \tag{4}$$

Adding the members of (3) and (4), we obtain

$$\mathbf{V}(\overrightarrow{AB}) + \mathbf{V}(\overrightarrow{DC}) = [\mathbf{V}(\overrightarrow{AE}) - \mathbf{V}(\overrightarrow{ED})] + 2\mathbf{V}(\overrightarrow{EF}) + [-\mathbf{V}(\overrightarrow{BF}) + \mathbf{V}(\overrightarrow{FC})] \tag{5}$$

Substituting from (1) and (2) into (5), we obtain

$$\mathbf{V}(\overrightarrow{AB}) + \mathbf{V}(\overrightarrow{DC}) = 2\mathbf{V}(\overrightarrow{EF})$$

Hence,

$$\mathbf{V}(\overrightarrow{EF}) = \frac{1}{2}[\mathbf{V}(\overrightarrow{AB}) + \mathbf{V}(\overrightarrow{DC})] \tag{6}$$

Because $\mathbf{V}(\overrightarrow{DC})$ is parallel to $\mathbf{V}(\overrightarrow{AB})$ and the vectors have the same direction, there is some positive scalar $k$, such that

$$\mathbf{V}(\overrightarrow{DC}) = k\,\mathbf{V}(\overrightarrow{AB}) \tag{7}$$

Thus,

$$\begin{aligned}\mathbf{V}(\overrightarrow{AB}) + \mathbf{V}(\overrightarrow{DC}) &= \mathbf{V}(\overrightarrow{AB}) + k\,\mathbf{V}(\overrightarrow{AB})\\ &= (1 + k)\mathbf{V}(\overrightarrow{AB})\end{aligned} \tag{8}$$

Substituting from (8) into (6), we obtain

$$\mathbf{V}(\overrightarrow{EF}) = \frac{1}{2}(1 + k)\mathbf{V}(\overrightarrow{AB}) \tag{9}$$

Because $\mathbf{V}(\overrightarrow{EF})$ is a scalar multiple of $\mathbf{V}(\overrightarrow{AB})$, then $\mathbf{V}(\overrightarrow{EF})$ is parallel to $\mathbf{V}(\overrightarrow{AB})$, and thus the line segment joining the midpoints of the nonparallel sides of the trapezoid is parallel to the parallel sides. Next, we show that the length of segment $EF$ is one-half the sum of the lengths of sides $AB$ and $DC$.

Because $k > 0$, from (7) we have

$$|\mathbf{V}(\overrightarrow{DC})| = k\,|\mathbf{V}(\overrightarrow{AB})| \tag{10}$$

Thus, from (6), (8), and (10), we obtain

$$\begin{aligned}
|\mathbf{V}(\overrightarrow{EF})| &= \frac{1}{2}|\mathbf{V}(\overrightarrow{AB}) + \mathbf{V}(\overrightarrow{DC})| \\
&= \frac{1}{2}|(1+k)\,\mathbf{V}(\overrightarrow{AB})| \\
&= \frac{1}{2}(1+k)|\mathbf{V}(\overrightarrow{AB})| \\
&= \frac{1}{2}(|\mathbf{V}(\overrightarrow{AB})| + k\,|\mathbf{V}(\overrightarrow{AB})|) \\
&= \frac{1}{2}(|\mathbf{V}(\overrightarrow{AB})| + |\mathbf{V}(\overrightarrow{DC})|)
\end{aligned}$$

Thus, the length of the segment joining the midpoints of the nonparallel sides of a trapezoid is one-half the sum of the lengths of the parallel sides.

## 17.4 VECTOR-VALUED FUNCTIONS AND PARAMETRIC EQUATIONS

**17.4.1 Definition** Let $f$ and $g$ be two real-valued functions of a real variable $t$. Then for every number $t$ in the domain common to $f$ and $g$, there is a vector $\mathbf{R}$ defined by

$$\mathbf{R}(t) = f(t)\mathbf{i} + g(t)\mathbf{j} \tag{1}$$

and $\mathbf{R}$ is called a *vector-valued function.*

The graph of the vector-valued function $\mathbf{R}$ of Definition 17.4.1 is the curve $C$ which is the set of all points $(x, y)$, such that

$$x = f(t) \quad \text{and} \quad y = g(t) \tag{2}$$

The equations (2) are called *parametric equations* of the curve $C$, and $t$ is called a *parameter*. If we eliminate $t$ from Eqs. (2), we have a Cartesian equation of the curve $C$. For this equation we have

$$\frac{dy}{dx} = \frac{\dfrac{dy}{dt}}{\dfrac{dx}{dt}} \qquad \text{if } \frac{dx}{dt} \neq 0$$

Thus if $dy/dt = 0$ and $dx/dt \neq 0$ at some point on the curve $C$, there is a horizontal tangent at that point. Moreover, if $dx/dt = 0$ and $dy/dt \neq 0$ at some point on $C$, there is a vertical tangent at that point. If $dy/dt = 0$ and $dx/dt = 0$ at a point, we may not have enough information to determine the inclination of the tangent line, if it exists.

### *Exercises 17.4*

4. Find the domain of the vector-valued function $\mathbf{R}$.

$$\mathbf{R}(t) = \ln(t+1)\mathbf{i} + (\tan^{-1}t)\mathbf{j}$$

SOLUTION: Let $f$ and $g$ be the functions defined by

$$f(t) = \ln(t+1) \quad \text{and} \quad g(t) = \tan^{-1}t$$

The domain of $\mathbf{R}$ is the intersection of the domains of $f$ and $g$. Because $\ln x$ is defined if and only if $x > 0$, the domain of $f$ is $\{t \mid t+1>0\} = (-1, +\infty)$. Because $\tan^{-1} x$ is defined for all real $x$, the domain of $g$ is $(-\infty, +\infty)$. Thus, the domain of $\mathbf{R}$ is $(-1, +\infty)$.

**10.** Find $dy/dx$ and $d^2y/dx^2$ without eliminating the parameter.

$$x = e^{2t} \qquad y = 1 + \cos t$$

SOLUTION: Because

$$\frac{dx}{dt} = 2e^{2t} \quad \text{and} \quad \frac{dy}{dt} = -\sin t$$

then

$$\frac{dy}{dx} = \frac{\frac{dy}{dt}}{\frac{dx}{dt}} = \frac{-\sin t}{2e^{2t}}$$

Let $y' = dy/dx$. Then

$$y' = \frac{-\sin t}{2e^{2t}}$$

and

$$\frac{d(y')}{dt} = \frac{-2e^{2t}\cos t + 4\sin t\, e^{2t}}{4e^{4t}} = \frac{-\cos t + 2\sin t}{2e^{2t}}$$

Thus,

$$\frac{d^2y}{dx^2} = \frac{d(y')}{dt} = \frac{\frac{d(y')}{dt}}{\frac{dx}{dt}} = \frac{\frac{-\cos t + 2\sin t}{2e^{2t}}}{2e^{2t}} = \frac{2\sin t - \cos t}{4e^{4t}}$$

**14.** Draw a sketch of the graph of the given vector equation and find a Cartesian equation of the graph.

$$\mathbf{R}(t) = \frac{4}{t^2}\mathbf{i} + \frac{4}{t}\mathbf{j}$$

SOLUTION: The graph is the curve $C$ which contains all points $(x, y)$, such that

$$x = \frac{4}{t^2} \quad \text{and} \quad y = \frac{4}{t} \tag{1}$$

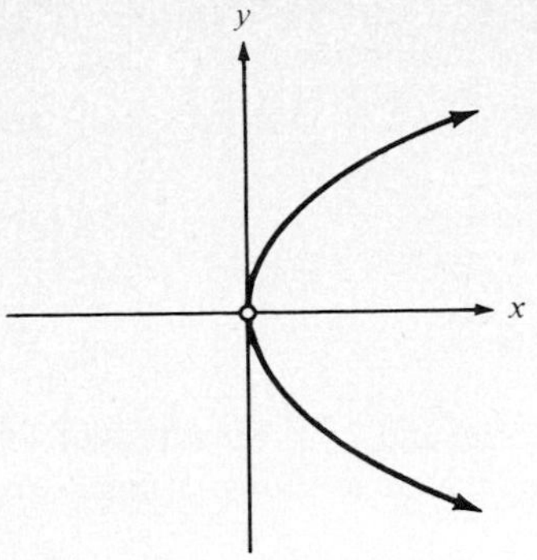

Figure 17.4.14

Thus, $t \neq 0$. Furthermore, $x > 0$ and $y \neq 0$.

We plot the points from Table 14 and draw the graph, shown in Fig. 17.4.14.

**Table 14**

| $t$ | $x$ | $y$ |
|---|---|---|
| $-4$ | $\frac{1}{4}$ | $-1$ |
| $-2$ | 1 | $-2$ |
| $-1$ | 4 | $-4$ |
| 1 | 4 | 4 |
| 2 | 1 | 2 |
| 4 | $\frac{1}{4}$ | 1 |

The graph appears to be a parabola with the vertex $(0, 0)$ deleted. Solving the equation $y = 4/t$ for $t$, we obtain $t = 4/y$.

Replacing $t$ by $4/y$ in the equation $x = 4/t^2$, we obtain $y^2 = 4x$, which is a Cartesian equation of the graph.

**18.** Find equations of the horizontal tangent lines by finding the values of $t$ for which $dy/dt = 0$, and find equations of the vertical tangent lines by finding the values of $t$ for which $dx/dt = 0$. Then draw a sketch of the graph of the given pair of parametric equations.

$$x = \frac{3at}{1+t^3} \qquad y = \frac{3at^2}{1+t^3}$$

SOLUTION:

$$\frac{dy}{dt} = \frac{(1+t^3)(6at) - (3at^2)(3t^2)}{(1+t^3)^2}$$

$$= \frac{-3at(t^3-2)}{(1+t^3)^2}$$

If $dy/dt = 0$, then either $t = 0$ or $t = \sqrt[3]{2}$. If $t = 0$, then $x = 0$ and $y = 0$. Thus, the line $y = 0$ is a horizontal tangent at the origin. If $t = \sqrt[3]{2}$, then $x = a\sqrt[3]{2}$ and $y = a\sqrt[3]{4}$. Thus, the line $y = a\sqrt[3]{4}$ is a horizontal tangent at the point $(a\sqrt[3]{2}, a\sqrt[3]{4})$. Moreover,

$$\frac{dx}{dt} = \frac{(1+t^3)(3a) - (3at)(3t^2)}{(1+t^3)^2}$$

$$= \frac{-3a(2t^3-1)}{(1+t^3)^2}$$

If $dx/dt = 0$, then $t = \frac{1}{2}\sqrt[3]{4}$ from which we get $x = a\sqrt[3]{4}$ and $y = a\sqrt[3]{2}$. Thus, the line $x = a\sqrt[3]{4}$ is a vertical tangent at the point $(a\sqrt[3]{4}, a\sqrt[3]{2})$.

We note that $(x, y)$ is not defined when $t = -1$. Moreover, if $a > 0$, then

$$\lim_{t \to -1^+} x = \lim_{t \to -1^+} \frac{3at}{1+t^3} = -\infty$$

and

$$\lim_{t \to -1^+} y = \lim_{t \to -1^+} \frac{3at^2}{1+t^3} = +\infty$$

Furthermore,

$$\lim_{t \to +\infty} x = \lim_{t \to +\infty} \frac{3at}{1+t^3} = 0$$

and

$$\lim_{t \to +\infty} y = \lim_{t \to +\infty} \frac{3at^2}{1+t^3} = 0$$

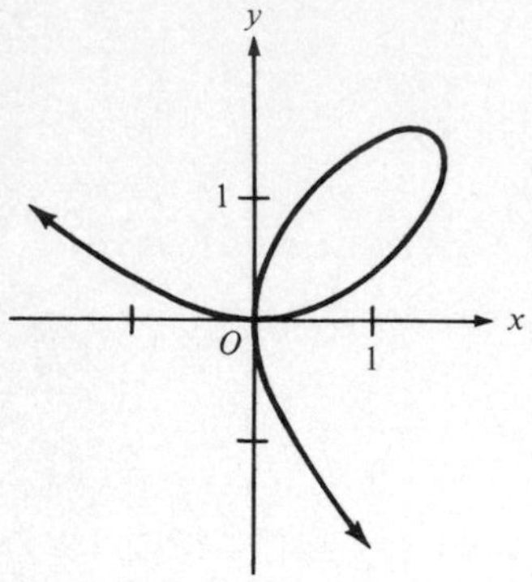

Figure 17.4.18

In Table 18 we show decimal approximations of $x/a$ and $y/a$ for values of $t \neq -1$. If $a > 0$, the graph for $t > -1$ is that part of the curve shown in Fig. 17.4.18 for which $y \geq 0$.

**Table 18**

| $t$ | $\frac{x}{a}$ | $\frac{y}{a}$ |
|---|---|---|
| 4 | 0.2 | 0.7 |
| 2 | 0.7 | 1.3 |
| $\sqrt[3]{2} =$ 1.3 | 1.3 | 1.6 (Horizontal tangent) |
| 1 | 1.5 | 1.5 |
| $\frac{1}{2}\sqrt[3]{4} =$ 0.8 | 1.6 | 1.3 (Vertical tangent) |
| 0.3 | 0.9 | 0.3 |
| 0 | 0 | 0 (Horizontal tangent) |
| –0.2 | –0.6 | 0.1 |
| –0.3 | –0.9 | 0.3 |
| –0.5 | –1.9 | 0.9 |
| –2 | 0.9 | –1.8 |
| –3 | 0.3 | –0.9 |
| –5 | 0.1 | –0.6 |

In $t < -1$, we note that

$$\lim_{t \to -1^-} x = +\infty \quad \text{and} \quad \lim_{t \to -1^-} y = -\infty$$

Furthermore,

$$\lim_{t \to -\infty} x = 0 \quad \text{and} \quad \lim_{t \to -\infty} y = 0$$

The graph for $t < -1$ is that part of the curve shown in Fig. 17.4.18 for which $y < 0$.

**22.** Show that the slope of the tangent line at $t = t_1$ to the cycloid having Equations (16) is $\cot(\frac{1}{2}t_1)$. Deduce then that the tangent line is vertical when $t = 2n\pi$, where $n$ is any integer.

SOLUTION: From Equations (16), we have

$$x = a(t - \sin t) \quad \text{and} \quad y = a(1 - \cos t)$$

Thus,

$$\frac{dx}{dt} = a(1 - \cos t) \quad \text{and} \quad \frac{dy}{dt} = a \sin t$$

Because

$$\frac{dy}{dx} = \frac{dy}{dt} \div \frac{dx}{dt}$$

we have

$$\begin{aligned} \frac{dy}{dx} &= \frac{a \sin t}{a(1 - \cos t)} \\ &= \frac{\sin t}{1 - \cos t} \end{aligned} \tag{1}$$

From trigonometry, we have

$$\cot \frac{1}{2}t = \frac{\sin t}{1 - \cos t} \tag{2}$$

From (1) and (2), we conclude that the slope of the tangent line at $t = t_1$ to the cycloid is $\cot(\frac{1}{2}t_1)$. For any integer $n$

$$\lim_{t \to 2n\pi} \cot\left(\frac{1}{2}t\right) = \lim_{t \to 2n\pi} \frac{\cos\left(\frac{1}{2}t\right)}{\sin\left(\frac{1}{2}t\right)} = \pm\infty$$

because

$$\lim_{t \to 2n\pi} \cos\left(\frac{1}{2}t\right) = \pm 1 \quad \text{and} \quad \lim_{t \to 2n\pi} \sin\left(\frac{1}{2}t\right) = 0$$

Thus, the tangent line is vertical when $t = 2n\pi$.

**26.** Parametric equations for the *tractrix* are:

$$x = t - a \tanh \frac{t}{a} \quad \text{and} \quad y = a \operatorname{sech} \frac{t}{a}$$

Draw a sketch of the curve for $a = 4$.

SOLUTION: If $a = 4$, we have

$$x = t - 4 \tanh \frac{1}{4}t \quad \text{and} \quad y = 4 \operatorname{sech} \frac{1}{4}t \tag{1}$$

Because $\tanh(-t) = -\tanh t$ and $\operatorname{sech}(-t) = \operatorname{sech} t$, we may replace $t$ by $-t$ and $x$ by $-x$ in Equations (1), and the resulting equations are equivalent. Thus, for every point $(x, y)$ on the curve that results from the replacement $t$ for the parameter, there is the point $(-x, y)$ which corresponds to the replacement $-t$ for the parameter. Hence, the graph is symmetric with respect to the $y$-axis, and we may plot points for $t \geq 0$ and use symmetry to complete the sketch. Furthermore,

$$\frac{dx}{dt} = 1 - \operatorname{sech}^2 \frac{1}{4}t \quad \text{and} \quad \frac{dy}{dt} = -\operatorname{sech} \frac{1}{4}t \tanh \frac{1}{4}t$$

Thus,

$$\begin{aligned} \frac{dy}{dx} &= \frac{-\operatorname{sech} \frac{1}{4}t \tanh \frac{1}{4}t}{1 - \operatorname{sech}^2 \frac{1}{4}t} \\ &= \frac{-\operatorname{sech} \frac{1}{4}t \tanh \frac{1}{4}t}{\tanh^2 \frac{1}{4}t} \end{aligned}$$

$$= \frac{-\text{sech}\, \frac{1}{4}t}{\tanh \frac{1}{4}t}$$

Because

$$\lim_{t \to 0} \frac{-\text{sech}\, \frac{1}{4}t}{\tanh \frac{1}{4}t} = \pm\infty$$

and when $t = 0$, then $x = 0$ and $y = 4$, we conclude that there is a vertical tangent at the point $(0, 4)$. Moreover, because

$$\lim_{t \to +\infty} x = \lim_{t \to +\infty} \left(t - 4 \tanh \frac{1}{4}t\right) = +\infty$$

and

$$\lim_{t \to +\infty} y = \lim_{t \to +\infty} 4\, \text{sech}\, \frac{1}{4}t = 0$$

the $x$-axis is a horizontal asymptote. We plot the points from Table 26 and draw a sketch of the graph, shown in Fig. 17.4.26.

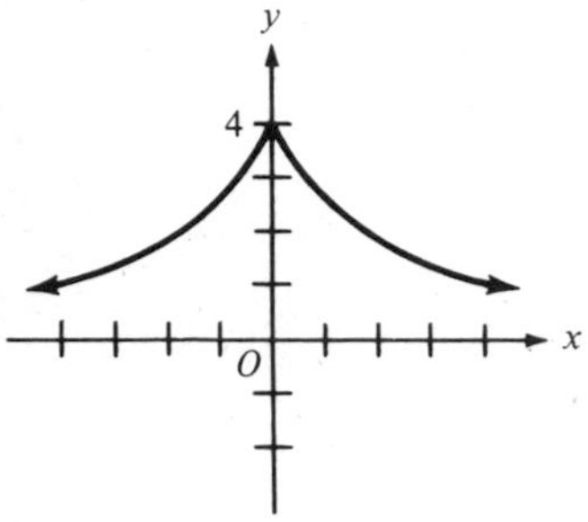

Figure 17.4.26

**Table 26**

| $t$ | $x$ | $y$ |
|---|---|---|
| 8 | 4.1 | 1.1 |
| 4 | 1.0 | 2.6 |
| 2 | 0.2 | 3.5 |
| 0 | 0 | 4 |

**30.** Find the centroid of the region bounded by the $x$-axis and one arch of the cycloid having Equations (16).

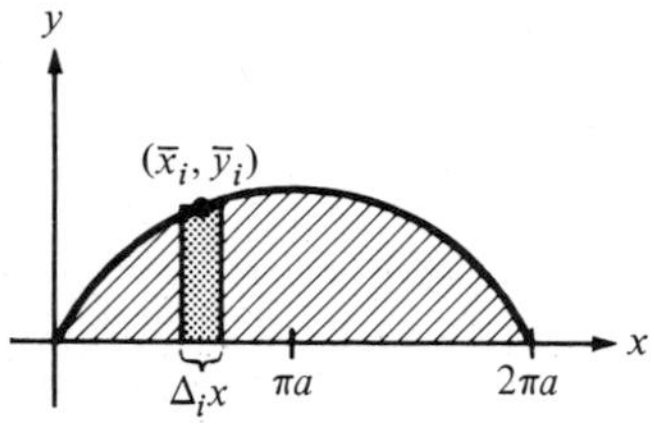

Figure 17.4.30

SOLUTION: From Equations (16) we have

$$x = a(t - \sin t) \quad \text{and} \quad y = a(1 - \cos t) \qquad \textbf{(1)}$$

In Fig. 17.4.30 we show a sketch of the region and an element of area which is $\Delta_i x$ units by $\bar{y}_i$ units. Thus, the number of square units in the area of the region is given by

$$A = \lim_{\|\Delta\| \to 0} \sum_{i=1}^{n} \bar{y}_i \, \Delta_i x$$

$$= \int_0^{2\pi a} y \, dx \qquad \textbf{(2)}$$

From (1), we have $y = a(1 - \cos t)$ and $dx = a(1 - \cos t)dt$. Moreover, when $x = 0$ and $y = 0$, then $t = 0$; when $x = 2\pi a$ and $y = 0$, then $t = 2\pi$. Substituting in (2), we obtain

$$A = a^2 \int_0^{2\pi} (1 - \cos t)^2 \, dt$$

$$= a^2 \int_0^{2\pi} (1 - 2\cos t + \cos^2 t)\, dt$$

$$= a^2 \int_0^{2\pi} \left(\frac{3}{2} - 2\cos t + \frac{1}{2}\cos 2t\right) dt$$

$$= a^2 \left[\frac{3}{2}t - 2\sin t + \frac{1}{4}\sin 2t\right]_0^{2\pi}$$

$$= 3\pi a^2 \tag{3}$$

Because the centroid of the element of area is at the point $(\bar{x}_i, \frac{1}{2}\bar{y}_i)$, we have

$$A\bar{y} = \lim_{\|\Delta\| \to 0} \sum_{i=1}^{n} \frac{1}{2}(\bar{y}_i)^2 \,\Delta_i x$$

$$= \frac{1}{2}\int_0^{2\pi a} y^2\, dx$$

$$= \frac{1}{2}a^3 \int_0^{2\pi} (1 - \cos t)^3\, dt$$

$$= \frac{1}{2}a^3 \int_0^{2\pi} (1 - 3\cos t + 3\cos^2 t + \cos^3 t)\, dt$$

$$= \frac{1}{2}a^3 \int_0^{2\pi} \left[1 - 3\cos t + \frac{3}{2}(1 + \cos 2t) + (1 - \sin^2 t)\cos t\right] dt$$

$$= \frac{1}{2}a^3 \int_0^{2\pi} \left(\frac{5}{2} - 2\cos t + \frac{3}{2}\cos 2t\right) dt - \frac{1}{2}a^3 \int_0^{2\pi} \sin^2 t \cos t\, dt$$

$$= \frac{1}{2}a^3 \left[\frac{5}{2}t - 2\sin t + \frac{3}{4}\sin 2t\right]_0^{2\pi} - \frac{1}{2}a^3\left[\frac{1}{3}\sin^3 t\right]_0^{2\pi}$$

$$= \frac{5}{2}\pi a^3 \tag{4}$$

Substituting the value for $A$ from (3) into (4), we have

$$3\pi a^2 \bar{y} = \frac{5}{2}\pi a^3$$

$$\bar{y} = \frac{5}{6}a$$

Because the region is symmetric about the line $x = \pi a$, we have $\bar{x} = \pi a$. Thus, the centroid is at $(\pi a, \frac{5}{6}a)$.

## 17.5 CALCULUS OF VECTOR-VALUED FUNCTIONS

The most important theorem in this section is the following.

**17.5.4 Theorem** If $\mathbf{R}$ is a vector-valued function defined by

$$\mathbf{R}(t) = f(t)\mathbf{i} + g(t)\mathbf{j} \tag{1}$$

then

$$\mathbf{R}'(t) = f'(t)\mathbf{i} + g'(t)\mathbf{j}$$

if $f'(t)$ and $g'(t)$ exist. For each replacement of $t$, the direction of the vector $\mathbf{R}'(t)$ is along the tangent line to the curve whose vector equation is (1).

The definitions of the limit of a vector-valued function, continuity of a vector-valued function, the derivative of a vector-valued function, differentiability of a vector-valued function, and the antiderivative of a vector-valued function are similar to the corresponding definitions for real-valued functions.

**17.5.1 Definition** Let $\mathbf{R}$ be a vector-valued function whose function values are given by

$$\mathbf{R}(t) = f(t)\mathbf{i} + g(t)\mathbf{j}$$

Then the *limit of* $\mathbf{R}(t)$ *as* $t$ *approaches* $t_1$ is defined by

$$\lim_{t \to t_1} \mathbf{R}(t) = [\lim_{t \to t_1} f(t)]\mathbf{i} + [\lim_{t \to t_1} g(t)]\mathbf{j}$$

if $\lim_{t \to t_1} f(t)$ and $\lim_{t \to t_1} g(t)$ both exist.

**17.5.2 Definition** The vector-valued function $\mathbf{R}$ is *continuous* at $t_1$ if and only if the following three conditions are satisfied:

(i) $\mathbf{R}(t_1)$ exists;
(ii) $\lim_{t \to t_1} \mathbf{R}(t)$ exists;
(iii) $\lim_{t \to t_1} \mathbf{R}(t) = \mathbf{R}(t_1)$.

**17.5.3 Definition** If $\mathbf{R}$ is a vector-valued function, then the *derivative* of $\mathbf{R}$ is another vector-valued function, denoted by $\mathbf{R}'$ and defined by

$$\mathbf{R}'(t) = \lim_{\Delta t \to 0} \frac{\mathbf{R}(t + \Delta t) - \mathbf{R}(t)}{\Delta t}$$

if this limit exists.

The notation $D_t\mathbf{R}(t)$ is sometimes used in place of $\mathbf{R}'(t)$.

**17.5.5 Definition** A vector-valued function $\mathbf{R}$ is said to be *differentiable* on an interval if $\mathbf{R}'(t)$ exists for all values of $t$ in the interval.

**17.5.10 Definition** If $\mathbf{Q}$ is the vector-valued function given by

$$\mathbf{Q}(t) = f(t)\mathbf{i} + g(t)\mathbf{j}$$

then the *indefinite integral of* $\mathbf{Q}(t)$ is defined by

$$\int \mathbf{Q}(t)\,dt = \mathbf{i}\int f(t)\,dt + \mathbf{j}\int g(t)\,dt$$

Furthermore, the differentiation formulas for a vector-valued function are similar to the corresponding differentiation formulas for a real-valued function.

**17.5.6 Theorem** If $\mathbf{R}$ and $\mathbf{Q}$ are differentiable vector-valued functions on an interval, then $\mathbf{R} + \mathbf{Q}$ is differentiable on the interval, and

$$D_t[\mathbf{R}(t) + \mathbf{Q}(t)] = D_t\mathbf{R}(t) + D_t\mathbf{Q}(t)$$

**17.5.7 Theorem** If $\mathbf{R}$ and $\mathbf{Q}$ are differentiable vector-valued functions on an interval, then $\mathbf{R} \cdot \mathbf{Q}$ is differentiable on the interval, and

$$D_t[\mathbf{R}(t) \cdot \mathbf{Q}(t)] = [D_t\mathbf{R}(t)] \cdot \mathbf{Q}(t) + \mathbf{R}(t) \cdot [D_t\mathbf{Q}(t)]$$

**17.5.8 Theorem** If **R** is a differentiable vector-valued function on an interval and $f$ is a differentiable real-valued function on the interval, then

$$D_t\{[f(t)][\mathbf{R}(t)]\} = [D_t f(t)]\mathbf{R}(t) + f(t)D_t\mathbf{R}(t)$$

**17.5.9 Theorem** Suppose that **F** is a vector-valued function, $h$ is a real-valued function such that $\phi = h(t)$, and $\mathbf{G}(t) = \mathbf{F}(h(t))$. If $h$ is continuous at $t$ and **F** is continuous at $h(t)$, then **G** is continuous at $t$. Furthermore, if $D_t\phi$ and $D_\phi \mathbf{G}(t)$ exist, then $D_t\mathbf{G}(t)$ exists and is given by

$$D_t\mathbf{G}(t) = [D_\phi \mathbf{G}(t)]\, D_t\phi$$

## *Exercises 17.5*

**4.** Find the indicated limit, if it exists.

$$\mathbf{R}(t) = \frac{t^2 - 2t - 3}{t - 3}\mathbf{i} + \frac{t^2 - 5t + 6}{t - 3}\mathbf{j};\ \lim_{t \to 3} \mathbf{R}(t)$$

SOLUTION: Let $f$ and $g$ be the real-valued functions defined by

$$f(t) = \frac{t^2 - 2t - 3}{t - 3} \quad \text{and} \quad g(t) = \frac{t^2 - 5t + 6}{t - 3}$$

Because

$$\lim_{t \to 3} f(t) = \lim_{t \to 3} \frac{(t-3)(t+1)}{t-3} = \lim_{t \to 3} (t+1) = 4$$

and

$$\lim_{t \to 3} g(t) = \lim_{t \to 3} \frac{(t-3)(t-2)}{t-3} = \lim_{t \to 3} (t-2) = 1$$

then

$$\begin{aligned} \lim_{t \to 3} \mathbf{R}(t) &= \lim_{t \to 3} [f(t)\mathbf{i} + g(t)\mathbf{j}] \\ &= \lim_{t \to 3} f(t)\mathbf{i} + \lim_{t \to 3} g(t)\mathbf{j} \\ &= 4\mathbf{i} + \mathbf{j} \end{aligned}$$

**8.** Find $\mathbf{R}'(t)$ and $\mathbf{R}''(t)$.

$$\mathbf{R}(t) = \cos 2t\,\mathbf{i} + \tan t\,\mathbf{j}$$

SOLUTION: By Theorem 17.5.4

$$\begin{aligned} \mathbf{R}'(t) &= D_t(\cos 2t)\mathbf{i} + D_t(\tan t)\mathbf{j} \\ &= -2 \sin 2t\,\mathbf{i} + \sec^2 t\,\mathbf{j} \end{aligned}$$

and

$$\begin{aligned} \mathbf{R}''(t) &= D_t(-2 \sin 2t)\mathbf{i} + D_t(\sec^2 t)\mathbf{j} \\ &= -4 \cos 2t\,\mathbf{i} + 2 \sec^2 t \tan t\,\mathbf{j} \end{aligned}$$

**12.** Find $D_t|\mathbf{R}(t)|$ if $\mathbf{R}(t) = (e^t + 1)\mathbf{i} + (e^t - 1)\mathbf{j}$.

SOLUTION:

$$|\mathbf{R}(t)| = \sqrt{(e^t + 1)^2 + (e^t - 1)^2}$$

$$= \sqrt{2}\sqrt{e^{2t}+1}$$

Thus,

$$D_t|\mathbf{R}(t)| = \frac{1}{2}\sqrt{2}\,(e^{2t}+1)^{-1/2}(2e^{2t})$$

$$= \frac{\sqrt{2}\,e^{2t}}{\sqrt{e^{2t}+1}}$$

**14.** Find $\mathbf{R}'(t)\cdot\mathbf{R}''(t)$ if $\mathbf{R}(t) = -\cos 2t\,\mathbf{i} + \sin 2t\,\mathbf{j}$.

SOLUTION:

$$\mathbf{R}'(t) = 2\sin 2t\,\mathbf{i} + 2\cos 2t\,\mathbf{j}$$

$$\mathbf{R}''(t) = 4\cos 2t\,\mathbf{i} - 4\sin 2t\,\mathbf{j}$$

Thus,

$$\begin{aligned}\mathbf{R}'(t)\cdot\mathbf{R}''(t) &= (2\sin 2t)(4\cos 2t) + (2\cos 2t)(-4\sin 2t)\\ &= 8\sin 2t\cos 2t - 8\sin 2t\cos 2t\\ &= 0\end{aligned}$$

**20.** Find the most general vector whose derivative has the given function value.

$$\frac{1}{4+t^2}\mathbf{i} - \frac{4}{1-t^2}\mathbf{j}$$

SOLUTION: We use Definition 17.5.10. Thus,

$$\int\left[\frac{1}{4+t^2}\mathbf{i} - \frac{4}{1-t^2}\mathbf{j}\right]dt = \mathbf{i}\int\frac{dt}{4+t^2} + \mathbf{j}\int\frac{4dt}{t^2-1}$$

$$= \frac{1}{2}\tan^{-1}\frac{1}{2}t\,\mathbf{i} + 2\ln\left|\frac{t-1}{t+1}\right|\mathbf{j} + C$$

**24.** If $\mathbf{R}'(t) = e^t\sin t\,\mathbf{i} + e^t\cos t\,\mathbf{j}$ and $\mathbf{R}(0) = \mathbf{i} - \mathbf{j}$, find $\mathbf{R}(t)$.

SOLUTION: Because $\mathbf{R}(t) = \int\mathbf{R}'(t)\,dt$, we have

$$\mathbf{R}(t) = \mathbf{i}\int e^t\sin t\,dt + \mathbf{j}\int e^t\cos t\,dt \tag{1}$$

Integrating by parts as in Example 4 of Section 11.2, we have

$$\int e^t\sin t\,dt = \frac{1}{2}e^t(\sin t - \cos t) + C_1 \tag{2}$$

and

$$\int e^t\cos t\,dt = \frac{1}{2}e^t(\sin t + \cos t) + C_2 \tag{3}$$

Substituting from (2) and (3) into (1), we obtain

$$\mathbf{R}(t) = \left[\frac{1}{2}e^t(\sin t - \cos t) + C_1\right]\mathbf{i} + \left[\frac{1}{2}e^t(\sin t + \cos t) + C_2\right]\mathbf{j} \tag{4}$$

Because $\mathbf{R}(0) = \mathbf{i} - \mathbf{j}$, we let $t = 0$ in (4) and get

$$\mathbf{i}-\mathbf{j}=\left[-\frac{1}{2}+C_1\right]\mathbf{i}+\left[\frac{1}{2}+C_2\right]\mathbf{j}$$

Thus, by equating the corresponding coefficients of **i** and *j*, we get

$$-\frac{1}{2}+C_1=1 \quad \text{and} \quad \frac{1}{2}+C_2=-1$$

$$C_1=\frac{3}{2} \qquad\qquad C_2=-\frac{3}{2}$$

Substituting the values for $C_1$ and $C_2$ into (4), we have

$$\mathbf{R}(t)=\frac{1}{2}[e^t(\sin t-\cos t)+3]\mathbf{i}+\frac{1}{2}[e^t(\sin t+\cos t)-3]\mathbf{j}$$

**26.** Given $\mathbf{R}(t)=2t\,\mathbf{i}+(t^2-1)\mathbf{j}$ and $\mathbf{Q}(t)=3t\,\mathbf{i}$. If $\alpha(t)$ is the radian measure of the angle between $\mathbf{R}(t)$ and $\mathbf{Q}(t)$, find $D_t\alpha(t)$.

SOLUTION: We use Theorem 17.3.5. Thus,

$$\begin{aligned}\cos\alpha(t)&=\frac{\mathbf{R}(t)\cdot\mathbf{Q}(t)}{|\mathbf{R}(t)||\mathbf{Q}(t)|}\\&=\frac{6t^2}{\sqrt{(2t)^2+(t^2-1)^2}\sqrt{(3t)^2}}\\&=\frac{6t^2}{\sqrt{(t^2+1)^2}\,|3t|}\\&=\frac{2|t|}{t^2+1}\end{aligned}$$

Thus,

$$\alpha(t)=\cos^{1}\frac{2|t|}{t^2+1}$$

$$\begin{aligned}\alpha'(t)&=\frac{-1}{\sqrt{1-\left(\frac{2|t|}{t^2+1}\right)^2}}\cdot\frac{(t^2+1)\frac{2t}{|t|}-2|t|(2t)}{(t^2+1)^2}\\&=\frac{-(t^2+1)}{\sqrt{(t^2-1)^2}}\cdot\frac{-2t(t^2-1)}{|t|(t^2+1)^2}\\&=\frac{2t(t^2-1)}{|t||t^2-1|(t^2+1)}\end{aligned}$$

**30.** Prove that if **A** and **B** are constant vectors and $f$ and $g$ are integrable functions, then

$$\int[\mathbf{A}f(t)+\mathbf{B}g(t)]\,dt=\mathbf{A}\int f(t)dt+\mathbf{B}\int g(t)dt$$

(*Hint*: Express **A** and **B** in terms of **i** and **j**.)

SOLUTION: Because **A** and **B** are constant vectors, then there are constants $a_1$, $a_2$, $b_1$, and $b_2$ such that

$$\mathbf{A}=a_1\mathbf{i}+a_2\mathbf{j} \quad \text{and} \quad \mathbf{B}=b_1\mathbf{i}+b_2\mathbf{j} \tag{1}$$

By (1), Definition 17.5.10, Theorem 17.2.1, and the properties of indefinite integrals for real-valued functions, we have

$$\begin{aligned}\int[\mathbf{A}f(t)+\mathbf{B}g(t)]\,dt &= \int[(a_1\mathbf{i}+a_2\mathbf{j})f(t)+(b_1\mathbf{i}+b_2\mathbf{j})g(t)]\,dt\\ &= \int[(a_1f(t)+b_1g(t))\mathbf{i}+(a_2f(t)+b_2g(t))\mathbf{j}]\,dt\\ &= \mathbf{i}\int[a_1f(t)+b_1g(t)]\,dt+\mathbf{j}\int[a_2f(t)+b_2g(t)]\,dt\\ &= \mathbf{i}\left[a_1\int f(t)dt+b_1\int g(t)dt\right]+\mathbf{j}\left[a_2\int f(t)dt+b_2\int g(t)dt\right]\\ &= (a_1\mathbf{i}+a_2\mathbf{j})\int f(t)dt+(b_1\mathbf{i}+b_2\mathbf{j})\int g(t)dt\\ &= \mathbf{A}\int f(t)dt+\mathbf{B}\int g(t)dt\end{aligned}$$

## 17.6 LENGTH OF ARC

In this section we have three formulas for calculating the number of units in the length of a curve.

**17.6.3 Theorem** Let the curve $C$ have parametric equations $x = f(t)$ and $y = g(t)$, and suppose that $f'$ and $g'$ are continuous on the closed interval $[a, b]$. Then the length of arc $L$ units of the curve $C$ from the point $(f(a), g(a))$ to the point $(f(b), g(b))$ is determined by

$$L = \int_a^b \sqrt{[f'(t)]^2 + [g'(t)]^2}\, dt$$

We may also write the formula of Theorem 17.6.3 as follows.

$$L = \int_a^b \sqrt{\left(\frac{dx}{dt}\right)^2 + \left(\frac{dy}{dt}\right)^2}\, dt$$

**17.6.4 Theorem** Let the curve $C$ have the vector equation $\mathbf{R}(t) = f(t)\mathbf{i} + g(t)\mathbf{j}$ and suppose that $f'$ and $g'$ are continuous on the closed interval $[a, b]$. Then the length of arc of $C$, traced by the terminal point of the position representation of $\mathbf{R}(t)$ as $t$ increases from $a$ to $b$, is determined by

$$L = \int_a^b |\mathbf{R}'(t)|\, dt$$

The formula for the number of units in the length of arc of a curve defined by a polar equation $r = f(\theta)$ from the point where $\theta = \alpha$ to the point where $\theta = \beta$ is given by

$$L = \int_\alpha^\beta \sqrt{\left(\frac{dr}{d\theta}\right)^2 + r^2}\, d\theta$$

When using the formulas for length of arc, you must remember that $\sqrt{z^2} = |z|$. Thus,

$$\sqrt{z^2} = \begin{cases} z & \text{if } z \geqslant 0 \\ -z & \text{if } z < 0 \end{cases}$$

*Exercises 17.6*

Find the length of the arc in each of the following exercises. Assume that $a > 0$.

4. $\mathbf{R}(t) = a(\cos t + t \sin t)\mathbf{i} + a(\sin t - t\cos t)\mathbf{j}$ from $t = 0$ to $t = \frac{1}{3}\pi$

SOLUTION: We use Theorem 17.6.4. Thus,

$$\begin{aligned}\mathbf{R}'(t) &= a(-\sin t + t\cos t + \sin t)\mathbf{i} + a(\cos t + t\sin t - \cos t)\mathbf{j} \\ &= at\cos t\,\mathbf{i} + at\sin t\,\mathbf{j}\end{aligned}$$

Hence,

$$\begin{aligned}|\mathbf{R}'(t)| &= \sqrt{(at\cos t)^2 + (at\sin t)^2} \\ &= \sqrt{(a^2t^2(\cos^2 t + \sin^2 t)} \\ &= a|t|\end{aligned}$$

Therefore,

$$\begin{aligned}L &= \int_0^{\pi/3} a|t|\,dt \\ &= a\int_0^{\pi/3} t\,dt \\ &= \frac{1}{2}at^2\Big]_0^{\pi/3} \\ &= \frac{1}{18}a\pi^2\end{aligned}$$

8. One arch of the cycloid: $x = a(t - \sin t), y = a(1 - \cos t)$

SOLUTION: We use Theorem 17.6.2. Thus,

$$\frac{dx}{dt} = a(1 - \cos t) \quad \text{and} \quad \frac{dy}{dt} = a\sin t$$

Hence,

$$\begin{aligned}\sqrt{\left(\frac{dx}{dt}\right)^2 + \left(\frac{dy}{dt}\right)^2} &= \sqrt{a^2(1 - 2\cos t + \cos^2 t) + a^2\sin^2 t} \\ &= a\sqrt{2 - 2\cos t} \qquad (1)\end{aligned}$$

Because,

$$\sin^2 \frac{1}{2}t = \frac{1}{2}(1 - \cos t)$$

then

$$4\sin^2 \frac{1}{2}t = 2 - 2\cos t \qquad (2)$$

Substituting from (2) into (1), we obtain

$$\begin{aligned}\sqrt{\left(\frac{dx}{dt}\right)^2 + \left(\frac{dy}{dt}\right)^2} &= a\sqrt{4\sin^2 \frac{1}{2}t} \\ &= 2a\left|\sin \frac{1}{2}t\right| \qquad (3)\end{aligned}$$

Because one arc of the cycloid is defined for $0 \leqslant t \leqslant 2\pi$, we have by (3)

$$\begin{aligned}L &= \int_0^{2\pi} \sqrt{\left(\frac{dx}{dt}\right)^2 + \left(\frac{dy}{dt}\right)^2}\,dt \\ &= 2a\int_0^{2\pi} \left|\sin \frac{1}{2}t\right|\,dt \qquad (4)\end{aligned}$$

Because $0 \leqslant t \leqslant 2\pi$, then $0 \leqslant \frac{1}{2}t \leqslant \pi$. Thus, $\sin \frac{1}{2}t \geqslant 0$. Hence, from (4) we have

$$\begin{aligned} L &= 2a \int_0^{2\pi} \sin \frac{1}{2}t \, dt \\ &= -4a \cos \frac{1}{2}t \Big]_0^{2\pi} \\ &= 8a \end{aligned}$$

**10.** The circumference of the circle: $r = a \sin \theta$

SOLUTION: We use the formula

$$L = \int_\alpha^\beta \sqrt{\left(\frac{dr}{d\theta}\right) + r^2} \, d\theta$$

Because $r = 0$ when $\theta = 0$; $r = 0$ when $\theta = \pi$; and $r > 0$ for $0 < \theta < \pi$, we have $\alpha = 0$ and $\beta = \pi$. That is, the entire circle is traced out when $\theta$ takes on all values in the closed interval $[0, \pi]$. Thus,

$$\begin{aligned} L &= \int_0^\pi \sqrt{(a \cos \theta)^2 + (a \sin \theta)^2} \, d\theta \\ &= a \int_0^\pi d\theta \\ &= a\theta \Big]_0^\pi \\ &= \pi a \end{aligned}$$

We note that the diameter of the circle is $a$ units. Thus, the measure of the radius is $r = \frac{1}{2}a$. The formula for the circumference of a circle, $C = 2\pi r$, gives $C = 2\pi(\frac{1}{2}a) = \pi a$, which agrees with our result for $L$.

**14.** $r = a\theta$ from $\theta = 0$ to $\theta = 2\pi$

SOLUTION:

$$\begin{aligned} L &= \int_0^{2\pi} \sqrt{\left(\frac{dr}{d\theta}\right)^2 + r^2} \, d\theta \\ &= \int_0^{2\pi} \sqrt{a^2 + (a\theta)^2} \, d\theta \\ &= a \int_0^{2\pi} \sqrt{1 + \theta^2} \, d\theta \qquad (1) \end{aligned}$$

We let $\theta = \tan z$. Because $0 \leqslant \theta \leqslant 2\pi$, we have $0 \leqslant z \leqslant \tan^{-1} 2\pi$. Also, $\sqrt{1 + \theta^2} = \sqrt{1 + \tan^2 z} = |\sec z|$. Because $0 \leqslant z \leqslant \tan^{-1} 2\pi$, then $0 \leqslant z \leqslant \frac{1}{2}\pi$. Thus, $\sec z > 0$, and hence $|\sec z| = \sec z$. Therefore, $\sqrt{1 + \theta^2} = \sec z$. Furthermore, $d\theta = \sec^2 z \, dz$. With these substitutions in (1), we have

$$L = a \int_0^{\tan^{-1} 2\pi} \sec^3 z \, dz \qquad (2)$$

We use the result of Example 5 in Section 11.2 for the integral in (2). Thus,

$$L = \frac{1}{2}a[\sec z \tan z + \ln|\sec z + \tan z|]_0^{\tan^{-1} 2\pi} \tag{3}$$

If $z = \tan^{-1}2\pi$, then $\tan z = 2\pi$ and $\sec z = \sqrt{1 + \tan^2 z} = \sqrt{1 + 4\pi^2}$. Thus, from (3) we have

$$L = \frac{1}{2}a[2\pi\sqrt{1 + 4\pi^2} + \ln(\sqrt{1 + 4\pi^2} + 2\pi)]$$

## 17.7 PLANE MOTION

**17.7.1 Definition** Let $C$ be the curve having parametric equations $x = f(t)$ and $y = g(t)$. If a particle is moving along $C$ so that its position at any time $t$ units is the point $(x, y)$, then the *instantaneous velocity* of the particle at time $t$ units is determined by the velocity vector

$$\mathbf{V}(t) = f'(t)\mathbf{i} + g'(t)\mathbf{j}$$

if $f'(t)$ and $g'(t)$ exist.

If a representation of the velocity vector $\mathbf{V}(t)$ is chosen so that its initial point is on the curve $C$, then the representation is tangent to the curve $C$ at all $t$. The measure of the *speed* of the particle at $t$ is a scalar and is the magnitude of the velocity vector. Because the number of units in the speed is $ds/dt$, we have

$$\frac{ds}{dt} = |\mathbf{V}(t)| = \sqrt{[f'(t)]^2 + [g'(t)]^2}$$

**17.7.2 Definition** The *instantaneous acceleration* at time $t$ units of a particle moving along a curve $C$ having parametric equations $x = f(t)$ and $y = g(t)$, is determined by the acceleration vector

$$\mathbf{A}(t) = \mathbf{V}'(t) = \mathbf{R}''(t)$$

where $\mathbf{R}(t) = f(t)\mathbf{i} + g(t)\mathbf{j}$ and $\mathbf{R}''(t)$ exists.

If a representation of the acceleration vector $\mathbf{A}(t)$ is chosen so that its initial point is on the curve $C$, then the representation is directed toward the "inside" or concave side of the curve at all $t$.

If a projectile is shot from a gun having an angle of elevation of radian measure $\alpha$ with a muzzle speed of $v_0$ ft/sec, then the position vector of the projectile at $t$ sec is $\mathbf{R}(t)$, given by

$$\mathbf{R}(t) = tv_0 \cos\alpha\,\mathbf{i} + \left(tv_0 \sin\alpha - \frac{1}{2}gt^2\right)\mathbf{j}$$

where $g$ ft/sec$^2$ is the acceleration due to the force of gravity ($g = 32$).

## *Exercises 17.7*

**4.** A particle is moving along the curve having the given parametric equations, where $t$ sec is the time. Find: **(a)** the velocity vector $\mathbf{V}(t)$, **(b)** the acceleration vector $\mathbf{A}(t)$, **(c)** the speed at $t = t_1$, and **(d)** the magnitude of the acceleration vector at $t = t_1$. Draw a sketch of the path of the particle and the representations of the velocity vector and the acceleration vector at $t = t_1$.

$$x = \frac{2}{t} \qquad y = -\frac{1}{4}t \qquad t_1 = 4$$

SOLUTION: A vector equation of the curve $C$ is

$$\mathbf{R}(t) = \frac{2}{t}\mathbf{i} + -\frac{1}{4}t\,\mathbf{j}$$

Thus,

(a) $\mathbf{V}(t) = \mathbf{R}'(t) = -\dfrac{2}{t^2}\mathbf{i} - \dfrac{1}{4}\mathbf{j}$

(b) $\mathbf{A}(t) = \mathbf{V}'(t) = \dfrac{4}{t^3}\mathbf{i}$

(c) $\mathbf{V}(4) = -\dfrac{1}{8}\mathbf{i} - \dfrac{1}{4}\mathbf{j}$

$$|\mathbf{V}(4)| = \sqrt{\left(-\frac{1}{8}\right)^2 + \left(-\frac{1}{4}\right)^2} = \frac{1}{8}\sqrt{5}$$

Thus, the speed is $\frac{1}{8}\sqrt{5}$ units per sec when $t = 4$.

(d) $\mathbf{A}(4) = \dfrac{1}{16}\mathbf{i}$

$$|\mathbf{A}(4)| = \frac{1}{16}$$

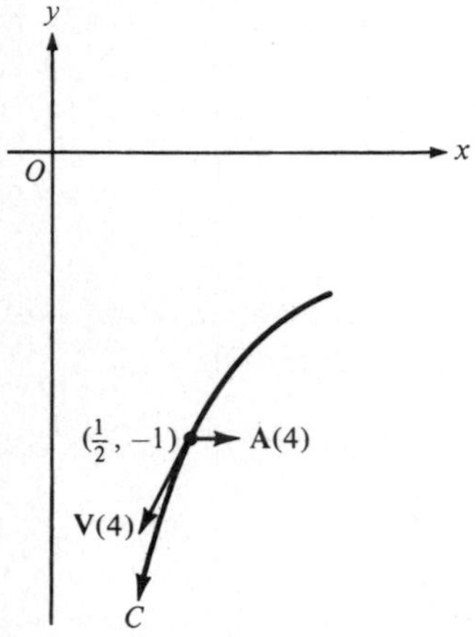

Figure 17.7.4

Thus, $\frac{1}{16}$ is the magnitude of the acceleration vector when $t = 4$.

In Fig. 17.7.4 we show a sketch of the curve $C$ for $t \geq 2$ and representations of the velocity vector and the acceleration vector at $t = 4$.

8. The position of a moving particle at $t$ sec is determined from a vector equation. Find: (a) $\mathbf{V}(t_1)$, (b) $\mathbf{A}(t_1)$, (c) $|\mathbf{V}(t_1)|$ and (d) $|\mathbf{A}(t_1)|$. Draw a sketch of a portion of the path of the particle containing the position of the particle at $t = t_1$, and draw the representations of $\mathbf{V}(t_1)$ and $\mathbf{A}(t_1)$ having initial position where $t = t_1$.

$$\mathbf{R}(t) = (t^2 + 3t)\mathbf{i} + (1 - 3t^2)\mathbf{j} \qquad t_1 = \frac{1}{2}$$

SOLUTION:

$$\mathbf{V}(t) = (2t + 3)\mathbf{i} - 6t\,\mathbf{j}$$
$$\mathbf{A}(t) = 2\mathbf{i} - 6\mathbf{j}$$

Thus,

(a) $\mathbf{V}\left(\dfrac{1}{2}\right) = 4\mathbf{i} - 3\mathbf{j}$

(b) $\mathbf{A}\left(\dfrac{1}{2}\right) = 2\mathbf{i} - 6\mathbf{j}$

(c) $\left|\mathbf{V}\left(\dfrac{1}{2}\right)\right| = \sqrt{4^2 + (-3)^2} = 5$

(d) $\left|\mathbf{A}\left(\dfrac{1}{2}\right)\right| = \sqrt{2^2 + (-6)^2} = 2\sqrt{10}$

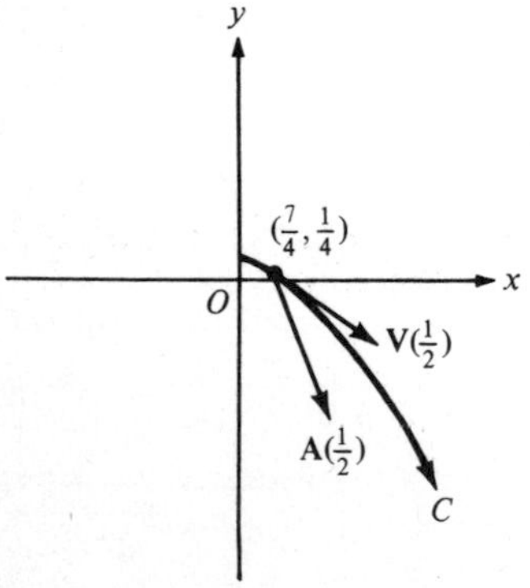

Figure 17.7.8

In Fig. 17.7.8 we show the required sketch.

14. Find the position vector $\mathbf{R}(t)$ if the acceleration vector

$$\mathbf{A}(t) = 2\cos 2t\,\mathbf{i} + 2\sin 2t\,\mathbf{j} \qquad \mathbf{V}(0) = \mathbf{i} + \mathbf{j} \qquad \mathbf{R}(0) = \frac{1}{2}\mathbf{i} - \frac{1}{2}\mathbf{j}$$

SOLUTION: Because $\mathbf{A}(t) = \mathbf{V}'(t)$, we have

$$\mathbf{V}'(t) = 2\cos 2t\,\mathbf{i} + 2\sin 2t\,\mathbf{j}$$

Thus,

$$\mathbf{V}(t) = \int (2 \cos 2t\, \mathbf{i} + 2 \sin 2t\, \mathbf{j})\, dt$$

$$= \sin 2t\, \mathbf{i} - \cos 2t\, \mathbf{j} + \mathbf{C}_1 \tag{1}$$

Because $\mathbf{V}(0) = \mathbf{i} + \mathbf{j}$, we replace $t$ by 0 in (1) and obtain

$$\mathbf{i} + \mathbf{j} = -\mathbf{j} + \mathbf{C}_1$$
$$\mathbf{C}_1 = \mathbf{i} + 2\mathbf{j}$$

Substituting for $\mathbf{C}_1$ in (1), we get

$$\mathbf{V}(t) = (1 + \sin 2t)\mathbf{i} + (2 - \cos 2t)\mathbf{j}$$

Because $\mathbf{V}(t) = \mathbf{R}'(t)$, we have

$$\mathbf{R}(t) = \int [(1 + \sin 2t)\mathbf{i} + (2 - \cos 2t)\mathbf{j}]\, dt$$

$$= \left(t - \frac{1}{2} \cos 2t\right)\mathbf{i} + \left(2t - \frac{1}{2} \sin 2t\right)\mathbf{j} + \mathbf{C}_2 \tag{2}$$

Because $\mathbf{R}(0) = \frac{1}{2}\mathbf{i} - \frac{1}{2}\mathbf{j}$. We let $t = 0$ in (2), obtaining

$$\frac{1}{2}\mathbf{i} - \frac{1}{2}\mathbf{j} = -\frac{1}{2}\mathbf{i} + \mathbf{C}_2$$

$$\mathbf{C}_2 = \mathbf{i} - \frac{1}{2}\mathbf{j}$$

Substituting for $\mathbf{C}_2$ in (2), we get

$$\mathbf{R}(t) = \left(1 + t - \frac{1}{2} \cos 2t\right)\mathbf{i} - \left(\frac{1}{2} - 2t + \frac{1}{2} \sin 2t\right)\mathbf{j}$$

**16.** A projectile is shot from a gun at an angle of elevation of 60°. The muzzle speed is 160 ft/sec. Find: **(a)** the position vector of the projectile at $t$ sec, **(b)** the time of flight, **(c)** the range, **(d)** the maximum height reached, **(e)** the velocity at impact, and **(f)** the speed at 4 sec.

SOLUTION: The position vector is given by

$$\mathbf{R}(t) = tv_0 \cos \alpha\, \mathbf{i} + \left(tv_0 \sin \alpha - \frac{1}{2} g t^2\right)\mathbf{j} \tag{1}$$

where $v_0$ ft/sec is the muzzle speed, $\alpha$ radians is the angle of elevation of the gun, and $g$ ft/sec$^2$ is the gravitational acceleration. Thus, we are given that

$$v_0 = 160 \qquad \alpha = \frac{1}{3}\pi \qquad g = 32$$

Substituting in (1), we have:

**(a)** $\mathbf{R}(t) = 80t\, \mathbf{i} + (80\sqrt{3}\, t - 16t^2)\mathbf{j}$ (2)

Thus, at any time $t$ sec, the projectile is at the point $(x, y)$ where

$$x = 80t \tag{3}$$
$$y = 80\sqrt{3}\, t - 16t^2 \tag{4}$$

**(b)** Because $y = 0$ at the end of the flight, we let $y = 0$ in Equation (4) and solve for $t$. The result is

$$t = 0 \quad \text{and} \quad t = 5\sqrt{3} \approx 8.66$$

Thus, the time of flight is approximately 8.66 seconds.

**(c)** If $t = 5\sqrt{3}$ in Equation (3), we get $x = 400\sqrt{3} \approx 693$. Thus, the range is approximately 693 feet.

**(d)** At the point of maximum height $dy/dt = 0$. From (4) we have

$$\frac{dy}{dt} = 80\sqrt{3} - 32t$$

Setting $dy/dt = 0$ and solving for $t$, we get $t = \frac{5}{2}\sqrt{3}$. With this replacement for $t$ in Equation (4), we obtain

$$\begin{aligned} y &= (80\sqrt{3})\left(\frac{5}{2}\sqrt{3}\right) - 16\left(\frac{5}{2}\sqrt{3}\right)^2 \\ &= 300 \end{aligned}$$

Thus, the maximum height reached is 300 feet.

**(e)** Differentiating on both sides of (2), we obtain

$$\mathbf{V}(t) = 80\,\mathbf{i} + (80\sqrt{3} - 32t)\mathbf{j} \tag{5}$$

Because the time of flight is $5\sqrt{3}$ sec, to find the velocity at impact we let $t = 5\sqrt{3}$ in Equation (5). Thus,

$$\mathbf{V}(5\sqrt{3}) = 80\,\mathbf{i} - 80\sqrt{3}\,\mathbf{j}$$

The velocity at impact is given by $80\,\mathbf{i} - 80\sqrt{3}\,\mathbf{j}$.

**(f)** From (5) we obtain

$$\begin{aligned} \mathbf{V}(4) &= 80\,\mathbf{i} + (80\sqrt{3} - 128)\mathbf{j} \\ |\mathbf{V}(4)| &= \sqrt{(80)^2 + (80\sqrt{3} - 128)^2} \\ &\approx 80.7 \end{aligned}$$

Thus, the speed at 4 sec is approximately 80.7 ft/sec.

**20.** The muzzle speed of a gun is 160 ft/sec. At what angle of elevation should the gun be fired so that a projectile will hit an object on the same level as the gun at a distance of 400 ft from it?

SOLUTION: We use Equation (1) of Exercise 16 with $v_0 = 160$ and $g = 32$. Thus, the position vector of the projectile is

$$\mathbf{R}(t) = 160t \cos\alpha\,\mathbf{i} + (160t \sin\alpha - 16t^2)\mathbf{j}$$

where $\alpha$ is the measure of the angle of elevation. At time $t$ sec, the projectile is at the point $(x, y)$ where

$$x = 160t\cos\alpha \quad \text{and} \quad y = 160t\sin\alpha - 16t^2 \tag{1}$$

Because we want to hit an object on the same level as the gun and at a distance of 400 ft from the gun and the gun is at the origin, we let $x = 400$ and $y = 0$ in (1). Thus, we have the system of equations

$$400 = 160t\cos\alpha \tag{2}$$

$$0 = 160t\sin\alpha - 16t^2 \tag{3}$$

We solve (3) for $t$. Thus,

$$16t(t - 10\sin\alpha) = 0$$

$$t = 0 \quad \text{and} \quad t = 10\sin\alpha$$

Substituting $t = 10\sin\alpha$ in (2), we obtain

$$400 = 1600\sin\alpha\cos\alpha$$

$$\frac{1}{2} = 2\sin\alpha\cos\alpha$$

$$\frac{1}{2} = \sin 2\alpha$$

Thus,

$$2\alpha = \frac{1}{6}\pi \quad \text{or} \quad 2\alpha = \frac{5}{6}\pi$$

$$\alpha = \frac{1}{12}\pi \qquad \alpha = \frac{5}{12}\pi$$

Thus, there are two possible angles of elevation: $\frac{1}{12}\pi$ radians or, equivalently $15°$, and $\frac{5}{12}\pi$ radians or, equivalently $75°$.

## 17.8 THE UNIT TANGENT AND UNIT NORMAL VECTORS AND ARC LENGTH AS A PARAMETER

**17.8.1 Definition** If $\mathbf{R}(t)$ is the position vector of curve $C$ at a point $P$ on $C$, then the *unit tangent vector* of $C$ at $P$, denoted by $\mathbf{T}(t)$, is the unit vector in the direction of $D_t\mathbf{R}(t)$ if $D_t\mathbf{R}(t) \neq 0$. Thus,

$$\mathbf{T}(t) = \frac{\mathbf{R}'(t)}{|\mathbf{R}'(t)|} = \frac{\mathbf{V}(t)}{|\mathbf{V}(t)|}$$

**17.8.2 Definition** If $\mathbf{T}(t)$ is the unit tangent vector of curve $C$ at a point $P$ on $C$, then the *unit normal vector,* denoted by $\mathbf{N}(t)$, is the unit vector in the direction of $D_t\mathbf{T}(t)$. Thus,

$$\mathbf{N}(t) = \frac{\mathbf{T}'(t)}{|\mathbf{T}'(t)|}$$

If $F, f,$ and $g$ are real-valued functions and $\mathbf{R}$ is the vector-valued function defined by

$$\mathbf{R}(t) = F(t)[f(t)\mathbf{i} + g(t)\mathbf{j}]$$

then the magnitude of $\mathbf{R}(t)$ is given by

$$|\mathbf{R}(t)| = |F(t)|\sqrt{[f(t)]^2 + [g(t)]^2}$$

where $|F(t)|$ represents the absolute value of the real function value $F(t)$. Thus, we may factor out real-valued functions when finding the magnitude of a vector-valued function, provided we also take the absolute value of the real-valued factor. We illustrate this technique in Exercise 10.

**17.8.3 Theorem** If the vector equation of a curve $C$ is $\mathbf{R}(s) = f(s)\mathbf{i} + g(s)\mathbf{j}$, where $s$ units is the length of arc measured from a particular point $P_0$ on $C$ to the point $P$, then the unit tangent vector of $C$ at $P$ is given by

$$\mathbf{T}(s) = D_s\mathbf{R}(s)$$

### *Exercises 17.8*

6. For the given curve, find $\mathbf{T}(t)$ and $\mathbf{N}(t)$, and at $t = t_1$ draw a sketch of a portion of the curve and draw the representations of $\mathbf{T}(t_1)$ and $\mathbf{N}(t_1)$ having initial point at $t = t_1$.

$$x = t - \sin t \qquad y = 1 - \cos t \qquad t_1 = \pi$$

SOLUTION: The position vector is

$$\mathbf{R}(t) = (t - \sin t)\mathbf{i} + (1 - \cos t)\mathbf{j} \tag{1}$$

Thus,

$$\mathbf{R}'(t) = (1 - \cos t)\mathbf{i} + \sin t\,\mathbf{j}$$
$$|\mathbf{R}'(t)| = \sqrt{(1-\cos t)^2 + \sin^2 t}$$
$$= \sqrt{2(1-\cos t)}$$

Therefore,

$$\mathbf{T}(t) = \frac{\mathbf{R}'(t)}{|\mathbf{R}'(t)|}$$
$$= \frac{1-\cos t}{\sqrt{2(1-\cos t)}}\mathbf{i} + \frac{\sin t}{\sqrt{2(1-\cos t)}}\mathbf{j} \qquad \textbf{(2)}$$

Because $\sin t = 2\sin\frac{1}{2}t\cos\frac{1}{2}t$ and $\sqrt{1-\cos t} = \sqrt{2}\sin\frac{1}{2}t$ if $0 < t < 2\pi$, we simplify equation (2) as follows for $0 < t < 2\pi$.

$$\mathbf{T}(t) = \frac{\sqrt{1-\cos t}}{\sqrt{2}}\mathbf{i} + \frac{2\sin\frac{1}{2}t\cos\frac{1}{2}t}{2\sin\frac{1}{2}t}\mathbf{j}$$
$$= \sin\frac{1}{2}t\,\mathbf{i} + \cos\frac{1}{2}t\,\mathbf{j} \qquad \textbf{(3)}$$

Hence,

$$\mathbf{T}'(t) = \frac{1}{2}\cos\frac{1}{2}t\,\mathbf{i} - \frac{1}{2}\sin\frac{1}{2}t\,\mathbf{j}$$
$$|\mathbf{T}'(t)| = \sqrt{\left(\frac{1}{2}\cos\frac{1}{2}t\right)^2 + \left(-\frac{1}{2}\sin\frac{1}{2}t\right)^2}$$
$$= \frac{1}{2}$$

Therefore,

$$\mathbf{N}(t) = \frac{\mathbf{T}'(t)}{|\mathbf{T}'(t)|}$$
$$= \frac{\frac{1}{2}\cos\frac{1}{2}t}{\frac{1}{2}}\mathbf{i} - \frac{\frac{1}{2}\sin\frac{1}{2}t}{\frac{1}{2}}\mathbf{j}$$
$$= \cos\frac{1}{2}t\,\mathbf{i} - \sin\frac{1}{2}t\,\mathbf{j} \qquad \textbf{(4)}$$

Substituting in (1), we obtain

$$\mathbf{R}(\pi) = (\pi - \sin\pi)\mathbf{i} + (1 - \cos\pi)\mathbf{j}$$
$$= \pi\mathbf{i} + 2\mathbf{j}$$

Substituting in (3), we obtain

$$\mathbf{T}(\pi) = \sin\frac{1}{2}\pi\mathbf{i} + \cos\frac{1}{2}\pi\mathbf{j}$$
$$= \mathbf{i}$$

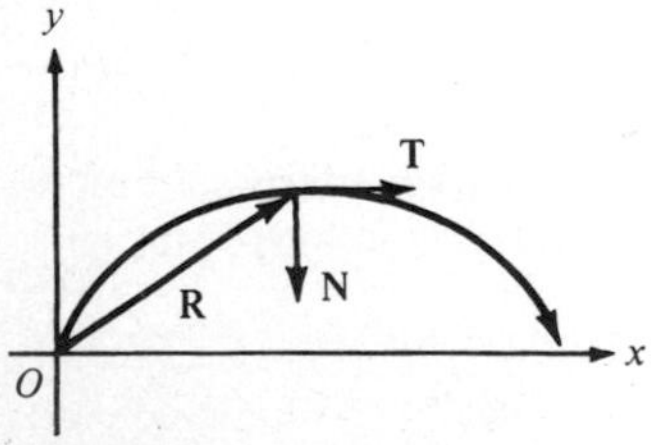

Figure 17.8.6

Substituting in (4), we obtain

$$\mathbf{N}(\pi) = \cos\frac{1}{2}\pi\mathbf{i} - \sin\frac{1}{2}\pi\mathbf{j}$$
$$= -\mathbf{j}$$

In Fig. 17.8.6 we show a portion of the curve with representations of $\mathbf{T}(\pi)$ and $\mathbf{N}(\pi)$ with initial point at the tip of the radius vector $\mathbf{R}(\pi)$.

**10.** If the vector equation of curve $C$ is $\mathbf{R}(t) = (4 - 3t^2)\mathbf{i} + (t^3 - 3t)\mathbf{j}$, find the radian measure of the angle between the vectors $\mathbf{N}(1)$ and $D_t^2\mathbf{R}(1)$.

SOLUTION: Differentiating on both sides of the equation for $\mathbf{R}(t)$, we obtain

$$\mathbf{R}'(t) = -6t\,\mathbf{i} + (3t^2 - 3)\mathbf{j} \tag{1}$$

$$\mathbf{R}''(t) = -6\,\mathbf{i} + 6t\,\mathbf{j}$$

Thus,

$$D_t^2\mathbf{R}(1) = \mathbf{R}''(1) = -6\,\mathbf{i} + 6\,\mathbf{j} \tag{2}$$

From (1), we have

$$\begin{aligned} |\mathbf{R}'(t)| &= 3\sqrt{(-2t)^2 + (t^2 - 1)^2} \\ &= 3\sqrt{(t^2 + 1)^2} \\ &= 3(t^2 + 1) \end{aligned}$$

Thus,

$$\begin{aligned} \mathbf{T}(t) &= \frac{\mathbf{R}'(t)}{|\mathbf{R}'(t)|} \\ &= \frac{-2t}{t^2 + 1}\mathbf{i} + \frac{t^2 - 1}{t^2 + 1}\mathbf{j} \end{aligned}$$

Differentiating, we obtain

$$\begin{aligned} \mathbf{T}'(t) &= \frac{2(t^2 - 1)}{(t^2 + 1)^2}\mathbf{i} + \frac{4t}{(t^2 + 1)^2}\mathbf{j} \\ &= \frac{2}{(t^2 + 1)^2}[(t^2 - 1)\mathbf{i} + 2t\,\mathbf{j}] \end{aligned}$$

Thus,

$$\begin{aligned} |\mathbf{T}'(t)| &= \frac{2}{(t^2 + 1)^2}\sqrt{(t^2 - 1)^2 + (2t)^2} \\ &= \frac{2\sqrt{(t^2 + 1)^2}}{(t^2 + 1)^2} \\ &= \frac{2}{t^2 + 1} \end{aligned}$$

Therefore,

$$\begin{aligned} \mathbf{N}(t) &= \frac{\mathbf{T}'(t)}{|\mathbf{T}'(t)|} \\ &= \frac{t^2 - 1}{t^2 + 1}\mathbf{i} + \frac{2t}{t^2 + 1}\mathbf{j} \end{aligned}$$

Hence,

$$\mathbf{N}(1) = \mathbf{j} \tag{3}$$

To find $\alpha$, the measure of the angle between $\mathbf{N}(1)$ and $D_t^2\mathbf{R}(1)$, we use Theorem 17.3.5. Thus, substituting from (2) and (3), we have

$$\cos\alpha = \frac{\mathbf{N}(1) \cdot D_t^2\mathbf{R}(1)}{|\mathbf{N}(1)||D_t^2(\mathbf{R}(1))|}$$

$$= \frac{\mathbf{j} \cdot (-6\mathbf{i} + 6\mathbf{j})}{\sqrt{1} \cdot \sqrt{(-6)^2 + 6^2}}$$

$$= \frac{6}{6\sqrt{2}}$$

$$= \frac{1}{2}\sqrt{2}$$

Therefore, $\alpha = \frac{1}{4}\pi$.

**14.** Find parametric equations of the curve having arc length $s$ as a parameter, where $s$ is measured from the point where $t = 0$. Check your result by using Equation (10).

$$x = 2(\cos t + t \sin t) \qquad y = 2(\sin t - t \cos t)$$

SOLUTION: We have $s = 0$ when $t = 0$, and for all $t > 0$

$$\frac{ds}{dt} = |\mathbf{V}(t)|$$

The position vector is

$$\mathbf{R}(t) = 2(\cos t + t \sin t)\mathbf{i} + 2(\sin t - t \cos t)\mathbf{j}$$

Differentiating with respect to $t$, we obtain

$$\mathbf{V}(t) = 2t \cos t\, \mathbf{i} + 2t \sin t\, \mathbf{j}$$
$$|\mathbf{V}(t)| = 2t$$

Thus,

$$\frac{ds}{dt} = 2t$$

$$s = t^2 + C$$

Because $s = 0$ when $t = 0$, we have $C = 0$. Thus,

$$s = t^2$$
$$t = \sqrt{s}$$

Substituting $t = \sqrt{s}$ into the original equation, we obtain

$$x = 2(\cos \sqrt{s} + \sqrt{s} \sin \sqrt{s}) \quad \text{and} \quad y = 2(\sin \sqrt{s} - \sqrt{s} \cos \sqrt{s}) \tag{1}$$

which are the required parametric equations.

Equation (10) from the text is:

$$(D_s x)^2 + (D_s y)^2 = 1 \tag{2}$$

Differentiating with respect to $s$ on both sides of Equations (1), we obtain

$$D_s x = 2\left[-\sin \sqrt{s}\left(\frac{1}{2}s^{-1/2}\right) + s^{1/2} \cos \sqrt{s}\left(\frac{1}{2}s^{-1/2}\right) + \sin \sqrt{s}\left(\frac{1}{2}s^{-1/2}\right)\right]$$
$$= \cos \sqrt{s}$$

and

$$D_s y = 2\left[\cos \sqrt{s}\left(\frac{1}{2}s^{-1/2}\right) + s^{1/2} \sin \sqrt{s}\left(\frac{1}{2}s^{-1/2}\right) - \cos \sqrt{s}\left(\frac{1}{2}s^{-1/2}\right)\right]$$
$$= \sin \sqrt{s}$$

Thus,

$$\begin{aligned}(D_s x)^2 + (D_s y)^2 &= \cos^2\sqrt{s} + \sin^2\sqrt{s}\\ &= 1\end{aligned}$$

which agrees with Equation (2).

**16.** Given the cycloid $x = 2(t - \sin t)$, $y = 2(1 - \cos t)$, express the arc length $s$ as a function of $t$, where $s$ is measured from the point where $t = 0$.

SOLUTION: We have

$$\mathbf{R}(t) = 2(t - \sin t)\mathbf{i} + 2(1 - \cos t)\mathbf{j}$$

Differentiating with respect to $t$, we get

$$\begin{aligned}\mathbf{V}(t) &- 2(1 - \cos t)\mathbf{i} + 2 \sin t\,\mathbf{j}\\ |\mathbf{V}(t)| &= 2\sqrt{(1 - \cos t)^2 + \sin^2 t}\\ &= 2\sqrt{2(1 - \cos t)}\\ &= 2\sqrt{4 \sin^2 \frac{1}{2}t}\\ &= 4 \sin \frac{1}{2}t \qquad \text{if } 0 \leqslant t \leqslant 2\pi\end{aligned}$$

Thus, if $0 \leqslant t \leqslant 2\pi$, we have

$$\begin{aligned}\frac{ds}{dt} &= |\mathbf{V}(t)|\\ &= 4 \sin \frac{1}{2}t\end{aligned}$$

$$ds = 4 \sin \frac{1}{2}t\,dt$$

$$\int ds = \int 4 \sin \frac{1}{2}t\,dt$$

$$s = -8 \cos \frac{1}{2}t + C$$

Because $s = 0$ when $t = 0$, we have $C = 8$. Therefore,

$$s = -8 \cos \frac{1}{2}t + 8 \qquad \text{if } 0 \leqslant t \leqslant 2\pi$$

## 17.9 CURVATURE

**17.9.1 Definition** If $\mathbf{T}(t)$ is the unit tangent vector to a curve $C$ at a point $P$, $s$ is the arc length measured from an arbitrarily chosen point on $C$ to $P$, and $s$ increases as $t$ increases, then the *curvature vector* of $C$ at $P$, denoted by $\mathbf{K}(t)$, is given by

$$\mathbf{K}(t) = D_s\mathbf{T}(t)$$

The *curvature* of $C$ at $P$, denoted by $K(t)$, is the magnitude of the curvature vector. Following is a formula for the curvature vector.

$$\mathbf{K}(t) = \frac{\mathbf{T}'(t)}{|\mathbf{R}'(t)|}$$

Thus, a formula for the curvature, which is a scalar, is

$$K(t) = \left|\frac{\mathbf{T}'(t)}{|\mathbf{R}'(t)|}\right| \tag{9}$$

We do not often use the above formulas to find the curvature. When $x$ and $y$ are expressed in terms of a parameter $t$, the following formula may be used to find the curvature.

$$K(t) = \frac{\left|\left(\frac{dx}{dt}\right)\left(\frac{d^2y}{dt^2}\right) - \left(\frac{dy}{dt}\right)\left(\frac{d^2x}{dt^2}\right)\right|}{\left[\left(\frac{dx}{dt}\right)^2 + \left(\frac{dy}{dt}\right)^2\right]^{3/2}} \tag{14}$$

If $y = f(x)$, we use the following formula to find the curvature.

$$K(x) = \frac{\left|\frac{d^2y}{dx^2}\right|}{\left[1 + \left(\frac{dy}{dx}\right)^2\right]^{3/2}}$$

And if $x = g(y)$, then the curvature may be found by the following formula.

$$K(y) = \frac{\left|\frac{d^2x}{dy^2}\right|}{\left[1 + \left(\frac{dx}{dy}\right)^2\right]^{3/2}}$$

**17.9.2 Definition** If $K(t)$ is the curvature of a curve $C$ at point $P$ and $K(t) \neq 0$, then the *radius of curvature* of $C$ at $P$, denoted by $\rho(t)$, is defined by

$$\rho(t) = \frac{1}{K(t)}$$

## *Exercises 17.9*

4. Find the curvature $K$ and the radius of curvature $\rho$ at the point where $t = t_1$. Use formula (9) to find $K$. Draw a sketch showing a portion of the curve, the unit tangent vector, and the circle of curvature at $t = t_1$.

$$\mathbf{R}(t) = \sin t\,\mathbf{i} + \sin 2t\,\mathbf{j} \qquad t_1 = \frac{1}{2}\pi$$

SOLUTION: We use the formula

$$K(t) = \left|\frac{\mathbf{T}'(t)}{|\mathbf{R}'(t)|}\right| \tag{1}$$

Thus,

$$\mathbf{R}'(t) = \cos t\,\mathbf{i} + 2\cos 2t\,\mathbf{j}$$
$$|\mathbf{R}'(t)| = \sqrt{\cos^2 t + 4\cos^2 2t} \tag{2}$$

Hence,

$$\mathbf{T}(t) = \frac{\mathbf{R}'(t)}{|\mathbf{R}'(t)|}$$
$$= \frac{\cos t}{\sqrt{\cos^2 t + 4\cos^2 2t}}\mathbf{i} + \frac{2\cos 2t}{\sqrt{\cos^2 t + 4\cos^2 2t}}\mathbf{j} \tag{3}$$

To find $\mathbf{T}'(t)$, we first express $T(t)$ as a product and then apply Theorem 17.5.8. Thus,

$$\mathbf{T}(t) = [\cos^2 t + 4\cos^2 2t]^{-1/2}[\cos t\,\mathbf{i} + 2\cos 2t\,\mathbf{j}]$$

and

$$\mathbf{T}'(t) = -\frac{1}{2}[\cos^2 t + 4\cos^2 2t]^{-3/2}[-2\cos t \sin t - 16\cos 2t \sin 2t][\cos t\,\mathbf{i} + 2\cos 2t\,\mathbf{j}]$$
$$+ [\cos^2 t + 4\cos^2 t]^{-1/2}[-\sin t\,\mathbf{i} - 4\sin 2t\,\mathbf{j}]$$

If $t = \frac{1}{2}\pi$, then $\cos t = 0$, $\cos 2t = -1$, $\sin t = 1$, and $\sin 2t = 0$. Substituting $t = \frac{1}{2}\pi$ into the formula for $\mathbf{T}'(t)$, we get

$$\mathbf{T}'\left(\frac{1}{2}\pi\right) = -\frac{1}{2}\mathbf{i} \tag{4}$$

From (2) with $t = \frac{1}{2}\pi$, we have

$$\left|\mathbf{R}'\left(\frac{1}{2}\pi\right)\right| = 2 \tag{5}$$

With $t = \frac{1}{2}\pi$ in (1) and substituting the values from (4) and (5), we obtain

$$K\left(\frac{1}{2}\pi\right) = \left|\frac{\mathbf{T}'\left(\frac{1}{2}\pi\right)}{\left|\mathbf{R}'\left(\frac{1}{2}\pi\right)\right|}\right|$$
$$= \left|-\frac{1}{4}\mathbf{i}\right|$$
$$= \frac{1}{4}$$

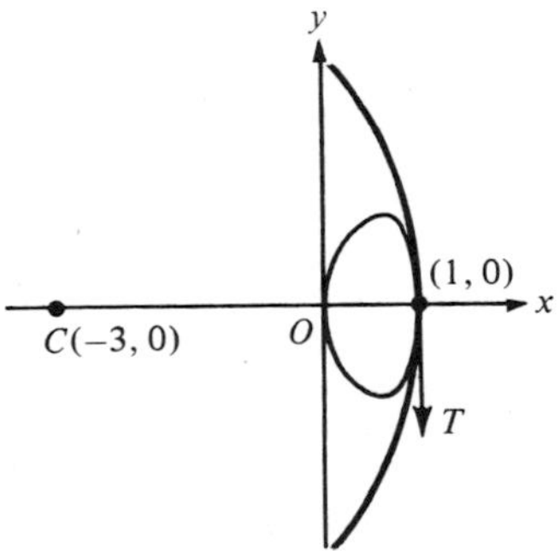

Figure 17.9.4

Thus, the curvature is $\frac{1}{4}$ at the point where $t = \frac{1}{2}\pi$. Because $\rho = 1/K = 4$, the radius of curvature is 4. In Fig. 17.9.4, we show a sketch of a portion of the curve containing the point where $t = \frac{1}{2}\pi$. The unit tangent vector is found from (3). Thus, $\mathbf{T}(\frac{1}{2}\pi) = -\mathbf{j}$. The circle of curvature at $t = \frac{1}{2}\pi$ has center on the line normal to the curve at $t = \frac{1}{2}\pi$; the curve has radius 4; and the circle is tangent to the curve. Because $\mathbf{R}(\frac{1}{2}\pi) = \mathbf{i}$, the point of tangency is $(1, 0)$. Thus, the center of the circle is at $(-3, 0)$.

6. Find the curvature $K$ by using formula (14). Then find $K$ and $\rho$ at the point where $t = t_1$ and draw a sketch showing a portion of the curve, the unit tangent vector, and the circle of curvature at $t = t_1$.

$$x = e^t + e^{-t} \qquad y = e^t - e^{-t} \qquad t_1 = 0$$

SOLUTION: We use the formula

$$K(t) = \frac{\left|\left(\frac{dx}{dt}\right)\left(\frac{d^2y}{dt^2}\right) - \left(\frac{dy}{dt}\right)\left(\frac{d^2x}{dt^2}\right)\right|}{\left[\left(\frac{dx}{dt}\right)^2 + \left(\frac{dy}{dt}\right)^2\right]^{3/2}} \tag{1}$$

Because $\cosh t = \frac{1}{2}(e^t + e^{-t})$ and $\sinh t = \frac{1}{2}(e^t - e^{-t})$, we have

$$x = 2\cosh t \quad \text{and} \quad y = 2\sinh t \tag{2}$$

Thus,

$$\frac{dx}{dt} = 2\sinh t \quad \text{and} \quad \frac{dy}{dt} = 2\cosh t$$

$$\frac{d^2x}{dt^2} = 2\cosh t \quad \text{and} \quad \frac{d^2y}{dt^2} = 2\sinh t$$

Substituting in (1), we get

$$K(t) = \frac{|4\sinh^2 t - 4\cosh^2 t|}{[4\sinh^2 t + 4\cosh^2 t]^{3/2}}$$

$$= \frac{|-4|}{(4 \cosh 2t)^{3/2}}$$

$$= \frac{1}{2(\cosh 2t)^{3/2}}$$

Thus,

$$K(0) = \frac{1}{2(\cosh 0)^{3/2}}$$

$$= \frac{1}{2}$$

and when $t = 0$

$$\rho = \frac{1}{K(0)} = 2$$

Furthermore, when $t = 0$, we have $x = 2$ and $y = 0$. Thus, the radius of curvature is 2 at the point $(2, 0)$. To find the unit tangent vector at this point, we use Equations (2) to write the position vector

$$\mathbf{R}(t) = 2 \cosh t\, \mathbf{i} + 2 \sinh t\, \mathbf{j}$$

Thus,

$$\mathbf{R}'(t) = 2 \sinh t\, \mathbf{i} + 2 \cosh t\, \mathbf{j}$$

$$\mathbf{R}'(0) = 2\, \mathbf{j}$$

$$|\mathbf{R}'(0)| = 2$$

$$\mathbf{T}(0) = \frac{\mathbf{R}'(0)}{|\mathbf{R}'(0)|} = \mathbf{j}$$

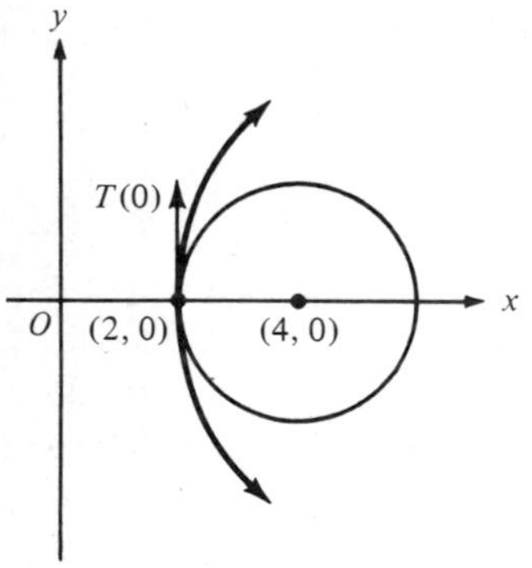

Figure 17.9.6

Thus, a representation of the unit tangent vector at the point $(2, 0)$ is a vertical line segment directed upward. In Fig. 17.9.6 we show a sketch of one branch of the curve, the unit tangent vector $\mathbf{T}(0)$, and the circle of curvature.

**10.** Find the curvature $K$ and the radius of curvature $\rho$ at the given point. Draw a sketch showing a portion of the curve, a piece of the tangent line, and the circle of curvature at the given point.

$$4x^2 + 9y^2 = 36; \qquad (0, 2)$$

SOLUTION: We use the curvature formula

$$K(x) = \frac{\left|\dfrac{d^2y}{dx^2}\right|}{\left[1 + \left(\dfrac{dy}{dx}\right)^2\right]^{3/2}} \tag{1}$$

Differentiating implicitly with respect to $x$ on both sides of the given equation, we have

$$8x + 18y\frac{dy}{dx} = 0$$

$$\frac{dy}{dx} = -\frac{4}{9}\left(\frac{x}{y}\right) \tag{2}$$

$$\left.\frac{dy}{dx}\right]_{\substack{x=0\\y=2}} = 0 \tag{3}$$

Differentiating implicitly with respect to $x$ on both sides of (2), we obtain

$$\frac{d^2y}{dx^2} = -\frac{4}{9}\left[\frac{y - x\dfrac{dy}{dx}}{y^2}\right] \tag{4}$$

From (3) we have $dy/dx = 0$, when $x = 0$ and $y = 2$. With these replacements in (4), we get

$$\left.\frac{d^2y}{dx^2}\right]_{\substack{x=0\\y=2}} = -\frac{2}{9} \tag{5}$$

Substituting from (3) and (5) into (1), we obtain

$$K = \frac{2}{9}$$

Thus,

$$\rho = \frac{9}{2}$$

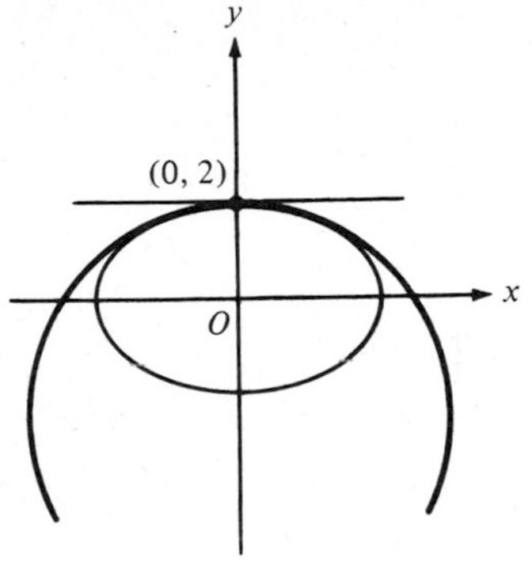

Figure 17.9.10

In Fig. 17.9.10 we show the ellipse $4x^2 + 9y^2 = 36$, the tangent line at the point $(0, 2)$, and the circle of curvature at the point $(0, 2)$.

**16.** Find the radius of curvature at any point on the given curve.

$\mathbf{R}(t) = e^t \sin t\,\mathbf{i} + e^t \cos t\,\mathbf{j}$

SOLUTION:

$$x = e^t \sin t \quad \text{and} \quad y = e^t \cos t$$

Then

$$\frac{dx}{dt} = e^t(\cos t + \sin t) \quad \text{and} \quad \frac{dy}{dt} = e^t(\cos t - \sin t)$$

$$\frac{d^2x}{dt^2} = 2e^t \cos t \quad \text{and} \quad \frac{d^2y}{dt^2} = -2e^t \sin t$$

Substituting into formula (14) for $K(t)$, we have

$$K(t) = \frac{|-2e^{2t} \sin t(\cos t + \sin t) - 2e^{2t} \cos t(\cos t - \sin t)|}{[e^{2t}(\cos t + \sin t)^2 + e^{2t}(\cos t - \sin t)^2]^{3/2}}$$

$$= \frac{2e^{2t}|\sin t \cos t + \sin^2 t + \cos^2 t - \sin t \cos t|}{e^{3t}[\cos^2 t + 2 \cos t \sin t + \sin^2 t + \cos^2 t - 2 \cos t \sin t + \sin^2 t]^{3/2}}$$

$$= \frac{2}{e^t(2)^{3/2}}$$

$$= \frac{1}{\sqrt{2}\,e^t}$$

Thus,

$$\rho(t) = \frac{1}{K(t)}$$

$$= \sqrt{2}\,e^t$$

**20.** Find a point on the given curve at which the curvature is an absolute maximum.

$y = e^x$

SOLUTION: We have

$$\frac{dy}{dx} = e^x \quad \text{and} \quad \frac{d^2y}{dx^2} = e^x$$

Thus,

$$K(x) = \frac{\left|\frac{d^2y}{dx^2}\right|}{\left[1 + \left(\frac{dy}{dx}\right)^2\right]^{3/2}}$$

$$= \frac{e^x}{(1 + e^{2x})^{3/2}}$$

$$= e^x(1 + e^{2x})^{-3/2}$$

Differentiating, we have

$$K'(x) = e^x\left(-\frac{3}{2}\right)(1 + e^{2x})^{-5/2}(2e^{2x}) + (1 + e^{2x})^{-3/2}e^x$$

$$= \frac{e^x(1 - 2e^{2x})}{(1 + e^{2x})^{5/2}} \qquad \textbf{(1)}$$

If $K'(x) = 0$, then from (1) we have

$$1 - 2e^{2x} = 0$$

$$e^{2x} = \frac{1}{2} \qquad \textbf{(2)}$$

$$2x = \ln\frac{1}{2}$$

$$x = -\frac{1}{2}\ln 2$$

Furthermore, if $x < -\frac{1}{2}\ln 2$, then by (2) we have $e^{2x} < \frac{1}{2}$. Thus, from (1) we have $K'(x) < 0$. And if $x > -\frac{1}{2}\ln 2$, then $K'(x) > 0$. Therefore, by the first-derivative test, $K$ has a relative maximum value at the critical number $-\frac{1}{2}\ln 2$. Because this is the only critical number, then $K$ has an absolute maximum value at $x = -\frac{1}{2}\ln 2$. If $x = -\frac{1}{2}\ln 2$, then $y = \frac{1}{2}\sqrt{2}$ because $y = e^x$, and from (2) we have $e^{2x} = \frac{1}{2}$. Thus, the point on the curve at which the curvature is an absolute maximum is the point $(-\frac{1}{2}\ln 2, \frac{1}{2}\sqrt{2})$.

**26.** Find the curvature $K$ and the radius of curvature $\rho$ at the indicated point. Use the formula of Exercise 24 to find $K$.

$$r = 1 - \sin\theta \qquad \theta = 0$$

SOLUTION: The formula of Exercise 24 is:

$$K = \frac{\left|r^2 + 2\left(\frac{dr}{d\theta}\right)^2 - r\left(\frac{d^2r}{d\theta^2}\right)\right|}{\left[r^2 + \left(\frac{dr}{d\theta}\right)^2\right]^{3/2}} \qquad \textbf{(1)}$$

We have

$$\frac{dr}{d\theta} = -\cos\theta \quad \text{and} \quad \frac{d^2r}{d\theta^2} = \sin\theta$$

When $\theta = 0$, we get

$$r = 1 \qquad \frac{dr}{d\theta} = -1 \qquad \frac{d^2r}{d\theta^2} = 0 \tag{2}$$

Substituting from (2) into (1), we obtain

$$K = \frac{|1^2 + 2(-1)^2 - 1(0)|}{[1^2 + (-1)^2]^{3/2}}$$

$$= \frac{3}{4}\sqrt{2}$$

Hence,

$$\rho = \frac{1}{K}$$

$$= \frac{2}{3}\sqrt{2}$$

32. Find the curvature $K$, the radius of curvature $\rho$, and the center of curvature at the given point. Draw a sketch of the curve and the circle of curvature.

$$y = \ln x; \qquad (1, 0)$$

SOLUTION: We have

$$\frac{dy}{dx} = \frac{1}{x} \quad \text{and} \quad \frac{d^2y}{dx^2} = -\frac{1}{x^2}$$

At the point $(1, 0)$, we get

$$\frac{dy}{dx} = 1 \quad \text{and} \quad \frac{d^2y}{dx^2} = -1 \tag{1}$$

Substituting from (1) into the curvature formula, we get

$$K = \frac{\left|\frac{d^2y}{dx^2}\right|}{\left[1 + \left(\frac{dy}{dx}\right)^2\right]^{3/2}}$$

$$= \frac{1}{2^{3/2}}$$

$$= \frac{1}{4}\sqrt{2}$$

Then,

$$\rho = \frac{1}{K}$$

$$= 2\sqrt{2}$$

To find the center of curvature at point $(1, 0)$ we use formulas from Exercise 29, and substitute the values from Equation (1). Thus,

$$x_C = x - \frac{\frac{dy}{dx}\left[1 + \left(\frac{dy}{dx}\right)^2\right]}{\frac{d^2y}{dx^2}}$$

$$= 1 - \frac{1[1 + 1^2]}{-1}$$

$$= 3$$

and

$$y_c = y + \frac{\left(\frac{dy}{dx}\right)^2 + 1}{\frac{d^2y}{dx^2}}$$

$$= 0 + \frac{1^2 + 1}{-1}$$

$$= -2$$

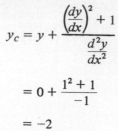

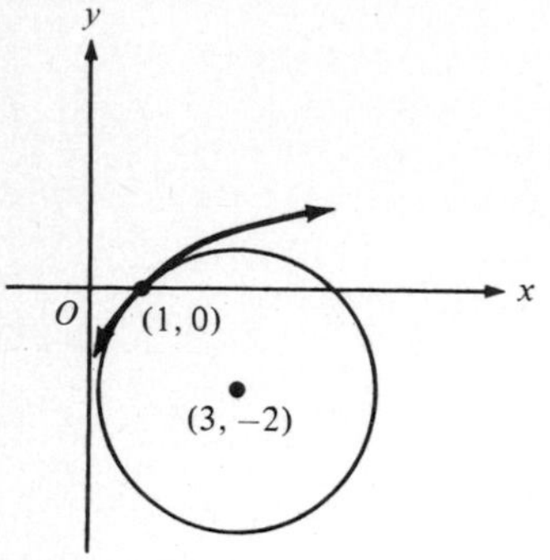

Figure 17.9.32

The center of curvature is $(3, -2)$. In Fig. 17.9.32, we show a sketch of the curve $y = \ln x$ and the circle of curvature at the point $(1, 0)$.

**36.** Find the coordinates of the center of curvature at any point.

$\mathbf{R}(t) = a \cos^3 t\,\mathbf{i} + a \sin^3 t\,\mathbf{j}$

SOLUTION: We use the formulas of Exercise 29. Let

$$x = a \cos^3 t \quad \text{and} \quad y = a \sin^3 t \tag{1}$$

Then

$$\frac{dx}{dt} = -3a \cos^2 t \sin t \quad \text{and} \quad \frac{dy}{dt} = 3a \sin^2 t \cos t \tag{2}$$

Thus,

$$\frac{dy}{dx} = \frac{\frac{dy}{dt}}{\frac{dx}{dt}}$$

$$= \frac{3a \sin^2 t \cos t}{-3a \cos^2 t \sin t}$$

$$= -\tan t \tag{3}$$

Let $dy/dx = y'$. Then

$$\frac{d^2y}{dx^2} = \frac{d(y')}{dx} = \frac{\frac{d(y')}{dt}}{\frac{dx}{dt}} \tag{4}$$

Differentiating on both sides of (3) with respect to $t$, we have

$$\frac{d(y')}{dt} = -\sec^2 t \tag{5}$$

Substituting from (2) and (5) into (4), we obtain

$$\frac{d^2y}{dx^2} = \frac{\sec^2 t}{3a \cos^2 t \sin t} \tag{6}$$

Substituting from (1), (3), and (6) into the formulas of Exercise 29, we obtain the coordinates.

$$x_c = x - \frac{\frac{dy}{dx}\left[1 + \left(\frac{dy}{dx}\right)^2\right]}{\frac{d^2y}{dx^2}}$$

$$= a\cos^3 t - \frac{\dfrac{-\tan t[1+\tan^2 t]}{\sec^2 t}}{3a\cos^2 t\sin t}$$

$$= a\cos^3 t + 3a\sin^2 t\cos t$$
$$= a\cos t(\cos^2 t + 3\sin^2 t)$$

and

$$y_c = y + \frac{\left(\dfrac{dy}{dx}\right)^2 + 1}{\dfrac{d^2y}{dx^2}}$$

$$= a\sin^3 t + \frac{\dfrac{\tan^2 t + 1}{\sec^2 t}}{3a\cos^2 t\sin t}$$

$$= a\sin^3 t + 3a\cos^2 t\sin t$$
$$= a\sin t(\sin^2 t + 3\cos^2 t)$$

## 17.10 TANGENTIAL AND NORMAL COMPONENTS OF ACCELERATION

The acceleration vector may be expressed as the sum of a scalar times the unit tangent vector and a scalar times the unit normal vector as follows.

$$\mathbf{A}(t) = A_T(t)\mathbf{T}(t) + A_N(t)\mathbf{N}(t) \qquad \text{(I)}$$

where

$$A_T(t) = \frac{d^2s}{dt^2} \qquad \text{(II)}$$

and

$$A_N(t) = K(t)\left(\frac{ds}{dt}\right)^2 \qquad \text{(III)}$$

The scalars $A_T$ and $A_N$ are called the tangential and normal components, respectively, of the acceleration vector. Because $A_T$ is the scalar projection of the acceleration vector onto the unit tangent vector and $A_N$ is the scalar projection of the acceleration vector onto the unit normal vector, we also have

$$A_T = \mathbf{A}\cdot\mathbf{T} \quad \text{and} \quad A_N = \mathbf{A}\cdot\mathbf{N}$$

It is no longer necessary to differentiate the unit vector $\mathbf{T}$ to find the unit vector $\mathbf{N}$. From (III) we see that $A_N \geqslant 0$. Because the unit vectors $\mathbf{T}$ and $\mathbf{N}$ are orthogonal, then $\mathbf{T}\cdot\mathbf{N} = 0$. Thus, if

$$\mathbf{T} = a\mathbf{i} + b\mathbf{j}$$

then we take either

$$\mathbf{N} = -b\mathbf{i} + a\mathbf{j}$$

if $\mathbf{A}\cdot\mathbf{N} \geqslant 0$, or we take

$$\overline{\mathbf{N}} = b\mathbf{i} - a\mathbf{j}$$

if $\mathbf{A}\cdot\overline{\mathbf{N}} \geqslant 0$. We illustrate the technique in the alternate solution to Exercise 2 and in Exercise 8.

## *Exercises 17.10*

In Exercises 1-6, a particle is moving along the curve having the given vector equation. In each problem find the vectors $\mathbf{V}(t)$, $\mathbf{A}(t)$, $\mathbf{T}(t)$, and $\mathbf{N}(t)$, and the following scalars for an arbitrary value of $t$: $|\mathbf{V}(t)|$, $A_T$, $A_N$, $K(t)$. Also find the particular values when

$t = t_1$. At $t = t_1$, draw a sketch of a portion of the curve and representations of the vectors $\mathbf{V}(t_1)$, $\mathbf{A}(t_1)$, $A_T\mathbf{T}(t_1)$, and $A_N\mathbf{N}(t_1)$.

2. $\mathbf{R}(t) = (t-1)\mathbf{i} + t^2\mathbf{j}$;  $t_1 = 1$

SOLUTION:

$$\mathbf{V}(t) = \mathbf{R}'(t) = \mathbf{i} + 2t\mathbf{j}$$

$$\mathbf{A}(t) = \mathbf{V}'(t) = 2\mathbf{j} \tag{1}$$

$$\mathbf{T}(t) = \frac{\mathbf{V}(t)}{|\mathbf{V}(t)|}$$

$$= \frac{1}{\sqrt{1+4t^2}}[\mathbf{i} + 2t\mathbf{j}] \tag{2}$$

Because

$$\frac{ds}{dt} = |\mathbf{V}(t)| = \sqrt{1+4t^2} \tag{3}$$

then

$$\frac{d^2s}{dt^2} = \frac{4t}{\sqrt{1+4t^2}}$$

Hence, from (II) we have

$$A_T(t) = D_t^2 s = \frac{4t}{\sqrt{1+4t^2}} \tag{4}$$

We have from (I)

$$\mathbf{A}(t) = A_T(t)\mathbf{T}(t) + A_N(t)\mathbf{N}(t) \tag{5}$$

Substituting from (1), (4), and (3) into (5), we have

$$2\mathbf{j} = \frac{4t}{\sqrt{1+4t^2}}\,\frac{1}{\sqrt{1+4t^2}}[\mathbf{i} + 2t\mathbf{j}] + A_N(t)\mathbf{N}(t)$$

Solving for $\mathbf{N}(t)$, we have

$$\mathbf{N}(t) = \frac{1}{A_N(t)}\left[\frac{-4t}{1+4t^2}\mathbf{i} + \left(2 - \frac{8t^2}{1+4t^2}\right)\mathbf{j}\right]$$

$$= \frac{2}{A_N(t)(1+4t^2)}[-2t\mathbf{i} + \mathbf{j}] \tag{6}$$

Because $\mathbf{N}(t)$ is a unit vector, then $|\mathbf{N}(t)| = 1$. By taking the magnitude of each vector in (6), we obtain

$$1 = \frac{2}{A_N(t)(1+4t^2)}\sqrt{4t^2+1}$$

Solving for $A_N(t)$, we have

$$A_N(t) = \frac{2}{\sqrt{1+4t^2}} \tag{7}$$

Substituting from (7) into (6), we obtain

$$\mathbf{N}(t) = \frac{1}{\sqrt{1+4t^2}}[-2t\mathbf{i} + \mathbf{j}] \tag{8}$$

From (III), we have

$$A_N(t) = K(t)(D_t s)^2 \tag{9}$$

Substituting from (7) and (3) into (9), we obtain

$$\frac{2}{\sqrt{1+4t^2}} = K(t)(\sqrt{1+4t^2})^2$$

Thus,

$$K(t) = \frac{2}{(1+4t^2)^{3/2}} \tag{10}$$

With $t = 1$ in (3), (4), (7), and (10), we have the particular values of the scalars. Thus,

$$|\mathbf{V}(1)| = \sqrt{5} \qquad A_T(1) = \frac{4}{5}\sqrt{5} \qquad A_N(1) = \frac{2}{5}\sqrt{5} \qquad K(1) = \frac{2}{25}\sqrt{5}$$

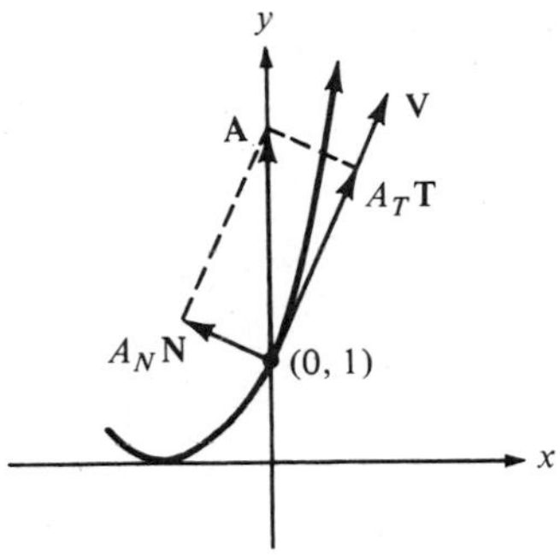

Figure 17.10.2

The particular values of the vectors required are found by letting $t = 1$ in the appropriate formula. Thus,

$$\mathbf{V}(1) = \mathbf{i} + 2\mathbf{j} \qquad \mathbf{A}(1) = 2\mathbf{j}$$

$$A_T\mathbf{T}(1) = \frac{4}{5}(\mathbf{i} + 2\mathbf{j}) \qquad A_N\mathbf{N}(1) = \frac{2}{5}(-2\mathbf{i} + \mathbf{j})$$

In Fig. 17.10.2 we show a sketch of the curve and representations of the vectors, which have initial point (0, 1), the terminal point of $\mathbf{R}(1) = \mathbf{j}$.

ALTERNATE SOLUTION: We find $\mathbf{V}(t)$, $|\mathbf{V}(t)|$, $\mathbf{T}(t)$, and $\mathbf{A}(t)$ as in the solution above. Then we proceed as follows.

$$\begin{aligned} A_T(t) &= \mathbf{A}(t) \cdot \mathbf{T}(t) \\ &= (2\mathbf{j}) \cdot \frac{1}{\sqrt{1+4t^2}}(\mathbf{i} + 2t\mathbf{j}) \\ &= \frac{4t}{\sqrt{1+4t^2}} \end{aligned}$$

Because

$$\mathbf{T}(t) = \frac{1}{\sqrt{1+4t^2}}(\mathbf{i} + 2t\mathbf{j})$$

and $\mathbf{T}(t) \cdot \mathbf{N}(t) = 0$, then

$$\mathbf{N}(t) = \frac{1}{\sqrt{1+4t^2}}(-2t\mathbf{i} + \mathbf{j})$$

and

$$\begin{aligned} A_N(t) &= \mathbf{A}(t) \cdot \mathbf{N}(t) \\ &= (2\mathbf{j}) \cdot \frac{1}{\sqrt{1+4t^2}}(-2t\mathbf{i} + \mathbf{j}) \\ &= \frac{2}{\sqrt{1+4t^2}} \end{aligned}$$

Note that we reject

$$\bar{\mathbf{N}}(t) = \frac{1}{\sqrt{1+4t^2}}(2t\mathbf{i} - \mathbf{j})$$

because

$$\mathbf{A}(t) \cdot \bar{\mathbf{N}}(t) = \frac{-2}{\sqrt{1+4t^2}} < 0$$

Next, we use formula (III) to find the curvature. Thus,

$$A_N(t) = K(t)\left(\frac{ds}{dt}\right)^2$$

$$\frac{2}{\sqrt{1+4t^2}} = K(t)(\sqrt{1+4t^2})^2$$

$$K(t) = \frac{2}{(1+4t^2)^{3/2}}$$

4. $\mathbf{R}(t) = \cos t^2\,\mathbf{i} + \sin t^2\,\mathbf{j};\quad t_1 = \frac{1}{2}\sqrt{\pi}$

SOLUTION: Because $t_1 > 0$, we assume $t > 0$. We have

$$\mathbf{V}(t) = \mathbf{R}'(t) = -2t \sin t^2\,\mathbf{i} + 2t \cos t^2\,\mathbf{j} \tag{1}$$
$$= -2t(\sin t^2\,\mathbf{i} - \cos t^2\,\mathbf{j})$$

$$\frac{ds}{dt} = |\mathbf{V}(t)| = |-2t|\sqrt{\sin^2 t^2 + \cos^2 t^2} = 2t \quad \text{if } t > 0 \tag{2}$$

$$A_T(t) = \frac{d^2s}{dt^2} = 2 \tag{3}$$

From (1) and (2) we obtain

$$\mathbf{T}(t) = \frac{\mathbf{V}(t)}{|\mathbf{V}(t)|} = \frac{1}{2t}[-2t \sin t^2\,\mathbf{i} + 2t \cos t^2\,\mathbf{j})]$$
$$= -\sin t^2\,\mathbf{i} + \cos t^2\,\mathbf{j} \tag{4}$$

From (1) we obtain

$$\mathbf{A}(t) = \mathbf{V}'(t) = (-4t^2 \cos t^2 - 2 \sin t^2)\mathbf{i} + (-4t^2 \sin t^2 + 2 \cos t^2)\mathbf{j} \tag{5}$$

For all $t$ we have from (I)

$$\mathbf{A}(t) = A_T(t)\mathbf{T}(t) + A_N(t)\mathbf{N}(t)$$

Solving for $\mathbf{N}(t)$, we get

$$\mathbf{N}(t) = \frac{1}{A_N(t)}[\mathbf{A}(t) - A_T(t)\mathbf{T}(t)] \tag{6}$$

Substituting from (5), (3), and (4) into (6), we obtain

$$\mathbf{N}(t) = \frac{1}{A_N(t)}[(-4t^2 \cos t^2 - 2 \sin t^2)\mathbf{i} + (-4t^2 \sin t^2 + 2 \cos t^2)\mathbf{j}$$
$$- 2(-\sin t^2\,\mathbf{i} + \cos t^2\,\mathbf{j})]$$
$$= \frac{1}{A_N(t)}[-4t^2 \cos t^2\,\mathbf{i} - 4t^2 \sin t^2\,\mathbf{j}]$$
$$= \frac{-4t^2}{A_N(t)}[\cos t^2\,\mathbf{i} + \sin t^2\,\mathbf{j}] \tag{7}$$

Taking the magnitude of both sides of (7), we obtain

$$|\mathbf{N}(t)| = \frac{4t^2}{A_N(t)}\sqrt{\cos^2 t^2 + \sin^2 t^2}$$

or, equivalently,

$$1 = \frac{4t^2}{A_N(t)}$$

Thus,

$$A_N(t) = 4t^2 \qquad (8)$$

Substituting from (8) into (7), we obtain

$$\mathbf{N}(t) = -\cos t^2\,\mathbf{i} - \sin t^2\,\mathbf{j} \qquad (9)$$

From (III), we have

$$A_N(t) = K(t)(D_t s)^2 \qquad (10)$$

Substituting from (8) and (2) into (10), we obtain

$$4t^2 = K(t)(2t)^2$$

Thus,

$$K(t) = 1 \qquad (11)$$

To find the particular values of the scalars when $t = \frac{1}{2}\sqrt{\pi}$, we substitute in (2), (3), (8), and (11). Thus,

$$\left|\mathbf{V}\left(\frac{1}{2}\sqrt{\pi}\right)\right| = \sqrt{\pi} \qquad A_T\left(\frac{1}{2}\sqrt{\pi}\right) = 2$$

$$A_N\left(\frac{1}{2}\sqrt{\pi}\right) = \pi \qquad K\left(\frac{1}{2}\sqrt{\pi}\right) = 1$$

Similarly, we find the particular values of the vectors when $t = \frac{1}{2}\sqrt{\pi}$. Thus,

$$\mathbf{V}\left(\frac{1}{2}\sqrt{\pi}\right) = -\frac{1}{2}\sqrt{2\pi}\,\mathbf{i} + \frac{1}{2}\sqrt{2\pi}\,\mathbf{j} \approx -1.25\,\mathbf{i} + 1.25\,\mathbf{j}$$

$$\mathbf{A}\left(\frac{1}{2}\sqrt{\pi}\right) = -\sqrt{2}\left(\frac{1}{2}\pi + 1\right)\mathbf{i} - \sqrt{2}\left(\frac{1}{2}\pi - 1\right)\mathbf{j} \approx -3.6\,\mathbf{i} - 0.8\,\mathbf{j}$$

$$A_T\mathbf{T}\left(\frac{1}{2}\sqrt{\pi}\right) = \sqrt{2}\,\mathbf{i} + \sqrt{2}\,\mathbf{j} \approx -1.4\,\mathbf{i} + 1.4\,\mathbf{j}$$

$$A_N\mathbf{N}\left(\frac{1}{2}\sqrt{\pi}\right) = -\frac{1}{2}\pi\sqrt{2}\,\mathbf{i} - \frac{1}{2}\pi\sqrt{2}\,\mathbf{j} \approx -2.2\,\mathbf{i} - 2.2\,\mathbf{j}$$

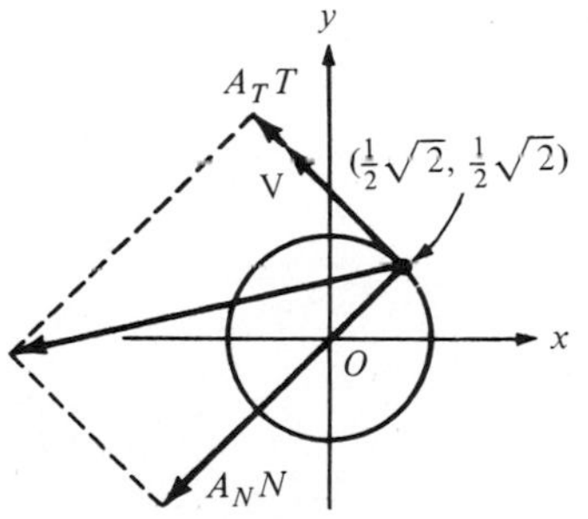

Figure 17.10.4

In Fig. 17.10.4 we show a sketch of the curve and representations of the vectors which have initial point $(\frac{1}{2}\sqrt{2}, \frac{1}{2}\sqrt{2})$, the terminal point of the radius vector $\mathbf{R}(\frac{1}{2}\pi)$.

8. A particle is moving along the top branch of the hyperbola $y^2 - x^2 = 9$, such that $D_t x$ is a positive constant. Find each of the following when the particle is at (4, 5): the position vector, the velocity vector, the acceleration vector, the unit tangent vector, the unit normal vector, $A_T$, and $A_N$.

SOLUTION: We express each of the required vectors and scalars in terms of $x$ and $y$, and then find the particular value when $x = 4$ and $y = 5$. Because $D_t x$ is a positive constant we let

$$D_t x = a \quad \text{where } a > 0 \qquad (1)$$

Differentiating with respect to $t$ on both sides of the equation $y^2 - x^2 = 9$, we obtain

$$2y\,D_t y - 2x\,D_t x = 0$$

Substituting $D_t x = a$ and solving for $D_t y$, we get

$$D_t y = \frac{ax}{y} \tag{2}$$

The position vector at the point $(4, 5)$ is

$$\mathbf{R} = 4\,\mathbf{i} + 5\,\mathbf{j}$$

If $\mathbf{R}$ is the position vector at the point $(x, y)$, then

$$\mathbf{R} = x\,\mathbf{i} + y\,\mathbf{j} \tag{3}$$

Differentiating with respect to $t$ on both sides of (3) and substituting from (1) and (2), we obtain the velocity vector in terms of $x$ and $y$. Thus,

$$\begin{aligned} \mathbf{V} = D_t\mathbf{R} &= D_t x\,\mathbf{i} + D_t y\,\mathbf{j} \\ &= a\,\mathbf{i} + \frac{ax}{y}\,\mathbf{j} \end{aligned} \tag{4}$$

To find the velocity vector at the point $(4, 5)$, we let $x = 4$ and $y = 5$ in Equation (4). Thus,

$$\mathbf{V} = a\,\mathbf{i} + \frac{4}{5}a\,\mathbf{j}$$

Differentiating on both sides of (4) with respect to $t$ and substituting from (1) and (2), we find the acceleration vector in terms of $x$ and $y$. Thus,

$$\begin{aligned} \mathbf{A} = D_t\mathbf{V} &= a\,\frac{yD_t x - xD_t y}{y^2}\,\mathbf{j} \\ &= \frac{a^2(y^2 - x^2)}{y^3}\,\mathbf{j} \end{aligned}$$

Substituting $x = 4$ and $y = 5$, we obtain the acceleration vector at the point $(4, 5)$. We have

$$\mathbf{A} = \frac{9}{125}a^2\,\mathbf{j} \tag{5}$$

From (4) we get

$$\frac{ds}{dt} = |\mathbf{V}| = \frac{a}{y}\sqrt{x^2 + y^2} \tag{6}$$

From (4) and (6) we find the unit tangent vector.

$$\mathbf{T} = \frac{\mathbf{V}}{|\mathbf{V}|} = \frac{1}{\sqrt{x^2 + y^2}}[y\,\mathbf{i} + x\,\mathbf{j}]$$

If $x = 4$ and $y = 5$, we get

$$\mathbf{T} = \frac{5}{\sqrt{41}}\,\mathbf{i} + \frac{4}{\sqrt{41}}\,\mathbf{j} \tag{7}$$

Furthermore, when $x = 4$ and $y = 5$ we have

$$\begin{aligned} A_T &= \mathbf{A} \cdot \mathbf{T} \\ &= \left(\frac{9}{125}a^2\mathbf{j}\right) \cdot \left(\frac{5}{\sqrt{41}}\mathbf{i} + \frac{4}{\sqrt{41}}\mathbf{j}\right) \\ &= \frac{36a^2}{125\sqrt{41}} \end{aligned}$$

Because $\mathbf{T} \cdot \mathbf{N} = 0$, from (7) we see that when $x = 4$ and $y = 5$, then

$$\mathbf{N} = -\frac{4}{\sqrt{41}}\mathbf{i} + \frac{5}{\sqrt{41}}\mathbf{j}$$

and

$$\begin{aligned} A_N &= \mathbf{A} \cdot \mathbf{N} \\ &= \left(\frac{9}{125}a^2\mathbf{j}\right) \cdot \left(-\frac{4}{\sqrt{41}}\mathbf{i} + \frac{5}{\sqrt{41}}\mathbf{j}\right) \\ &= \frac{9a^2}{25\sqrt{41}} \end{aligned}$$

Note that we reject the vector

$$\bar{\mathbf{N}} = \frac{4}{\sqrt{41}}\mathbf{i} - \frac{5}{\sqrt{41}}\mathbf{j}$$

because

$$\mathbf{A} \cdot \bar{\mathbf{N}} = -\frac{9a^2}{25\sqrt{41}} < 0$$

## *Review Exercises*

In Exercises 1-12, let

$$\mathbf{A} = 4\mathbf{i} - 6\mathbf{j}, \quad \mathbf{B} = \mathbf{i} + 7\mathbf{j}, \text{ and } \mathbf{C} = 9\mathbf{i} - 5\mathbf{j}$$

**6.** Find $(\mathbf{A} \cdot \mathbf{B})\mathbf{C}$.

SOLUTION:

$$\begin{aligned} \mathbf{A} \cdot \mathbf{B} &= (4\mathbf{i} - 6\mathbf{j}) \cdot (\mathbf{i} + 7\mathbf{j}) \\ &= 4 \cdot 1 + (-6)7 \\ &= -38 \end{aligned}$$

Thus,

$$\begin{aligned} (\mathbf{A} \cdot \mathbf{B})\mathbf{C} &= -38(9\mathbf{i} - 5\mathbf{j}) \\ &= -342\mathbf{i} + 190\mathbf{j} \end{aligned}$$

**10.** Find the vector projection of $\mathbf{C}$ onto $\mathbf{A}$.

SOLUTION: The vector projection of $\mathbf{C}$ onto $\mathbf{A}$ is given by

$$\begin{aligned} \mathbf{C}_{\mathbf{A}} &= \frac{\mathbf{C} \cdot \mathbf{A}}{|\mathbf{A}|^2}\mathbf{A} \\ &= \frac{(9\mathbf{i} - 5\mathbf{j}) \cdot (4\mathbf{i} - 6\mathbf{j})}{4^2 + (-6)^2}(4\mathbf{i} - 6\mathbf{j}) \\ &= \frac{33}{26}(4\mathbf{i} - 6\mathbf{j}) \\ &= \frac{66}{13}\mathbf{i} - \frac{99}{13}\mathbf{j} \end{aligned}$$

16. Find equations of the horizontal and vertical tangent lines, and then draw a sketch of the graph of the given pair of parametric equations.

$$x = \frac{2at^2}{1+t^2}; \quad y = \frac{2at^3}{1+t^2}; \quad a > 0$$

(the cissoid of Diocles).

SOLUTION: Differentiating with respect to $t$, we obtain

$$\frac{dx}{dt} = \frac{4at}{(1+t^2)^2} \quad \text{and} \quad \frac{dy}{dt} = \frac{2at^2(t^2+3)}{(1+t^2)^2} \qquad \textbf{(1)}$$

Thus,

$$\frac{dy}{dx} = \frac{\frac{dy}{dt}}{\frac{dx}{dt}} = \frac{\frac{2at^2(t^2+3)}{(1+t^2)^2}}{\frac{4at}{(1+t^2)^2}}$$

$$= \frac{1}{2}t(t^2+3)$$

Because $dy/dx = 0$ when $t = 0$, there is a horizontal tangent at $t = 0$. Moreover,

$$0 \leqslant \frac{2at^2}{1+t^2} < 2a$$

for all $t$. Thus, $x$ is in the interval $[0, 2a]$. Furthermore,

$$\lim_{t \to +\infty} x = \lim_{t \to +\infty} \frac{2at^2}{1+t^2} = 2a$$

$$\lim_{t \to +\infty} y = \lim_{t \to +\infty} \frac{2at^3}{1+t^2} = +\infty$$

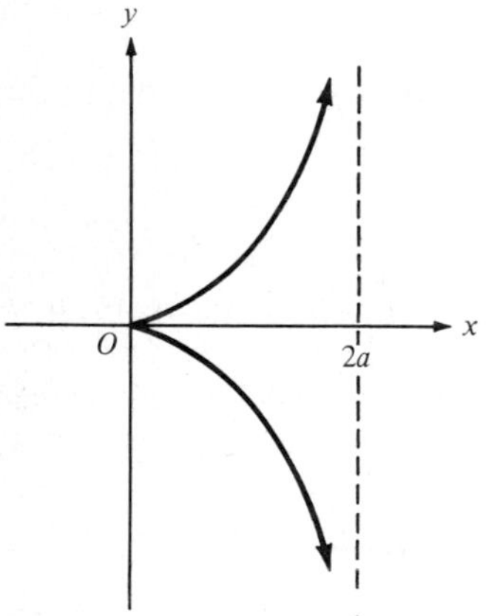


Figure 17.16R

Therefore, the line $x = 2a$ is a vertical asymptote. If $t$ is replaced by $-t$, then $x$ is unchanged and $y$ is replaced by $-y$. Thus, for every point $(x, y)$ on the curve we have the point $(x, -y)$, and hence, the graph is symmetric with respect to the $x$-axis.

We use the above information to draw a sketch of the graph, as shown in Fig. 17.16R.

18. Find the length of the arc of the curve $r = 3 \sec \theta$ from $\theta = 0$ to $\theta = \frac{1}{4}\pi$.

SOLUTION: We use the formula

$$L = \int_\alpha^\beta \sqrt{\left(\frac{dr}{d\theta}\right)^2 + r^2}\, d\theta$$

We have

$$\sqrt{\left(\frac{dr}{d\theta}\right)^2 + r^2} = \sqrt{(3 \sec \theta \tan \theta)^2 + (3 \sec \theta)^2}$$

$$= 3\sqrt{\sec^2 \theta(\tan^2 \theta + 1)}$$

$$= 3\sqrt{\sec^4 \theta}$$

$$= 3 \sec^2 \theta$$

Thus,

$$L = 3 \int_0^{\pi/4} \sec^2 \theta \, d\theta$$

$$= 3 \tan \theta \Big]_0^{\pi/4}$$
$$= 3$$

22. Find the curvature at any point of the branch of the hyperbola defined by $x = a \cosh t,\ y = b \sinh t$. Also show that the curvature is an absolute maximum at the vertex.

SOLUTION: We use the curvature formula (14) of Section 17.9. We have

$$\frac{dx}{dt} = a \sinh t \qquad \frac{dy}{dt} = b \cosh t$$

$$\frac{d^2x}{dt^2} = a \cosh t \qquad \frac{d^2y}{dt^2} = b \sinh t$$

Substituting in the formula, we obtain

$$K(t) = \frac{|(a \sinh t)(b \sinh t) - (b \cosh t)(a \cosh t)|}{(a^2 \sinh^2 t + b^2 \cosh^2 t)^{3/2}}$$

$$= \frac{|ab|}{(a^2 \sinh^2 t + b^2 \cosh^2 t)^{3/2}}$$

Differentiating with respect to $t$, we get

$$K'(t) = -\frac{3}{2}|ab|(a^2 \sinh^2 t + b^2 \cosh^2 t)^{-5/2}(2a^2 \sinh t \cosh t + 2b^2 \cosh t \sinh t)$$

$$= \frac{-3|ab|(a^2 + b^2) \sinh t \cosh t}{(a^2 \sinh^2 t + b^2 \cosh^2 t)^{5/2}} \tag{1}$$

Because $\cosh t > 0$ for all $t$, from (1) we see that the only critical number of $K$ is when $\sinh t = 0$. Thus, $t = 0$ is the only critical number. Furthermore, if $t < 0$, then $K'(t) > 0$; if $t > 0$, then $K'(t) < 0$. Thus, $K$ has an absolute maximum value when $t = 0$. Moreover, when $t = 0$, we have $x = a$ and $y = 0$. Because $(a, 0)$ is the vertex, then the curvature is an absolute maximum at the vertex.

28. If a particle is moving along a curve, under what conditions will the acceleration vector and the unit tangent vector have the same or opposite directions?

SOLUTION: At all times the acceleration vector may be expressed in terms of the unit tangent vector and the unit normal vector as follows:

$$\mathbf{A}(t) = A_{\mathbf{T}}(t)\mathbf{T}(t) + A_{\mathbf{N}}(t)\mathbf{N}(t)$$

Thus, if $A_{\mathbf{N}}(t) = 0$, the acceleration vector is parallel to the unit tangent vector because $\mathbf{A}$ is a scalar times $\mathbf{T}$. Furthermore,

$$A_{\mathbf{N}}(t) = K(t)\left(\frac{ds}{dt}\right)^2$$

Then if $A_{\mathbf{N}}(t) = 0$, we have $K(t) = 0$, and thus the motion is in a straight line. Furthermore, if $A_{\mathbf{N}}(t) = 0$, we have

$$\mathbf{A}(t) = A_{\mathbf{T}}(t)\mathbf{T}(t)$$

$$= \frac{d^2s}{dt^2}\mathbf{T}(t)$$

If $d^2s/dt^2 > 0$, then the speed $ds/dt$ is increasing, and the acceleration vector is in the same direction as the unit tangent vector. If $d^2s/dt^2 < 0$, then the speed is decreasing, and the acceleration vector is in the opposite direction as the unit tangent vector.

32. Find the radian measure of the angle of elevation at which a gun should be fired in order to obtain the maximum range for a given muzzle speed.

SOLUTION: Let

$\alpha =$ the number of radians in the angle of elevation of the gun
$t =$ the number of units of time that have elapsed since the gun was fired
$z =$ the number of units of distance in the range

From Equation (14) of Section 17.7, at $t$ units of time the projectile is at point $(x, y)$, where

$$x = tv_0 \cos \alpha \quad \text{and} \quad y = tv_0 \sin \alpha - \frac{1}{2}gt^2$$

with $v_0$ the number of units in the muzzle speed and $g$ a constant. If $y = 0$, then either $t = 0$ or

$$t_1 = \frac{2v_0 \sin \alpha}{g}$$

Substituting $t = t_1$ into the equation for $x$, we obtain

$$x_1 = \frac{2v_0^2 \sin \alpha \cos \alpha}{g}$$

where $x_1$ is the number of units in the range. Thus $z = x_1$, or, equivalently,

$$z = \frac{v_0^2}{g} \sin 2\alpha$$

Differentiating with respect to $\alpha$, we obtain

$$D_\alpha z = \frac{2v_0^2}{g} \cos 2\alpha$$

If $D_\alpha z = 0$, then $\cos 2\alpha = 0$, and thus $\alpha = \frac{1}{4}\pi$ is the only critical number in the interval $[0, \frac{1}{2}\pi]$. Furthermore, we have

$$D_\alpha^2 z = \frac{-4v_0^2}{g} \sin 2\alpha$$

If $\alpha = \frac{1}{4}\pi$, then $D_\alpha^2 z < 0$. Thus, $z$ has an absolute maximum value at $\alpha = \frac{1}{4}\pi$. The angle of elevation has measure $\frac{1}{4}\pi$ for maximum range.

34. Find the position vector $\mathbf{R}(t)$ if the acceleration vector

$$\mathbf{A}(t) = t^2\mathbf{i} - \frac{1}{t^2}\mathbf{j}, \quad \mathbf{V}(1) = \mathbf{j}, \quad \text{and} \quad \mathbf{R}(1) = -\frac{1}{4}\mathbf{i} + \frac{1}{2}\mathbf{j}$$

SOLUTION:

$$\mathbf{V}(t) = \int \mathbf{A}(t)\, dt$$

$$= \int (t^2\mathbf{i} - t^{-2}\mathbf{j})\, dt$$

$$= \frac{1}{3}t^3\mathbf{i} + t^{-1}\mathbf{j} + \mathbf{C}_1 \tag{1}$$

Because $\mathbf{V}(1) = \mathbf{j}$, we have from (1)

$$\mathbf{j} = \frac{1}{3}\mathbf{i} + \mathbf{j} + \mathbf{C}_1$$

Thus, $\mathbf{C}_1 = -\frac{1}{3}\mathbf{i}$, and substituting in (1), we obtain

$$\mathbf{V}(t) = \frac{1}{3}(t^3 - 1)\mathbf{i} + t^{-1}\mathbf{j}$$

Then,

$$\mathbf{R}(t) = \int \mathbf{V}(t)\, dt$$

$$= \int \left(\frac{1}{3}(t^3 - 1)\mathbf{i} + t^{-1}\mathbf{j}\right) dt$$

$$= \left(\frac{1}{12}t^4 - \frac{1}{3}t\right)\mathbf{i} + \ln|t|\mathbf{j} + \mathbf{C}_2 \qquad (2)$$

Because $\mathbf{R}(1) = -\frac{1}{4}\mathbf{i} + \frac{1}{2}\mathbf{j}$, we have from (2)

$$-\frac{1}{4}\mathbf{i} + \frac{1}{2}\mathbf{j} = -\frac{1}{4}\mathbf{i} + \mathbf{C}_2$$

Thus $\mathbf{C}_2 = \frac{1}{2}\mathbf{j}$. Substituting in (2), we obtain

$$\mathbf{R}(t) = \left(\frac{1}{12}t^4 - \frac{1}{3}t\right)\mathbf{i} + \left(\ln|t| + \frac{1}{2}\right)\mathbf{j}$$

**38.** Find the velocity and acceleration vectors, the speed, and the tangental and normal components of acceleration.

$$\mathbf{R}(t) = (2\tan^{-1} t - t)\mathbf{i} + \ln(1 + t^2)\mathbf{j}$$

SOLUTION:

$$\mathbf{V}(t) = \mathbf{R}'(t) = \frac{1 - t^2}{1 + t^2}\mathbf{i} + \frac{2t}{1 + t^2}\mathbf{j}$$

$$\mathbf{A}(t) = \mathbf{V}'(t) = \frac{-4t}{(1 + t^2)^2}\mathbf{i} + \frac{2(1 - t^2)}{(1 + t^2)^2}\mathbf{j} \qquad (1)$$

$$\frac{ds}{dt} = |\mathbf{V}(t)| = \frac{1}{1 + t^2}\sqrt{(1 - t^2)^2 + (2t)^2}$$

$$= 1$$

$$A_T(t) = \frac{d^2s}{dt^2} = 0$$

Because

$$\mathbf{A}(t) = A_T(t)\mathbf{T}(t) + A_N(t)\mathbf{N}(t)$$

and $A_T(t) = 0$, then

$$\mathbf{A}(t) = A_N(t)\mathbf{N}(t)$$

Thus,

$$|\mathbf{A}(t)| = |A_N(t)\mathbf{N}(t)|$$

Because $|\mathbf{N}(t)| = 1$, the above is equivalent to

$$A_N(t) = |\mathbf{A}(t)|$$

$$= \frac{2}{(1 + t^2)^2}\sqrt{(-2t)^2 + (1 - t^2)^2}$$

$$= \frac{2}{1 + t^2}$$

# 18

# Vectors in three-dimensional space and solid analytic geometry

## 18.1 $R^3$–THE THREE-DIMENSIONAL NUMBER SPACE

**18.1.1 Definition** The set of all ordered triples of real numbers is called the *three-dimensional number space* and is denoted by $R^3$. Each ordered triple $(x, y, z)$ is called a *point* in the three-dimensional number space.

**18.1.2 Theorem**

(i) A line is parallel to the $yz$ plane if and only if all points on the line have equal $x$ coordinates.
(ii) A line is parallel to the $xz$ plane if and only if all points on the line have equal $y$ coordinates.
(iii) A line is parallel to the $xy$ plane if and only if all points on the line have equal $z$ coordinates.

**18.1.3 Theorem**

(i) A line is parallel to the $x$-axis if and only if all points on the line have equal $y$ coordinates and equal $z$ coordinates.
(ii) A line is parallel to the $y$-axis if and only if all points on the line have equal $x$ coordinates and equal $z$ coordinates.
(iii) A line is parallel to the $z$-axis if and only if all points on the line have equal $x$ coordinates and equal $y$ coordinates.

The remaining theorems in this section are the extensions of the corresponding theorems in two-dimensional space.

**18.1.5 Theorem** The undirected distance between the two points $P_1(x_1, y_1, z_1)$ and $P_2(x_2, y_2, z_2)$ is given by

$$|\overline{P_1P_2}| = \sqrt{(x_2 - x_1)^2 + (y_2 - y_1)^2 + (z_2 - z_1)^2}$$

**18.1.6 Theorem** The coordinates of the midpoint of the line segment having endpoints $P_1(x_1, y_1, z_1)$ and $P_2(x_2, y_2, z_2)$ are given by

$$\bar{x} = \frac{x_1 + x_2}{2} \qquad \bar{y} = \frac{y_1 + y_2}{2} \qquad \bar{z} = \frac{z_1 + z_2}{2}$$

**18.1.9 Theorem** An equation of the sphere of radius $r$ and center at $(h, k, l)$ is

$$(x - h)^2 + (y - k)^2 + (z - l)^2 = r^2$$

**18.1.10 Theorem** The graph of any second-degree equation in $x, y,$ and $z$, of the form

$$x^2 + y^2 + z^2 + Gx + Hy + Iz + J = 0$$

is either a sphere, a point-sphere, or the empty set.

## *Exercises 18.1*

**4.** The given points $A$ and $B$ are opposite vertices of a rectangular parallelepiped, having its faces parallel to the coordinate planes.

(a) Draw a sketch of the figure.
(b) Find the coordinates of the other six vertices.
(c) Find the length of the diagonal $AB$.

$A(2, -1, -3) \quad B(4, 0, -1)$

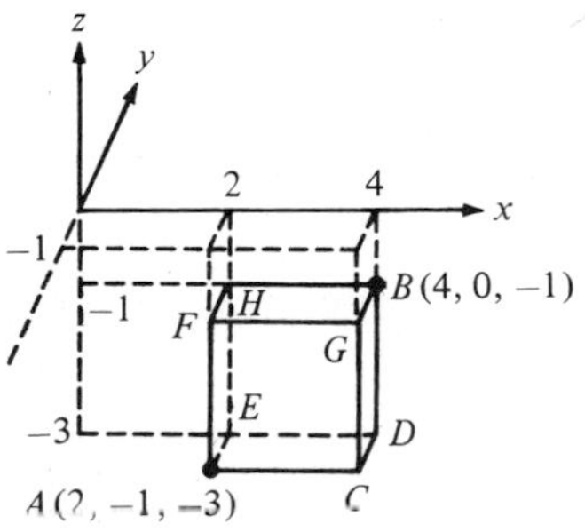

Figure 18.1.4

SOLUTION:

(a) In Fig. 18.1.4 we show a sketch of the parallelepiped. Rectangles $ACDE$ and $FGBH$ are parallel faces.
(b) We use Theorems 18.1.2 and 18.1.3 to find the coordinates of vertex $C$. Because line $BC$ is parallel to the $yz$ plane, then points $B$ and $C$ have equal $x$ coordinates. Because line $AC$ is parallel to the $x$-axis, then points $A$ and $C$ have equal $y$ coordinates and equal $z$ coordinates. Thus, $C = (4, -1, -3)$. By similar reasoning, we find the coordinates of the remaining vertices. Thus, $D = (4, 0, -3)$; $E = (2, 0, -3)$; $F = (2, -1, -1)$; $G = (4, -1, -1)$; and $H = (2, 0, -1)$.
(c) We use the distance formula (18.1.5). Thus,

$$|\overline{AB}| = \sqrt{(4 - 2)^2 + (0 + 1)^2 + (-1 + 3)^2}$$
$$= 3$$

**8.** Find: **(a)** the undirected distance between the points $A$ and $B$ and **(b)** the midpoint of the line segment joining $A$ and $B$.

$A(4, -3, 2) \quad B(-2, 3, -5)$

SOLUTION:

(a) Applying the distance formula (18.1.5), we obtain

$$|\overline{AB}| = \sqrt{(-2 - 4)^2 + (3 + 3)^2 + (-5 - 2)^2}$$
$$= 11$$

(b) We apply Theorem 18.1.6 to find $(\bar{x}, \bar{y}, \bar{z})$, the midpoint of segment $AB$. Thus,

$$\bar{x} = \frac{4 - 2}{2} = 1 \qquad \bar{y} = \frac{-3 + 3}{2} = 0 \qquad \bar{z} = \frac{2 - 5}{2} = -\frac{3}{2}$$

Hence, the midpoint of segment $AB$ is $(1, 0, -\frac{3}{2})$.

**10.** Prove that the three points $(1, -1, 3)$, $(2, 1, 7)$, and $(4, 2, 6)$ are the vertices of a right triangle, and find its area.

SOLUTION: Let $A = (1, -1, 3)$, $B = (2, 1, 7)$, and $C = (4, 2, 6)$. Applying the distance formula (18.1.5), we get

$$|\overline{AB}| = \sqrt{(2-1)^2 + (1+1)^2 + (7-3)^2} = \sqrt{21}$$
$$|\overline{BC}| = \sqrt{(4-2)^2 + (2-1)^2 + (6-7)^2} = \sqrt{6}$$
$$|\overline{AC}| = \sqrt{(4-1)^2 + (2+1)^2 + (6-3)^2} = \sqrt{27}$$

Because $|\overline{AB}|^2 + |\overline{BC}|^2 = |\overline{AC}|^2$, then triangle $ABC$ is a right triangle with hypoteneuse side $AC$. To find the area, we use the area formula $\frac{1}{2}bh$. We take $b = |\overline{AB}| = \sqrt{21}$ and $h = |\overline{BC}| = \sqrt{6}$. Thus, the number of square units in the area is $\frac{1}{2}\sqrt{21}\sqrt{6} = \frac{3}{2}\sqrt{14}$.

**14.** Find the vertices of the triangle whose sides have midpoints at $(3, 2, 3)$, $(-1, 1, 5)$, and $(0, 3, 4)$.

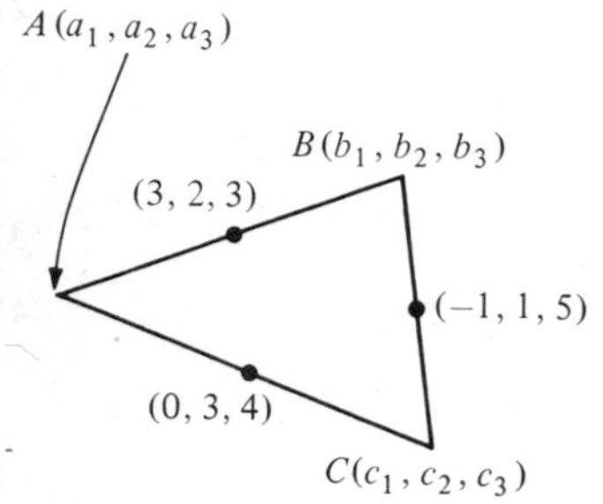


Figure 18.1.14

SOLUTION: Let $A(a_1, a_2, a_3)$, $B(b_1, b_2, b_3)$, and $C(c_1, c_2, c_3)$ be the vertices of the triangle with $(3, 2, 3)$ the midpoint of side $AB$, $(-1, 1, 5)$ the midpoint of side $BC$, and $(0, 3, 4)$ the midpoint of side $AC$, as shown in Fig. 18.1.14. Because $(3, 2, 3)$ is the midpoint of side $AB$, we have

$$\frac{a_1 + b_1}{2} = 3 \qquad \frac{a_2 + b_2}{2} = 2 \qquad \frac{a_3 + b_3}{2} = 3 \tag{1}$$

Because $(-1, 1, 5)$ is the midpoint of side $BC$, we have

$$\frac{b_1 + c_1}{2} = -1 \qquad \frac{b_2 + c_2}{2} = 1 \qquad \frac{b_3 + c_3}{2} = 5 \tag{2}$$

Because $(0, 3, 4)$ is the midpoint of side $AC$, we have

$$\frac{a_1 + c_1}{2} = 0 \qquad \frac{a_2 + c_2}{2} = 3 \qquad \frac{a_3 + c_3}{2} = 4 \tag{3}$$

We solve the first equations in (1), (2), and (3) simultaneously. The result is

$$a_1 = 4 \qquad b_1 = 2 \qquad c_1 = -4$$

Solving the second equations in (1), (2), and (3) simultaneously, we obtain

$$a_2 = 4 \qquad b_2 = 0 \qquad c_2 = 2$$

Solving the third equations in (1), (2), and (3) simultaneously, we obtain

$$a_3 = 2 \qquad b_3 = 4 \qquad c_3 = 6$$

Thus,

$$A = (a_1, a_2, a_3) = (4, 4, 2)$$
$$B = (b_1, b_2, b_3) = (2, 0, 4)$$
$$C = (c_1, c_2, c_3) = (-4, 2, 6)$$

**18.** Determine the graph of the given equation.

$$x^2 + y^2 + z^2 - 8y + 6z - 25 = 0$$

SOLUTION: We complete the squares on the terms in $y$ and $z$. Thus,

$$x^2 + (y^2 - 8y) \qquad + (z^2 + 6z) \qquad = 25$$
$$x^2 + (y^2 - 8y + 16) + (z^2 + 6z + 9) = 25 + 16 + 9$$
$$x^2 + (y - 4)^2 \qquad + (z + 3)^2 \qquad = 50 \tag{1}$$

By comparison of (1) with Theorem 18.1.9, we determine that the graph is a sphere with center at $(0, 4, -3)$ and radius $\sqrt{50} = 5\sqrt{2}$ units.

**24.** Find an equation of the sphere that contains the points $(0, 0, 4)$, $(2, 1, 3)$, and $(0, 2, 6)$ and has its center in the $yz$ plane.

SOLUTION: Let $(h, k, l)$ be the center of the sphere and let $r$ be the radius. Because the center is in the $yz$ plane, we have $h = 0$. Thus, by Theorem 18.1.9 with $h = 0$, an equation of the sphere is of the form

$$x^2 + (y - k)^2 + (z - l)^2 = r^2 \tag{1}$$

Because the sphere contains the point $(0, 0, 4)$, then Eq. (1) holds if $x = 0, y = 0$, and $z = 4$. Thus,

$$k^2 + (4 - l)^2 = r^2 \tag{2}$$

Similarly, we may substitute the coordinates of $(2, 1, 3)$ and $(0, 2, 6)$ into (1) and obtain

$$4 + (1 - k)^2 + (3 - l)^2 = r^2 \tag{3}$$

$$(2 - k)^2 + (6 - l)^2 = r^2 \tag{4}$$

We solve (2), (3), and (4) simultaneously. Elimintaing $r^2$ from (2) and (3), we get

$$k^2 + (4 - l)^2 = 4 + (1 - k)^2 + (3 - l)^2$$

$$k - l + 1 = 0 \tag{5}$$

Eliminating $r^2$ from (2) and (4), we get

$$k + l - 6 = 0 \tag{6}$$

Solving (5) and (6), we obtain $k = \frac{5}{2}$ and $l = \frac{7}{2}$. Substituting these values into (2), we obtain $r^2 = \frac{13}{2}$. With these replacements for $k, l$, and $r^2$ in Equation (1), we have

$$x^2 + \left(y - \frac{5}{2}\right)^2 + \left(z - \frac{7}{2}\right)^2 = \frac{13}{2}$$

$$x^2 + y^2 + z^2 - 5y - 7z + 12 = 0$$

## 18.2 VECTORS IN THREE-DIMENSIONAL SPACE

**18.2.1 Definition** A *vector in three-dimensional space* is an ordered triple of real numbers $\langle x, y, z\rangle$. The numbers $x, y$, and $z$ are called the *components* of the vector $\langle x, y, z\rangle$. The magnitude of the vector $\mathbf{A} = \langle a_1, a_2, a_3\rangle$ is given by

$$|\mathbf{A}| = \sqrt{a_1{}^2 + a_2{}^2 + a_3{}^2}$$

**18.2.2 Definition** The *direction angles* of a nonzero vector are the three angles that have the smallest nonnegative radian measures $\alpha, \beta$, and $\gamma$ measured from the positive $x, y$, and $z$ axes, respectively, to the position representation of the vector.

Thus the direction angles for the vector $\mathbf{A} = \langle a_1, a_2, a_3\rangle$ are $\alpha, \beta$, and $\gamma$ with $0 \leqslant \alpha \leqslant \pi, 0 \leqslant \beta \leqslant \pi$, and $0 \leqslant \gamma \leqslant \pi$, such that

$$\cos\alpha = \frac{a_1}{|\mathbf{A}|} \qquad \cos\beta = \frac{a_2}{|\mathbf{A}|} \qquad \cos\gamma = \frac{a_3}{|\mathbf{A}|}$$

**18.2.3 Theorem** If $\cos\alpha, \cos\beta$, and $\cos\gamma$ are the direction cosines of a vector, then

$$\cos^2\alpha + \cos^2\beta + \cos^2\gamma = 1$$

The remaining definitions and theorems in this section are extensions of the corresponding definitions and theorems given in Section 17.2 for vectors in $V_2$. We define the unit vectors

$$\mathbf{i} = \langle 1, 0, 0 \rangle \qquad \mathbf{j} = \langle 0, 1, 0 \rangle \qquad \mathbf{k} = \langle 0, 0, 1 \rangle$$

Thus, if $\mathbf{A} = \langle a_1, a_2, a_3 \rangle$, then

$$\mathbf{A} = a_1\mathbf{i} + a_2\mathbf{j} + a_3\mathbf{k}$$

and

$$\mathbf{A} = |\mathbf{A}|(\cos\alpha\,\mathbf{i} + \cos\beta\,\mathbf{j} + \cos\gamma\,\mathbf{k})$$

## *Exercises 18.2*

**4.** Prove Theorem 18.2.8(v).

SOLUTION: We want to prove that if $c$ and $d$ are any scalars and $\mathbf{A}$ is any vector in $V_3$, then

$$(cd)\mathbf{A} = c(d\mathbf{A})$$

Let $\mathbf{A} = \langle a_1, a_2, a_3 \rangle$. By the associative property of real numbers

$$(cd)a_1 = c(da_1) \qquad (cd)a_2 = c(da_2) \qquad (cd)a_3 = c(da_3)$$

Thus,

$$\begin{aligned}(cd)\mathbf{A} &= (cd)\langle a_1, a_2, a_3 \rangle \\ &= \langle (cd)a_1, (cd)a_2, (cd)a_3 \rangle \\ &= \langle c(da_1), c(da_2), c(da_3) \rangle \\ &= c\langle da_1, da_2, da_3 \rangle \\ &= c(d\langle a_1, a_2, a_3 \rangle) \\ &= c(d\mathbf{A})\end{aligned}$$

In Exercises 7-18, let $\mathbf{A} = \langle 1, 2, 3 \rangle$, $\mathbf{B} = \langle 4, -3, -1 \rangle$, $\mathbf{C} = \langle -5, -3, 5 \rangle$, $\mathbf{D} = \langle -2, 1, 6 \rangle$.

**8.** Find $2\mathbf{A} - \mathbf{C}$.

SOLUTION:

$$\begin{aligned}2\mathbf{A} - \mathbf{C} &= 2\langle 1, 2, 3 \rangle - \langle -5, -3, 5 \rangle \\ &= \langle 2, 4, 6 \rangle + \langle 5, 3, -5 \rangle \\ &= \langle 7, 7, 1 \rangle\end{aligned}$$

**12.** Find $|4\mathbf{B}| + |6\mathbf{C}| - |2\mathbf{D}|$.

SOLUTION:

$$\begin{aligned}|4\mathbf{B}| &= 4|\mathbf{B}| \\ &= 4\sqrt{4^2 + (-3)^2 + (-1)^2} \\ &= 4\sqrt{26}\end{aligned}$$

$$\begin{aligned}|6\mathbf{C}| &= 6|\mathbf{C}| \\ &= 6\sqrt{(-5)^2 + (-3)^2 + 5^2} \\ &= 6\sqrt{59}\end{aligned}$$

$$\begin{aligned}|2\mathbf{D}| &= 2\sqrt{(-2)^2 + 1^2 + 6^2} \\ &= 2\sqrt{41}\end{aligned}$$

Thus,

$$|4\mathbf{B}| + |6\mathbf{C}| - |2\mathbf{D}| = 4\sqrt{26} + 6\sqrt{59} - 2\sqrt{41}$$

**16.** Find $|\mathbf{A}|\mathbf{C} - |\mathbf{B}|\mathbf{D}$.

SOLUTION:

$$|\mathbf{A}| = \sqrt{1^2 + 2^2 + 3^2} = \sqrt{14}$$
$$|\mathbf{B}| = \sqrt{4^2 + (-3)^2 + (-1)^2} = \sqrt{26}$$

Thus,

$$\begin{aligned}|\mathbf{A}|\mathbf{C} - |\mathbf{B}|\mathbf{D} &= \sqrt{14}\langle -5, -3, 5\rangle - \sqrt{26}\langle -2, 1, 6\rangle \\ &= \langle -5\sqrt{14}, -3\sqrt{14}, 5\sqrt{14}\rangle + \langle 2\sqrt{26}, -\sqrt{26}, -6\sqrt{26}\rangle \\ &= \langle -5\sqrt{14} + 2\sqrt{26}, -3\sqrt{14} - \sqrt{26}, 5\sqrt{14} - 6\sqrt{26}\rangle\end{aligned}$$

**18.** Find scalars $a$, $b$, and $c$ such that

$$a\mathbf{A} + b\mathbf{B} + c\mathbf{C} = \mathbf{D}$$

SOLUTION: Substituting for **A**, **B**, **C**, and **D**, in the given equation, we have

$$a\langle 1, 2, 3\rangle + b\langle 4, -3, -1\rangle + c\langle -5, -3, 5\rangle = \langle -2, 1, 6\rangle$$

or, equivalently,

$$\langle a + 4b - 5c, 2a - 3b - 3c, 3a - b + 5c\rangle = \langle -2, 1, 6\rangle$$

Thus,

$$a + 4b - 5c = -2 \quad (1)$$
$$2a - 3b - 3c = 1 \quad (2)$$
$$3a - b + 5c = 6 \quad (3)$$

We solve equations (1), (2), and (3) simultaneously as follows.

$$a + 4b - 5c = -2 \qquad [\text{Eq. (1)}]$$
$$-11b + 7c = 5 \qquad [-2 \text{ times Eq. (1)} + \text{Eq. (2)}] \quad (4)$$
$$-13b + 20c = 12 \qquad [-3 \text{ times Eq. (1)} + \text{Eq. (3)}] \quad (5)$$

We use determinants with Cramer's rule to solve Eqs. (4) and (5). Let $\Delta$ be the determinant of the coefficient matrix. Then

$$\Delta = \begin{vmatrix} -11 & 7 \\ -13 & 20 \end{vmatrix} = (-11)(20) - (7)(-13) = -129$$

$$\mathbf{b} \cdot \Delta = \begin{vmatrix} 5 & 7 \\ 12 & 20 \end{vmatrix} = 5 \cdot 20 - 7 \cdot 12 = 16$$

$$\mathbf{c} \cdot \Delta = \begin{vmatrix} -11 & 5 \\ -13 & 12 \end{vmatrix} = (-11)(12) - (5)(-13) = -67$$

Thus,

$$b = -\frac{16}{129} \quad \text{and} \quad c = \frac{67}{129}$$

Substituting these values into (1), we obtain

$$a = \frac{141}{129}$$

22. Find the direction cosines of the vector $\mathbf{V}(\overrightarrow{P_1P_2})$ and check the answers by verifying that the sum of their squares is 1.

$P_1(-2, 6, 5) \quad P_2(2, 4, 1)$

SOLUTION: We have

$$\begin{aligned}\mathbf{V} &= \langle 2 + 2, 4 - 6, 1 - 5\rangle \\ &= \langle 4, -2, -4\rangle \\ &= 2\langle 2, -1, -2\rangle\end{aligned}$$

Thus,

$$\begin{aligned}|\mathbf{V}| &= 2\sqrt{2^2 + (-1)^2 + (-2)^2} \\ &= 6\end{aligned}$$

We have

$$\cos\alpha = \frac{a_1}{|\mathbf{V}|} = \frac{4}{6} = \frac{2}{3}$$

$$\cos\beta = \frac{a_2}{|\mathbf{V}|} = \frac{-2}{6} = -\frac{1}{3}$$

$$\cos\gamma = \frac{a_3}{|\mathbf{V}|} = \frac{-4}{6} = -\frac{2}{3}$$

Furthermore,

$$\cos^2\alpha + \cos^2\beta + \cos^2\gamma = \left(\frac{2}{3}\right)^2 + \left(-\frac{1}{3}\right)^2 + \left(-\frac{2}{3}\right)^2 = 1$$

24. Find the point $R$ such that $\mathbf{V}(\overrightarrow{P_1R}) = -2\mathbf{V}(\overrightarrow{P_2R})$ for $P_1(1, 3, 5)$ and $P_2(2, -1, 4)$.

SOLUTION: Let $R = (r_1, r_2, r_3)$. Then

$$\mathbf{V}(\overrightarrow{P_1R}) = \langle r_1 - 1, r_2 - 3, r_3 - 5\rangle$$

$$\mathbf{V}(\overrightarrow{P_2R}) = \langle r_1 - 2, r_2 + 1, r_3 - 4\rangle$$

Because $\mathbf{V}(\overrightarrow{P_1R}) = -2\mathbf{V}(\overrightarrow{P_2R})$, we have

$$\langle r_1 - 1, r_2 - 3, r_3 - 5\rangle = -2\langle r_2 - 2, r_2 + 1, r_3 - 4\rangle$$

or, equivalently,

$$\langle r_1 - 1, r_2 - 3, r_3 - 5\rangle = \langle -2r_1 + 4, -2r_2 - 2, -2r_3 + 8\rangle$$

Thus,

$$r_1 - 1 = -2r_1 + 4 \quad r_2 - 3 = -2r_2 - 2 \quad r_3 - 5 = -2r_3 + 8$$

$$r_1 = \frac{5}{3} \qquad r_2 = \frac{1}{3} \qquad r_3 = \frac{13}{3}$$

Therefore, $R = (\frac{5}{3}, \frac{1}{3}, \frac{13}{3})$.

28. Express the given vector in terms of its magnitude and direction cosines.

$3\mathbf{i} + 4\mathbf{j} - 5\mathbf{k}$

SOLUTION:

$$|3\mathbf{i} + 4\mathbf{j} - 5\mathbf{k}| = \sqrt{3^2 + 4^2 + (-5)^2} = 5\sqrt{2}$$

As in Exercise 22, we have

$$\cos \alpha = \frac{3}{5\sqrt{2}} = \frac{3}{10}\sqrt{2}$$

$$\cos \beta = \frac{4}{5\sqrt{2}} = \frac{2}{5}\sqrt{2}$$

$$\cos \gamma = \frac{-5}{5\sqrt{2}} = -\frac{1}{2}\sqrt{2}$$

Therefore,

$$3\mathbf{i} + 4\mathbf{j} - 5\mathbf{k} = 5\sqrt{2}\left(\frac{3}{10}\sqrt{2}\,\mathbf{i} + \frac{2}{5}\sqrt{2}\,\mathbf{j} - \frac{1}{2}\sqrt{2}\,\mathbf{k}\right)$$

**32.** Find the unit vector having the same direction as $\mathbf{V}(\overrightarrow{P_1P_2})$.

$P_1(-8, -5, 2) \quad P_2(-3, -9, 4)$

SOLUTION:

$$\mathbf{V} = \langle -3 + 8,\ -9 + 5,\ 4 - 2\rangle$$
$$= \langle 5, -4, 2\rangle$$

Thus,

$$|\mathbf{V}| = \sqrt{5^2 + (-4)^2 + 2^2} = 3\sqrt{5}$$

The required unit vector is $\mathbf{U}$, where

$$\mathbf{U} = \frac{\mathbf{V}}{|\mathbf{V}|} = \frac{1}{3\sqrt{5}}(5\mathbf{i} - 4\mathbf{j} + 2\mathbf{k})$$

$$= \frac{1}{3}\sqrt{5}\,\mathbf{i} - \frac{4}{15}\sqrt{5}\,\mathbf{j} + \frac{2}{15}\sqrt{5}\,\mathbf{k}$$

## 18.3 THE DOT PRODUCT IN $V_3$

The definitions and theorems in this section are all extensions of the corresponding definitions and theorems given in Section 17.3 for vectors in $V_2$.

## *Exercises 18.3*

In Exercises 5-14, let $\mathbf{A} = \langle -4, -2, 4\rangle$, $\mathbf{B} = \langle 2, 7, -1\rangle$, $\mathbf{C} = \langle 6, -3, 0\rangle$, and $\mathbf{D} = \langle 5, 4, -3\rangle$.

**10.** Find $(2\mathbf{A} + 3\mathbf{B}) \cdot (4\mathbf{C} - \mathbf{D})$.

SOLUTION:

$$2\mathbf{A} + 3\mathbf{B} = 2\langle -4, -2, 4\rangle + 3\langle 2, 7, -1\rangle$$
$$= \langle -8, -4, 8\rangle + \langle 6, 21, -3\rangle$$
$$= \langle -2, 17, 5\rangle$$

And

$$4\mathbf{C} - \mathbf{D} = 4\langle 6, -3, 0\rangle - \langle 5, 4, -3\rangle$$
$$= \langle 24, -12, 0\rangle + \langle -5, -4, 3\rangle$$
$$= \langle 19, -16, 3\rangle$$

Thus,

$$(2\mathbf{A} + 3\mathbf{B}) \cdot (4\mathbf{C} - \mathbf{D}) = \langle -2, 17, 5\rangle \cdot \langle 19, -16, 3\rangle$$
$$= -2 \cdot 19 + 17(-16) + 5 \cdot 3$$
$$= -295$$

12. Find the cosine of the measure of the angle between **C** and **D**.

SOLUTION: We use Theorem 18.3.4. Thus

$$\cos\theta = \frac{\mathbf{C}\cdot\mathbf{D}}{|\mathbf{C}||\mathbf{D}|}$$

$$= \frac{\langle 6, -3, 0\rangle \cdot \langle 5, 4, -3\rangle}{\sqrt{6^2 + (-3)^2 + 0^2}\sqrt{5^2 + 4^2 + (-3)^2}}$$

$$= \frac{6\cdot 5 + (-3)4 + 0(-3)}{\sqrt{45}\sqrt{50}}$$

$$= \frac{3}{25}\sqrt{10}$$

14. Find: **(a)** the component of **B** in the direction of **D** and **(b)** the vector projection of **B** onto **D**,

SOLUTION:

**(a)** The component of **B** in the direction of **D** is the scalar projection of **B** **D**, onto **D**, represented by $B_D$. We have

$$B_D = \frac{\mathbf{B}\cdot\mathbf{D}}{|\mathbf{D}|}$$

$$= \frac{\langle 2, 7, -1\rangle \cdot \langle 5, 4, -3\rangle}{\sqrt{5^2 + 4^2 + (-3)^2}}$$

$$= \frac{2\cdot 5 + 7\cdot 4 + (-1)(-3)}{\sqrt{50}}$$

$$= \frac{41}{5\sqrt{2}}$$

**(b)** The vector projection of **B** onto **D** is represented by $\mathbf{B_D}$ and is given by

$$\mathbf{B_D} = \frac{\mathbf{B}\cdot\mathbf{D}}{|\mathbf{D}||\mathbf{D}|}\mathbf{D} = B_D\frac{\mathbf{D}}{|\mathbf{D}|}$$

Substituting the values for $B_D$ and $|\mathbf{D}|$ which we found in part (a), we obtain

$$\mathbf{B_D} = \frac{41}{5\sqrt{2}}\cdot\frac{1}{5\sqrt{2}}\langle 5, 4, -3\rangle$$

$$= \frac{41}{50}\langle 5, 4, -3\rangle$$

$$= \left\langle \frac{41}{10}, \frac{82}{25}, -\frac{123}{50}\right\rangle$$

18. Prove by using vectors that the points $(2, 2, 2)$, $(0, 1, 2)$, $(-1, 3, 3)$, and $(3, 0, 1)$ are the vertices of a parallelogram.

SOLUTION: Let $A = (2, 2, 2,)$, $B = (0, 1, 2)$, $C = (-1, 3, 3)$, and $D = (3, 0, 1)$. Then

$$\mathbf{V}(\overrightarrow{AC}) = \mathbf{V}(\overrightarrow{OC}) - \mathbf{V}(\overrightarrow{OA}) = \langle -1, 3, 3\rangle - \langle 2, 2, 2\rangle = \langle -3, 1, 1\rangle$$

$$\mathbf{V}(\overrightarrow{DB}) = \mathbf{V}(\overrightarrow{OB}) - \mathbf{V}(\overrightarrow{OD}) = \langle 0, 1, 2\rangle - \langle 3, 0, 1\rangle = \langle -3, 1, 1\rangle$$

Because $\mathbf{V}(\overrightarrow{AC}) = \mathbf{V}(\overrightarrow{DB})$, then line segment $AC$ is parallel to line segment $DB$ and

$|\overline{AC}| = |\overline{DB}|$. If one pair of sides in a quadrilateral are both parallel and have the same length, then the quadrilateral is a parallelogram. Thus, the quadrilateral $ACBD$ is a parallelogram.

**22.** If a force has the vector representation $\mathbf{F} = 5\mathbf{i} - 3\mathbf{k}$, find the work done by the force in moving an object from the point $P_1(4, 1, 3)$ along a straight line to the point $P_2(-5, 6, 2)$. The magnitude of the force is measured in pounds, and distance is measured in feet. (*Hint*: Review Sec. 17.3.)

SOLUTION: In Sec. 17.3 we reasoned that the number of units in the work done by a force $\mathbf{F}$ which causes a displacement $\mathbf{D}$ is given by

$$W = \mathbf{F} \cdot \mathbf{D}$$

We have

$$\mathbf{D} = \mathbf{V}(\overrightarrow{P_1P_2}) = \langle -5, 6, 2\rangle - \langle 4, 1, 3\rangle = \langle -9, 5, -1\rangle$$

and

$$\mathbf{F} = 5\mathbf{i} - 3\mathbf{k} = \langle 5, 0, -3\rangle$$

Thus,

$$\begin{aligned} W &= \langle 5, 0, -3\rangle \cdot \langle -9, 5, -1\rangle \\ &= 5(-9) + 0 \cdot 5 + (-3)(-1) \\ &= -42 \end{aligned}$$

The work done is $-42$ foot-pounds.

**24.** If $\mathbf{A}$ and $\mathbf{B}$ are nonzero vectors, prove that the vector $\mathbf{A} - c\mathbf{B}$ is orthogonal to $\mathbf{B}$ if $c = \mathbf{A} \cdot \mathbf{B}/|\mathbf{B}|^2$.

SOLUTION: By Definition 18.3.7, the vectors $\mathbf{A} - c\mathbf{B}$ and $\mathbf{B}$ are orthogonal if and only if

$$(\mathbf{A} - c\mathbf{B}) \cdot \mathbf{B} = 0 \tag{1}$$

By Theorem 18.3.2, Eq. (1) is true if and only if

$$\begin{aligned} \mathbf{A} \cdot \mathbf{B} - c\mathbf{B} \cdot \mathbf{B} &= 0 \\ \mathbf{A} \cdot \mathbf{B} - c(\mathbf{B} \cdot \mathbf{B}) &= 0 \\ \mathbf{A} \cdot \mathbf{B} - c|\mathbf{B}|^2 &= 0 \end{aligned} \tag{2}$$

Because $|\mathbf{B}| \neq 0$, we may solve Eq. (2), which contains all scalars, for $c$. Thus, (1) is true if and only if

$$c = \frac{\mathbf{A} \cdot \mathbf{B}}{|\mathbf{B}|^2}$$

**28.** Prove that if $\mathbf{A}$ and $\mathbf{B}$ are any nonzero vectors and $\mathbf{C} = |\mathbf{B}|\mathbf{A} + |\mathbf{A}|\mathbf{B}$, then the angle between $\mathbf{A}$ and $\mathbf{C}$ has the same measure as the angle between $\mathbf{B}$ and $\mathbf{C}$.

SOLUTION: Let $\theta_1$ be the angle between $\mathbf{A}$ and $\mathbf{C}$. Then applying Theorems 18.3.4 and 18.3.2, we obtain

$$\begin{aligned} \cos\theta_1 &= \frac{\mathbf{A} \cdot \mathbf{C}}{|\mathbf{A}||\mathbf{C}|} \\ &= \frac{\mathbf{A} \cdot (|\mathbf{B}|\mathbf{A} + |\mathbf{A}|\mathbf{B})}{|\mathbf{A}||\mathbf{C}|} \\ &= \frac{\mathbf{A} \cdot (|\mathbf{B}|\mathbf{A}) + \mathbf{A} \cdot (|\mathbf{A}|\mathbf{B})}{|\mathbf{A}||\mathbf{C}|} \end{aligned}$$

$$= \frac{|\mathbf{B}|(\mathbf{A} \cdot \mathbf{A}) + |\mathbf{A}|(\mathbf{A} \cdot \mathbf{B})}{|\mathbf{A}||\mathbf{C}|}$$

$$= \frac{|\mathbf{B}||\mathbf{A}|^2 + |\mathbf{A}|(\mathbf{A} \cdot \mathbf{B})}{|\mathbf{A}||\mathbf{C}|}$$

$$= \frac{|\mathbf{A}|(|\mathbf{A}||\mathbf{B}| + \mathbf{A} \cdot \mathbf{B})}{|\mathbf{A}||\mathbf{C}|}$$

$$= \frac{|\mathbf{A}||\mathbf{B}| + \mathbf{A} \cdot \mathbf{B}}{|\mathbf{C}|} \tag{1}$$

Let $\theta_2$ be the angle between **B** and **C**. Then

$$\cos \theta_2 = \frac{\mathbf{B} \cdot \mathbf{C}}{|\mathbf{B}||\mathbf{C}|}$$

$$= \frac{\mathbf{B} \cdot (|\mathbf{B}|\mathbf{A} + |\mathbf{A}|\mathbf{B})}{|\mathbf{B}||\mathbf{C}|}$$

$$= \frac{|\mathbf{B}|(\mathbf{A} \cdot \mathbf{B}) + |\mathbf{A}|(\mathbf{B} \cdot \mathbf{B})}{|\mathbf{B}||\mathbf{C}|}$$

$$= \frac{|\mathbf{B}|(\mathbf{A} \cdot \mathbf{B}) + |\mathbf{A}||\mathbf{B}|^2}{|\mathbf{B}||\mathbf{C}|}$$

$$= \frac{|\mathbf{B}|(\mathbf{A} \cdot \mathbf{B} + |\mathbf{A}||\mathbf{B}|)}{|\mathbf{B}||\mathbf{C}|}$$

$$= \frac{\mathbf{A} \cdot \mathbf{B} + |\mathbf{A}||\mathbf{B}|}{|\mathbf{C}|} \tag{2}$$

From (1) and (2) we have $\cos \theta_1 = \cos \theta_2$. Thus, the angle between **A** and **C** has the same measure as the angle between **B** and **C**.

## 18.4 PLANES

**18.4.1 Definition** If **N** is a given nonzero vector and $P_0$ is a given point, then the set of all points $P$ for which $\mathbf{V}(\overrightarrow{P_0P})$ and **N** are orthogonal is defined to be a *plane* through $P_0$ having **N** as a *normal vector.*

**18.4.2 Theorem** If $P_0(x_0, y_0, z_0)$ is a point in a plane and a normal vector to the plane is $\mathbf{N} = \langle a, b, c \rangle$, then an equation of the plane is

$$a(x - x_0) + b(y - y_0) + c(z - z_0) = 0$$

**18.4.3 Theorem** If $a, b,$ and $c$ are not all zero, the graph of an equation of the form

$$ax + by + cz + d = 0$$

is a plane, and $\langle a, b, c \rangle$ is a normal vector to the plane.

If one of the coefficients $a, b,$ and $c$ in Theorem 18.4.3 is zero, then the plane is parallel to a coordinate axis. If two of the coefficients $a, b,$ and $c$ are zero, then the plane is parallel to two coordinate axes and, hence, perpendicular to one of the coordinate axes. We summarize the possible cases in Table A.

**18.4.4 Definition** The *angle between two planes* is defined to be the angle between the normal vectors of the two planes.

**18.4.5 Definition** Two planes are *parallel* if and only if their normal vectors are parallel.

**18.4.6 Definition** Two planes are *perpendicular* if and only if their normal vectors are orthogonal.

**Table A**
$ax + by + cz + d = 0$

| | |
|---|---|
| $a = 0$ | Plane is parallel to the $x$-axis. |
| $b = 0$ | Plane is parallel to the $y$-axis. |
| $c = 0$ | Plane is parallel to the $z$-axis. |
| $a = 0$ and $b = 0$ | Plane is perpendicular to the $z$-axis. |
| $a = 0$ and $c = 0$ | Plane is perpendicular to the $y$-axis. |
| $b = 0$ and $c = 0$ | Plane is perpendicular to the $x$-axis. |

Let $\Phi$ be a plane with normal vector **N**, and let $P$ be a point. If $Q$ is any point in the plane $\Phi$, then the undirected distance between the point $P$ and the plane $\Phi$ is the absolute value of the scalar projection of vector $\mathbf{V}(\overrightarrow{PQ})$ on vector **N**. In symbols

$$d = \left| \frac{\mathbf{N} \cdot \mathbf{V}(\overrightarrow{PQ})}{|\mathbf{N}|} \right| \tag{I}$$

We illustrate the above formula in Exercise 20.

## *Exercises 18.4*

**2.** Find an equation of the plane containing the given point $P$ and having the given vector **N** as a normal vector.

$P(-1, 8, 3) \qquad \mathbf{N} = \langle -7, -1, 1 \rangle$

SOLUTION: We apply Theorem 18.4.2. Thus, an equation of the plane is

$$-7(x + 1) - (y - 8) + (z - 3) = 0$$
$$7x + y - z + 2 = 0$$

**6.** Find an equation of the plane containing the given three points.

$(0, 0, 2) \qquad (2, 4, 1) \qquad (-2, 3, 3)$

SOLUTION: By Theorem 18.4.3, an equation of the plane is of the form

$$ax + by + cz + d = 0 \tag{1}$$

Because the plane contains the point $(0, 0, 2)$, then Eq. (1) holds when $x = 0$, $y = 0$, and $z = 2$. Thus

$$2c + d = 0 \tag{2}$$

Similarly, we may substitute the coordinates of $(2, 4, 1)$ and $(-2, 3, 3)$ into (1). Thus,

$$2a + 4b + c + d = 0 \tag{3}$$

and

$$-2a + 3b + 3c + d = 0 \tag{4}$$

We solve Eqs. (2), (3), and (4) for $a, b,$ and $c$ in terms of $d$. From (2), we have

$$c = -\frac{1}{2}d \tag{5}$$

Substituting from (5) into (3), we get

$$2a + 4b - \frac{1}{2}d + d = 0$$

$$2a + 4b = -\frac{1}{2}d \qquad (6)$$

If $c = -\frac{1}{2}d$ in Eq. (4), we get

$$-2a + 3b - \frac{3}{2}d + d = 0$$

$$-2a + 3b = \frac{1}{2}d \qquad (7)$$

Adding the corresponding terms in Eq. (6) and (7), we get

$$7b = 0$$
$$b = 0 \qquad (8)$$

Substituting $b = 0$ into Eq. (6), we obtain

$$a = -\frac{1}{4}d \qquad (9)$$

Substituting the values of $a$, $b$, and $c$ from Eqs. (9), (8), and (5) into (1), we obtain

$$-\frac{1}{4}dx + 0 \cdot y - \frac{1}{2}dz + d = 0$$

Dividing by $-\frac{1}{4}d$, we get

$$x + 2z - 4 = 0$$

**10.** Draw a sketch of the given plane and find two unit vectors that are normal to the plane.

$$y + 2z - 4 = 0$$

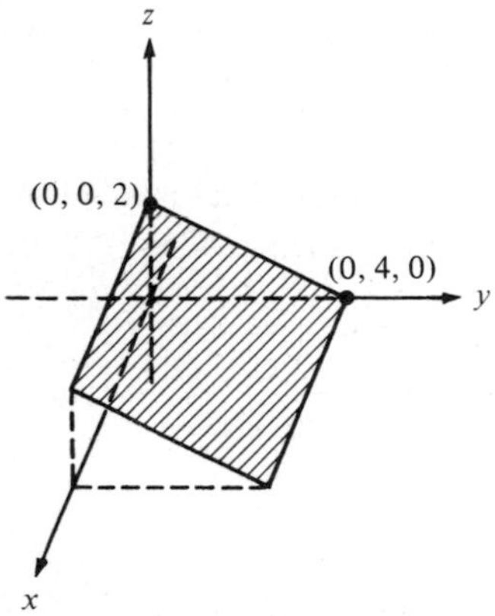

Figure 18.4.10

SOLUTION: Because the coefficient of $x$ is zero, the plane is parallel to the $x$-axis and perpendicular to the $yz$ plane. The $yz$ trace is the line in the $yz$ plane determined by $y + 2z - 4 = 0$. This line intersects the $y$-axis at $(0, 4, 0)$ and the $z$-axis at $(0, 0, 2)$. In Fig. 18.4.10 we show a sketch of the plane. By Theorem 18.4.3, a normal vector of the plane is $\mathbf{N} = \langle 0, 1, 2\rangle$. Because $|\mathbf{N}| = \sqrt{5}$, unit normal vectors are

$$\mathbf{U} = \frac{\mathbf{N}}{|\mathbf{N}|}$$
$$= \frac{1}{\sqrt{5}}\langle 0, 1, 2\rangle$$
$$= \left\langle 0, \frac{1}{5}\sqrt{5}, \frac{2}{5}\sqrt{5}\right\rangle$$

and

$$-\mathbf{U} = \left\langle 0, -\frac{1}{5}\sqrt{5}, -\frac{2}{5}\sqrt{5}\right\rangle$$

**14.** Find an equation of the plane parallel to the plane $4x - 2y + z - 1 = 0$ and containing the point $(2, 6, -1)$.

SOLUTION: Let $\Phi$ be the required plane and let $\Phi_1$ be the plane $4x - 2y + z - 1 = 0$. By Theorem 18.4.3, a normal vector of the plane $\Phi_1$ is $\mathbf{N}_1 = \langle 4, -2, 1\rangle$. Because $\Phi$ is parallel to $\Phi_1$, then $\mathbf{N}_1$ is also a normal vector of $\Phi$. We apply Theorem 18.4.2 to write an equation of $\Phi$. Thus, we have

$$4(x-2)-2(y-6)+(z+1)=0$$
$$4x-2y+z+5=0$$

**20.** Find the distance from the plane $2x+2y-z-6=0$ to the point $(2,2,-4)$.

SOLUTION: Let $P=(2,2,-4)$, and let $\Phi$ be the plane $2x+2y-z-6=0$. Then the distance between the point $P$ and the plane $\Phi$ is given by

$$d=\left|\frac{\mathbf{N}\cdot\mathbf{V}(\overrightarrow{PQ})}{|\mathbf{N}|}\right| \tag{1}$$

where $\mathbf{N}$ is a normal vector of plane $\Phi$ and $Q$ is any point in plane $\Phi$. By Theorem 18.4.3, we have

$$\mathbf{N}=\langle 2,2,-1\rangle \tag{2}$$

To find $Q$ we let $x=0$ and $y=0$ in the equation of $\Phi$ and solve for $z$. Thus, $Q=(0,0,-6)$. We have

$$\begin{aligned}\mathbf{V}(\overrightarrow{PQ})&=\langle 0,0,-6\rangle-\langle 2,2,-4\rangle\\&=\langle -2,-2,-2\rangle\end{aligned} \tag{3}$$

Substituting from (2) and (3) into (1), we obtain

$$\begin{aligned}d&=\left|\frac{\langle 2,2,-1\rangle\cdot\langle -2,-2,-2\rangle}{\sqrt{2^2+2^2+(-1)^2}}\right|\\&=\left|\frac{-6}{3}\right|\\&=2\end{aligned}$$

**22.** Find the perpendicular distance between the parallel planes

$$4x-8y-z+9=0 \quad\text{and}\quad 4x-8y-z-6=0$$

SOLUTION: Let $\Phi_1$ be the plane $4x-8y-z+9=0$ and let $\Phi_2$ be the plane $4x-8y-z-6=0$. If $P$ is a point in plane $\Phi_1$, then the distance between the planes is also the distance between the point $P$ and the plane $\Phi_2$. If $x=0$ and $y=0$ in the equation $4x-8y-z+9=0$, we get $z=9$. Thus, we take $P=(0,0,9)$. To find the distance between $P$ and plane $\Phi_2$, we use the formula

$$d=\left|\frac{\mathbf{N}\cdot\mathbf{V}(\overrightarrow{PQ})}{|\mathbf{N}|}\right| \tag{1}$$

where $\mathbf{N}$ is a normal vector of plane $\Phi_2$ and $Q$ is any point in $\Phi_2$. We have

$$\mathbf{N}=\langle 4,-8,-1\rangle \tag{2}$$

If $x=0$ and $y=0$ in the equation $4x-8y-z-6=0$, we get $z=-6$. Thus we take $Q=(0,0,-6)$, and so

$$\mathbf{V}(\overrightarrow{PQ})=\langle 0,0,-15\rangle \tag{3}$$

Substituting from (2) and (3) into (1), we get

$$\begin{aligned}d&=\left|\frac{\langle 4,-8,-1\rangle\cdot\langle 0,0,-15\rangle}{\sqrt{4^2+(-8)^2+(-1)^2}}\right|\\&=\frac{5}{3}\end{aligned}$$

**24.** Prove that the undirected distance from the plane $ax+by+cz+d=0$ to the point $(x_0,y_0,z_0)$ is given by

$$\frac{|ax_0 + by_0 + cz_0 + d|}{\sqrt{a^2 + b^2 + c^2}}$$

SOLUTION: We use formula (I). Thus, the undirected distance between the point $P_0(x_0, y_0, z_0)$ and the plane $ax + by + cz + d = 0$ is given by

$$\text{distance} = \left| \frac{\mathbf{N} \cdot \mathbf{V}(\overrightarrow{P_0Q})}{|\mathbf{N}|} \right| \tag{1}$$

where $\mathbf{N}$ is a normal vector of the plane and $Q$ is any point in the plane. We have

$$\mathbf{N} = \langle a, b, c \rangle \tag{2}$$

As in Exercise 20, we may take $Q = (0, 0, -d/c)$. Thus,

$$\mathbf{V}(\overrightarrow{P_0Q}) = \left\langle -x_0, -y_0, -\frac{d}{c} - z_0 \right\rangle \tag{3}$$

Substituting from (2) and (3) into (1), we have

$$\begin{aligned} \text{distance} &= \left| \frac{\langle a, b, c \rangle \cdot \left\langle -x_0, -y_0, -\frac{d}{c} - z_0 \right\rangle}{\sqrt{a^2 + b^2 + c^2}} \right| \\ &= \left| \frac{-ax_0 - by_0 - d - cz_0}{\sqrt{a^2 + b^2 + c^2}} \right| \\ &= \frac{|ax_0 + by_0 + cz_0 + d|}{\sqrt{a^2 + b^2 + c^2}} \end{aligned}$$

## 18.5 LINES IN $R^3$

A line $L$ that contains the point $(x_0, y_0, z_0)$ and is parallel to representations of the vector $\langle a, b, c \rangle$ has *parametric equations*

$$x = x_0 + at \qquad y = y_0 + bt \qquad z = z_0 + ct \tag{I}$$

The vector $\langle a, b, c \rangle$ is called a *direction vector* of the line $L$, and the set of numbers $[a, b, c]$ is called a set of *direction numbers* for $L$. Any set of numbers that is proportional to the set $[a, b, c]$ is also a set of direction numbers for $L$. Thus, parametric equations are not unique. If two lines are parallel, a set of direction numbers for one line is also a set of direction numbers for the other line. If $\langle a_1, b_1, c_1 \rangle$ is a direction vector of line $L_1$ and $\langle a_2, b_2, c_2 \rangle$ is a direction vector of line $L_2$, then $L_1$ is perpendicular to $L_2$ if and only if

$$\langle a_1, b_1, c_1 \rangle \cdot \langle a_2, b_2, c_2 \rangle = 0$$

Let $\Phi$ be the plane $ax + by + cz + d = 0$ with normal vector $\langle a, b, c \rangle$. If line $L$ is perpendicular to plane $\Phi$, then $\langle a, b, c \rangle$ is a direction vector for line $L$. If line $L$ lies in plane $\Phi$ or if $L$ is parallel to $\Phi$, then $\langle a', b', c' \rangle$ is a direction vector of line $L$ if and only if

$$\langle a, b, c \rangle \cdot \langle a', b', c' \rangle = 0$$

A line $L$ that contains the point $(x_0, y_0, z_0)$ and has direction vector $\langle a, b, c \rangle$ with $a \neq 0, b \neq 0$, and $c \neq 0$, has *symmetric equations*

$$\frac{x - x_0}{a} = \frac{y - y_0}{b} = \frac{z - z_0}{c} \tag{II}$$

If $a \neq 0, b \neq 0$, and $c = 0$, line $L$ has symmetric equations

$$\frac{x - x_0}{a} = \frac{y - y_0}{b} \quad \text{and} \quad z = z_0$$

If $a \neq 0, b = 0$, and $c \neq 0$, line $L$ has symmetric equations

$$\frac{x - x_0}{a} = \frac{z - z_0}{c} \quad \text{and} \quad y = y_0$$

If $a = 0$, $b \neq 0$, and $c \neq 0$, line $L$ has symmetric equations

$$x = x_0 \quad \text{and} \quad \frac{y - y_0}{b} = \frac{z - z_0}{c}$$

## *Exercises 18.5*

**4.** Find parametric and symmetric equations for the line through the origin and perpendicular to the lines having direction numbers $[4, 2, 1]$ and $[-3, -2, 1]$.

SOLUTION: Let $[a, b, c]$ be a set of direction numbers of the required line $L$. Because line $L$ is perpendicular to the line with direction numbers $[4, 2, 1]$, the vector $\langle a, b, c\rangle$ is orthogonal to the vector $\langle 4, 2, 1\rangle$. Thus,

$$\langle a, b, c\rangle \cdot \langle 4, 2, 1\rangle = 0$$
$$4a + 2b + c = 0 \tag{1}$$

Because line $L$ is perpendicular to the line with direction numbers $[-3, -2, 1]$, we have

$$\langle a, b, c\rangle \cdot \langle -3, -2, 1\rangle = 0$$
$$-3a - 2b + c = 0 \tag{2}$$

We solve Eqs. (1) and (2) for $a$ and $b$ in terms of $c$. Thus,

$$a = -2c \quad \text{and} \quad b = \frac{7}{2}c$$

Therefore $[-2c, \frac{7}{2}c, c]$ is a set of direction numbers for the line $L$. We take $c = 2$ (to eliminate the fraction) and obtain $[-4, 7, 2]$ as a set of direction numbers for $L$. Because $L$ contains the origin, we use parametric equations (I) with $(x_0, y_0, z_0) = (0, 0, 0)$. Thus, parametric equations for $L$ are

$$x = -4t \qquad y = 7t \qquad z = 2t$$

We use symmetric equations (II) to obtain

$$\frac{x}{-4} = \frac{y}{7} = \frac{z}{2}$$

**8.** Prove that the line $x + 1 = -\frac{1}{2}(y - 6) = z$ lies in the plane $3x + y - z = 3$.

SOLUTION: We write the equations of the line in symmetric form. Thus, we have

$$\frac{x+1}{1} = \frac{y-6}{-2} = \frac{z}{1}$$

Hence, the vector $\mathbf{A} = \langle 1, -2, 1\rangle$ is a direction vector of the line and the point $P(-1, 6, 0)$ lies on the line. A normal vector of the plane is $\mathbf{N} = \langle 3, 1, -1\rangle$. Because

$$\mathbf{A} \cdot \mathbf{N} = \langle 1, -2, 1\rangle \cdot \langle 3, 1, -1\rangle = 0$$

the vectors $\mathbf{A}$ and $\mathbf{N}$ are orthogonal. Thus, the line is either parallel to the plane or lies in the plane. Because the coordinates of $P(-1, 6, 0)$ satisfy the equation $3x + y - z = 3$, the point $P$ lies in the plane. Because $P$ also lies on the line, then the line and the plane are not parallel. Therefore, the line lies in the plane.

**12.** The planes through a line that are perpendicular to the coordinate planes are called the *projecting planes* of the line. Find equations of the projecting planes of the given line and draw a sketch of the line.

$$2x - y + z - 7 = 0 \tag{1}$$

$$4x - y + 3z - 13 = 0 \tag{2}$$

SOLUTION: Let $L$ be the given line. Eliminating $z$ from Eqs. (1) and (2), we obtain

$$x - y - 4 = 0 \tag{3}$$

Every point that lies on line $L$ satisfies Eq. (3). Thus, the graph of Eq. (3) contains the line $L$. By Theorem 18.4.3, the graph of (3) is a plane. Furthermore, because the coefficient of $z$ in Eq. (3) is zero, the plane is perpendicular to the $xy$ plane. Thus, the graph of (3) is a projecting plane of the line $L$. Next, we eliminate $y$ from Eqs. (1) and (2). The result is

$$x + z - 3 = 0 \tag{4}$$

which is the projecting plane perpendicular to the $xz$ plane. Eliminating $x$ from Eqs. (1) and (2), we obtain

$$y + z + 1 = 0$$

which is the projecting plane perpendicular to the $yz$ plane. The line $L$ is the intersection of any two of its projecting planes. Thus, we draw sketches of planes (3) and (4), as shown in Fig. 18.5.12. The intersection of the planes is line $L$.

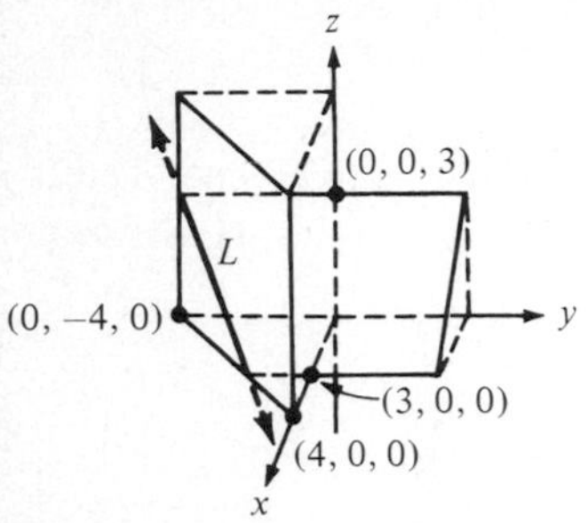

Figure 18.5.12

**16.** Find an equation of the plane containing the given intersecting lines.

$$\frac{x}{2} = \frac{y-2}{3} = \frac{z-1}{1} \quad \text{and} \quad \frac{x}{1} = \frac{y-2}{-1} = \frac{z-1}{1}$$

SOLUTION: The lines intersect in the point $P(0, 2, 1)$ because the coordinates of $P$ satisfy the equations for both lines. Let $\mathbf{N} = \langle a, b, c\rangle$ be a normal vector of the required plane. A direction vector of the first line is $\mathbf{A} = \langle 2, 3, 1\rangle$ and a direction vector of the second line is $\mathbf{B} = \langle 1, -1, 1\rangle$. Because the plane contains the lines, $\mathbf{N}$ is orthogonal to both $\mathbf{A}$ and $\mathbf{B}$. Thus

$$\langle 2, 3, 1\rangle \cdot \langle a, b, c\rangle = 0$$
$$2a + 3b + c = 0 \tag{1}$$

and

$$\langle 1, -1, 1\rangle \cdot \langle a, b, c\rangle = 0$$
$$a - b + c = 0 \tag{2}$$

Solving Eqs. (1) and (2) for $a$ and $b$ in terms of $c$, we get

$$a = -\frac{4}{5}c \quad \text{and} \quad b = \frac{1}{5}c$$

Thus, $\mathbf{N} = \langle -\frac{4}{5}c, \frac{1}{5}c, c\rangle$ is a normal vector of the plane. We take $c = 5$ and obtain $\mathbf{N} = \langle -4, 1, 5\rangle$. Applying Theorem 18.4.3, with $\mathbf{N}$ and the point $P = (0, 2, 1)$, we get an equation of the plane

$$-4x + (y - 2) + 5(z - 1) = 0$$
$$4x - y - 5z + 7 = 0$$

**20.** Find equations of the line through the origin, perpendicular to the line $x = y - 5$, $z = 2y - 3$, and intersecting the line $y = 2x + 1$, $z = x + 2$.

SOLUTION: Let $L$ be the required line and let $L_1$ be the line $y = 2x + 1, z = x + 2$ that intersects $L$. If $P(a, b, c)$ is the point of intersection of lines $L$ and $L_1$, then the coordinates of $P$ satisfy the equations of $L_1$. Thus, we let $x = a$, $y = b$, and $z = c$ in the equations of $L_1$, obtaining

$$b = 2a + 1 \tag{1}$$

$$c = a + 2 \tag{2}$$

Because line $L$ contains the origin and point $P$, then the vector $\mathbf{V}(\overrightarrow{OP}) = \langle a, b, c\rangle$ is a direction vector of line $L$. Let $L_2$ be the line $x = y - 5, z = 2y - 3$. We find symmetric equations of $L_2$. Solving each equation for $y$, we have

$$x + 5 = y \quad \text{and} \quad y = \frac{1}{2}(z + 3)$$

Thus, symmetric equations of $L_2$ are

$$\frac{x+5}{1} = \frac{y}{1} = \frac{z+3}{2}$$

Hence, a direction vector of line $L_2$ is $\langle 1, 1, 2\rangle$. Because line $L$ is perpendicular to line $L_2$, any direction vector of $L$ is orthogonal to a direction vector of $L_2$. Thus,

$$\langle 1, 1, 2\rangle \cdot \langle a, b, c\rangle = 0$$

$$a + b + 2c = 0 \tag{3}$$

Solving Eqs. (1), (2), and (3) simultaneously, we obtain

$$a = -1 \qquad b = -1 \qquad c = 1$$

Thus, $[-1, -1, 1]$ is a set of direction numbers of line $L$. We use these direction numbers and the origin to write symmetric equations of $L$, which are

$$\frac{x}{-1} = \frac{y}{-1} = \frac{z}{1}$$

**22.** Find the perpendicular distance from the origin to the line

$$x = -2 + \frac{6}{7}t \qquad y = 7 - \frac{2}{7}t \qquad z = 4 + \frac{3}{7}t \tag{1}$$

SOLUTION: Let $L$ be the given line and let $P(a, b, c)$ be the point on line $L$ such that line $OP$ is perpendicular to $L$. The required distance is given by $|\mathbf{V}(\overrightarrow{OP})|$. Because $[\frac{6}{7}, -\frac{2}{7}, \frac{3}{7}]$ are direction numbers for line $L$, then the vector $\mathbf{A} = \langle 6, -2, 3\rangle$ is a direction vector of $L$. Because $\mathbf{A}$ is orthogonal to $\mathbf{V}(\overrightarrow{OP}) = \langle a, b, c\rangle$, we have

$$\langle 6, -2, 3\rangle \cdot \langle a, b, c\rangle = 0$$

$$6a - 2b + 3c = 0 \tag{2}$$

Because line $L$ contains point $P$, the coordinates of $P$ must satisfy all equations of $L$. Thus, we let $x = a$, $y = b$, and $z = c$ in (1), the given parametric equations of $L$, resulting in

$$a = -2 + \frac{6}{7}t \qquad b = 7 - \frac{2}{7}t \qquad c = 4 + \frac{3}{7}t \tag{3}$$

Substituting from Eqs. (3) into (2), we obtain

$$6\left(-2 + \frac{6}{7}t\right) - 2\left(7 - \frac{2}{7}t\right) + 3\left(4 + \frac{3}{7}t\right) = 0$$

Solving for $t$, we get $t = 2$. Substituting $t = 2$ into Eqs. (3), we obtain

$$a = -\frac{2}{7} \qquad b = \frac{45}{7} \qquad c = \frac{34}{7}$$

Hence, the required distance is given by

$$|\mathbf{V}(\overrightarrow{OP})| = \sqrt{\left(-\frac{2}{7}\right)^2 + \left(\frac{45}{7}\right)^2 + \left(\frac{34}{7}\right)^2} = \sqrt{65}$$

## 18.6 CROSS PRODUCT

**18.6.1 Definition** If $\mathbf{A} = \langle a_1, a_2, a_3 \rangle$ and $\mathbf{B} = \langle b_1, b_2, b_3 \rangle$, then the *cross product* of **A** and **B**, denoted by $\mathbf{A} \times \mathbf{B}$, is given by

$$\mathbf{A} \times \mathbf{B} = \langle a_2b_3 - a_3b_2, a_3b_1 - a_1b_3, a_1b_2 - a_2b_1 \rangle$$

This is an operation for vectors in $V_3$ that is not defined for vectors in $V_2$. It is not necessary to memorize the definition for the cross product. If $\mathbf{A} = \langle a_1, a_2, a_3 \rangle$ and $\mathbf{B} = \langle b_1, b_2, b_3 \rangle$, we may use the rules for evaluating a third order determinant to find $\mathbf{A} \times \mathbf{B}$. That is,

$$\mathbf{A} \times \mathbf{B} = \begin{vmatrix} \mathbf{i} & \mathbf{j} & \mathbf{k} \\ a_1 & a_2 & a_3 \\ b_1 & b_2 & b_3 \end{vmatrix}$$

$$= \begin{vmatrix} a_2 & a_3 \\ b_2 & b_3 \end{vmatrix} \mathbf{i} - \begin{vmatrix} a_1 & a_3 \\ b_1 & b_3 \end{vmatrix} \mathbf{j} + \begin{vmatrix} a_1 & a_2 \\ b_1 & b_2 \end{vmatrix} \mathbf{k}$$

We illustrate the use of determinants to find $\mathbf{A} \times \mathbf{B}$ in Exercise 4.

The operation of cross product for vectors is neither commutative nor associative. However, cross multiplication of vectors is distributive with respect to vector addition. Furthermore, scalar multiplication and cross multiplication are associative. We formally state these properties in the following theorems.

**18.6.4 Theorem** If **A**, **B**, and **C** are any vectors in $V_3$, then

$$\mathbf{A} \times (\mathbf{B} + \mathbf{C}) = \mathbf{A} \times \mathbf{B} + \mathbf{A} \times \mathbf{C}$$

**18.6.5 Theorem** If **A** and **B** are any two vectors in $V_3$ and $c$ is a scalar, then

(i) $(c\mathbf{A}) \times \mathbf{B} = \mathbf{A} \times (c\mathbf{B})$
(ii) $(c\mathbf{A}) \times \mathbf{B} = c(\mathbf{A} \times \mathbf{B})$

In Exercise 8 we illustrate how to use Theorems 18.6.4 and 18.6.5 together with the following results for cross multiplying the unit vectors **i**, **j**, and **k** to find the cross product to any two vectors in $V_3$.

$$\mathbf{i} \times \mathbf{i} = \mathbf{j} \times \mathbf{j} = \mathbf{k} \times \mathbf{k} = \mathbf{0}$$
$$\mathbf{i} \times \mathbf{j} = \mathbf{k} \qquad \mathbf{j} \times \mathbf{k} = \mathbf{i} \qquad \mathbf{k} \times \mathbf{i} = \mathbf{j}$$
$$\mathbf{j} \times \mathbf{i} = -\mathbf{k} \qquad \mathbf{k} \times \mathbf{j} = -\mathbf{i} \qquad \mathbf{i} \times \mathbf{k} = -\mathbf{j}$$

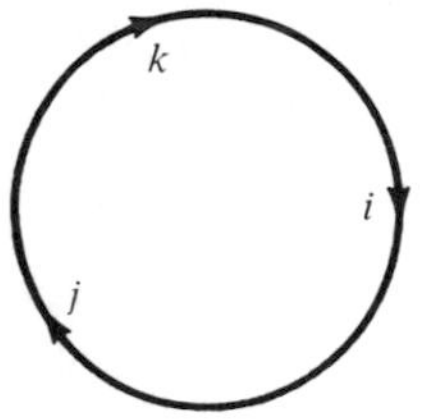

Figure 18.6.1

Note that the cross product of each unit vector with itself is the vector **O**. In Figure 18.6.1 we illustrate a method to remember the remaining six cross products of the unit vectors. The cross product of two consecutive vectors, in the clockwise direction, is the next vector; the cross product of two consecutive vectors, in the counterclockwise direction, is the negative of the next vector.

The following two theorems give a geometric interpretation of the cross product of two vectors.

**18.6.6 Theorem** If **A** and **B** are two vectors in $V_3$ and $\theta$ is the radian measure of the angle between **A** and **B**, then

$$|\mathbf{A} \times \mathbf{B}| = |\mathbf{A}||\mathbf{B}| \sin \theta$$

**18.6.9 Theorem** If **A** and **B** are two vectors in $V_3$, then the vector $\mathbf{A} \times \mathbf{B}$ is orthogonal to both **A** and **B**. Thus,

$$(\mathbf{A} \times \mathbf{B}) \cdot \mathbf{A} = 0 \quad \text{and} \quad (\mathbf{A} \times \mathbf{B}) \cdot \mathbf{B} = 0$$

We may use this fact to check the calculation of the cross product of vectors **A** and **B**.

If representations of the vectors **A** and **B** are two adjacent sides of a parallelogram, then the number of square units in the area of the parallelogram is given by $|\mathbf{A} \times \mathbf{B}|$. If representations of the vectors **A** and **B** are two line segments that lie in a plane, then the vector $\mathbf{A} \times \mathbf{B}$ is a normal vector of the plane. This fact can be used to more easily solve some of the exercises in Section 18.5. If representations of the vectors **A, B,** and **C** are three nonparallel edges of a parallelepiped, then the number of cubic units in the volume of the parallelepiped is given by $|\mathbf{A} \times \mathbf{B} \cdot \mathbf{C}|$.

Let $L$ be a line with direction vector **L**, and let $P$ be a point. If $Q$ is any point on the line $L$, then the undirected distance between the point $P$ and the line $L$ is given by

$$d = \frac{|\mathbf{L} \times \mathbf{V}(\overrightarrow{PQ})|}{|\mathbf{L}|}$$

Following are some additional theorems for cross products.

**18.6.2 Theorem** If **A** is any vector in $V_3$, then

(i) $\mathbf{A} \times \mathbf{A} = \mathbf{0}$
(ii) $\mathbf{0} \times \mathbf{A} = \mathbf{0}$
(iii) $\mathbf{A} \times \mathbf{0} = \mathbf{0}$

**18.6.3 Theorem** If **A** and **B** are any vectors in $V_3$,

$$\mathbf{A} \times \mathbf{B} = -(\mathbf{B} \times \mathbf{A})$$

**18.6.7 Theorem** If **A** and **B** are two vectors in $V_3$, **A** and **B** are parallel if and only if $\mathbf{A} \times \mathbf{B} = \mathbf{0}$.

**18.6.8 Theorem** If **A, B,** and **C** are vectors in $V_3$, then

$$\mathbf{A} \cdot \mathbf{B} \times \mathbf{C} = \mathbf{A} \times \mathbf{B} \cdot \mathbf{C}$$

## *Exercises 18.6*

In Exercises 1-12, let $\mathbf{A} = \langle 1, 2, 3 \rangle$, $\mathbf{B} = \langle 4, -3, -1 \rangle$, $\mathbf{C} = \langle -5, -3, 5 \rangle$, $\mathbf{D} = \langle -2, 1, 6 \rangle$, $\mathbf{E} = \langle 4, 0, -7 \rangle$, and $\mathbf{F} = \langle 0, 2, 1 \rangle$.

**4.** Find $(\mathbf{C} \times \mathbf{E}) \cdot (\mathbf{D} \times \mathbf{F})$.

SOLUTION: We use the method of determinants.

$$\mathbf{C} \times \mathbf{E} = \begin{vmatrix} \mathbf{i} & \mathbf{j} & \mathbf{k} \\ -5 & -3 & 5 \\ 4 & 0 & -7 \end{vmatrix}$$

$$= \begin{vmatrix} -3 & 5 \\ 0 & -7 \end{vmatrix} \mathbf{i} - \begin{vmatrix} -5 & 5 \\ 4 & -7 \end{vmatrix} \mathbf{j} + \begin{vmatrix} -5 & -3 \\ 4 & 0 \end{vmatrix} \mathbf{k}$$

$$= [(-3)(-7) - 0 \cdot 5]\mathbf{i} - [(-5)(-7) - 4 \cdot 5]\mathbf{j} + [-5 \cdot 0 - 4(-3)]\mathbf{k}$$

$$= 21\mathbf{i} - 15\mathbf{j} + 12\mathbf{k}$$

$$\mathbf{D} \times \mathbf{F} = \begin{vmatrix} \mathbf{i} & \mathbf{j} & \mathbf{k} \\ -2 & 1 & 6 \\ 0 & 2 & 1 \end{vmatrix}$$

$$= \begin{vmatrix} 1 & 6 \\ 2 & 1 \end{vmatrix} \mathbf{i} - \begin{vmatrix} -2 & 6 \\ 0 & 1 \end{vmatrix} \mathbf{j} + \begin{vmatrix} -2 & 1 \\ 0 & 2 \end{vmatrix} \mathbf{k}$$

$$= -11\mathbf{i} + 2\mathbf{j} - 4\mathbf{k}$$

Thus,

$$\begin{aligned}(\mathbf{C} \times \mathbf{E}) \cdot (\mathbf{D} \times \mathbf{F}) &= (21\mathbf{i} - 15\mathbf{j} + 12\mathbf{k}) \cdot (-11\mathbf{i} + 2\mathbf{j} - 4\mathbf{k}) \\ &= (21)(-11) + (-15)(2) + 12(-4) \\ &= -309\end{aligned}$$

8. Verify Theorem 18.6.5(ii) for vectors **A** and **B** and $c = 3$.

SOLUTION: We use the method of unit vectors.

$$\begin{aligned}(c\mathbf{A}) \times \mathbf{B} &= 3\langle 1, 2, 3\rangle \times \langle 4, -3, -1\rangle \\ &= (3\mathbf{i} + 6\mathbf{j} + 9\mathbf{k}) \times (4\mathbf{i} - 3\mathbf{j} - \mathbf{k}) \\ &= (3\mathbf{i} \times 4\mathbf{i}) + (3\mathbf{i} \times -3\mathbf{j}) + (3\mathbf{i} \times -\mathbf{k}) + (6\mathbf{j} \times 4\mathbf{i}) + (6\mathbf{j} \times -3\mathbf{j}) \\ &\quad + (6\mathbf{j} \times -\mathbf{k}) + (9\mathbf{k} \times 4\mathbf{i}) + (9\mathbf{k} \times -3\mathbf{j}) + (9\mathbf{k} \times -\mathbf{k}) \\ &= 12(\mathbf{i} \times \mathbf{i}) - 9(\mathbf{i} \times \mathbf{j}) - 3(\mathbf{i} \times \mathbf{k}) + 24(\mathbf{j} \times \mathbf{i}) - 18(\mathbf{j} \times \mathbf{j}) - 6(\mathbf{j} \times \mathbf{k}) \\ &\quad + 36(\mathbf{k} \times \mathbf{i}) - 27(\mathbf{k} \times \mathbf{j}) - 9(\mathbf{k} \times \mathbf{k}) \\ &= 12(\mathbf{0}) - 9(\mathbf{k}) - 3(-\mathbf{j}) + 24(-\mathbf{k}) - 18(\mathbf{0}) - 6(\mathbf{i}) + 36(\mathbf{j}) - 27(-\mathbf{i}) \\ &\quad - 9(\mathbf{0}) \\ &= 21\mathbf{i} + 39\mathbf{j} - 33\mathbf{k} \qquad (1)\end{aligned}$$

Furthermore,

$$\begin{aligned}\mathbf{A} \times \mathbf{B} &= (\mathbf{i} + 2\mathbf{j} + 3\mathbf{k}) \times (4\mathbf{i} - 3\mathbf{j} - \mathbf{k}) \\ &= -3(\mathbf{i} \times \mathbf{j}) - (\mathbf{i} \times \mathbf{k}) + 8(\mathbf{j} \times \mathbf{i}) - 2(\mathbf{j} \times \mathbf{k}) + 12(\mathbf{k} \times \mathbf{i}) - 9(\mathbf{k} \times \mathbf{j}) \\ &= -3(\mathbf{k}) - (-\mathbf{j}) + 8(-\mathbf{k}) - 2(\mathbf{i}) + 12(\mathbf{j}) - 9(-\mathbf{i}) \\ &= 7\mathbf{i} + 13\mathbf{j} - 11\mathbf{k}\end{aligned}$$

Thus,

$$\begin{aligned}c(\mathbf{A} \times \mathbf{B}) &= 3(7\mathbf{i} + 13\mathbf{j} - 11\mathbf{k}) \\ &= 21\mathbf{i} + 39\mathbf{j} - 33\mathbf{k} \qquad (2)\end{aligned}$$

Comparing (1) and (2), we see that

$$(c\mathbf{A}) \times \mathbf{B} = c(\mathbf{A} \times \mathbf{B})$$

12. Find $|\mathbf{A} \times \mathbf{B}||\mathbf{C} \times \mathbf{D}|$.

SOLUTION: We illustrate the two methods for cross products.

$$\begin{aligned}\mathbf{A} \times \mathbf{B} &= (\mathbf{i} + 2\mathbf{j} + 3\mathbf{k}) \times (4\mathbf{i} - 3\mathbf{j} - \mathbf{k}) \\ &= -3(\mathbf{k}) - (-\mathbf{j}) + 8(-\mathbf{k}) - 2(\mathbf{i}) + 12(\mathbf{j}) - 9(-\mathbf{i}) \\ &= 7\mathbf{i} + 13\mathbf{j} - 11\mathbf{k}\end{aligned}$$

$$\mathbf{C} \times \mathbf{D} = \begin{vmatrix} \mathbf{i} & \mathbf{j} & \mathbf{k} \\ -5 & -3 & 5 \\ -2 & 1 & 6 \end{vmatrix}$$

$$= \begin{vmatrix} -3 & 5 \\ 1 & 6 \end{vmatrix}\mathbf{i} - \begin{vmatrix} -5 & 5 \\ -2 & 6 \end{vmatrix}\mathbf{j} + \begin{vmatrix} -5 & -3 \\ -2 & 1 \end{vmatrix}\mathbf{k}$$

$$= -23\mathbf{i} + 20\mathbf{j} - 11\mathbf{k}$$

Hence,

$$\begin{aligned}|\mathbf{A} \times \mathbf{B}||\mathbf{C} \times \mathbf{D}| &= \sqrt{7^2 + 13^2 + (-11)^2} \cdot \sqrt{(-23)^2 + 20^2 + (-11)^2} \\ &= \sqrt{339} \cdot \sqrt{1050} \\ &= 15\sqrt{1582}\end{aligned}$$

**18.** Given the two unit vectors

$$\mathbf{A} = \frac{4}{9}\mathbf{i} + \frac{7}{9}\mathbf{j} - \frac{4}{9}\mathbf{k} \quad \text{and} \quad \mathbf{B} = -\frac{2}{3}\mathbf{i} + \frac{2}{3}\mathbf{j} + \frac{1}{3}\mathbf{k}$$

If $\theta$ is the radian measure of the angle between **A** and **B**, find $\sin\theta$ in two ways: **(a)** by using the cross product (formula (3) of this section), and **(b)** by using the dot product and a trigonometric identity.

SOLUTION:

**(a)** From Theorem 18.6.6 we have

$$\sin\theta = \frac{|\mathbf{A} \times \mathbf{B}|}{|\mathbf{A}||\mathbf{B}|}$$

$$= |\mathbf{A} \times \mathbf{B}|$$

because $|\mathbf{A}| = 1$ and $|\mathbf{B}| = 1$. We have

$$\mathbf{A} \times \mathbf{B} = \frac{1}{9}(4\mathbf{i} + 7\mathbf{j} - 4\mathbf{k}) \times \frac{1}{3}(-2\mathbf{i} + 2\mathbf{j} + \mathbf{k})$$

$$= \frac{1}{27}(8\mathbf{k} - 4\mathbf{j} + 14\mathbf{k} + 7\mathbf{i} + 8\mathbf{j} + 8\mathbf{i})$$

$$= \frac{1}{27}(15\mathbf{i} + 4\mathbf{j} + 22\mathbf{k})$$

Therefore,

$$\sin\theta = |\mathbf{A} \times \mathbf{B}|$$

$$= \frac{1}{27}\sqrt{15^2 + 4^2 + 22^2}$$

$$= \frac{5}{27}\sqrt{29}$$

**(b)** From Theorem 18.3.4

$$\cos\theta = \frac{\mathbf{A} \cdot \mathbf{B}}{|\mathbf{A}||\mathbf{B}|}$$

$$= \mathbf{A} \cdot \mathbf{B}$$

because $|\mathbf{A}| = |\mathbf{B}| = 1$. We have

$$\mathbf{A} \cdot \mathbf{B} = \frac{1}{9}(4\mathbf{i} + 7\mathbf{j} - 4\mathbf{k}) \cdot \frac{1}{3}(-2\mathbf{i} + 2\mathbf{j} + \mathbf{k})$$

$$= \frac{1}{27}[4(-2) + 7 \cdot 2 + (-4) \cdot 1]$$

$$= \frac{2}{27}$$

Thus,

$$\cos\theta = \frac{2}{27}$$

Because $0 \leqslant \theta \leqslant \pi$, we have

$$\sin\theta = \sqrt{1 - \cos^2\theta}$$

$$= \sqrt{1 - \left(\frac{2}{27}\right)^2}$$

$$= \frac{5}{27}\sqrt{29}$$

**24.** Find the area of the triangle having vertices at $(4, 5, 6)$, $(4, 4, 5)$, and $(3, 5, 5)$.

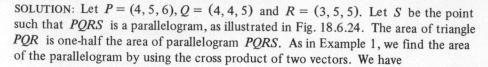

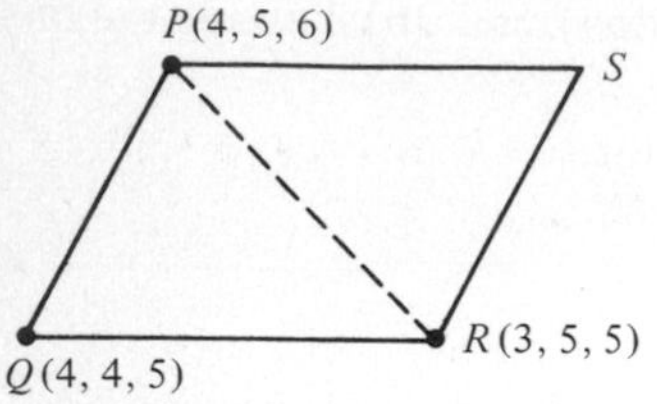

Figure 18.6.24

SOLUTION: Let $P = (4, 5, 6)$, $Q = (4, 4, 5)$ and $R = (3, 5, 5)$. Let $S$ be the point such that $PQRS$ is a parallelogram, as illustrated in Fig. 18.6.24. The area of triangle $PQR$ is one-half the area of parallelogram $PQRS$. As in Example 1, we find the area of the parallelogram by using the cross product of two vectors. We have

$$\mathbf{V}(\overrightarrow{QP}) = \mathbf{j} + \mathbf{k}$$
$$\mathbf{V}(\overrightarrow{QR}) = -\mathbf{i} + \mathbf{j}$$

Thus,

$$\begin{aligned}\mathbf{V}(\overrightarrow{QP}) \times \mathbf{V}(\overrightarrow{QR}) &= (\mathbf{j} + \mathbf{k}) \times (-\mathbf{i} + \mathbf{j}) \\ &= -(\mathbf{j} \times \mathbf{i}) - (\mathbf{k} \times \mathbf{i}) + (\mathbf{k} \times \mathbf{j}) \\ &= \mathbf{k} - \mathbf{j} - \mathbf{i} \\ &= -\mathbf{i} - \mathbf{j} + \mathbf{k}\end{aligned}$$

Therefore, the number of square units in the area of the parallelogram is

$$|\mathbf{V}(\overrightarrow{QP}) \times \mathbf{V}(\overrightarrow{QR})| = \sqrt{(-1)^2 + (-1)^2 + 1^2} = \sqrt{3}$$

Hence, the area of the triangle $PQR$ is $\frac{1}{2}\sqrt{3}$ square units.

**26.** Find a unit vector whose representations are perpendicular to the plane containing $\overrightarrow{PQ}$ and $\overrightarrow{PR}$ if $\overrightarrow{PQ}$ is a representation of the vector $\mathbf{i} + 3\mathbf{j} - 2\mathbf{k}$ and $\overrightarrow{PR}$ is a representation of the vector $2\mathbf{i} - \mathbf{j} - \mathbf{k}$.

SOLUTION: Let

$$\begin{aligned}\mathbf{N} &= (\mathbf{i} + 3\mathbf{j} - 2\mathbf{k}) \times (2\mathbf{i} - \mathbf{j} - \mathbf{k}) \\ &= -5\mathbf{i} - 3\mathbf{j} - 7\mathbf{k}\end{aligned}$$

By Theorem 18.6.9 the vector $\mathbf{N}$ is a normal vector to the plane that contains $\overrightarrow{PQ}$ and $\overrightarrow{PR}$, and hence the required unit vector is parallel to $\mathbf{N}$. We take

$$\mathbf{U} = \frac{\mathbf{N}}{|\mathbf{N}|} = \frac{1}{\sqrt{(-5)^2 + (-3)^2 + (-7)^2}}(-5\mathbf{i} - 3\mathbf{j} - 7\mathbf{k})$$
$$= \frac{1}{\sqrt{83}}(-5\mathbf{i} - 3\mathbf{j} - 7\mathbf{k})$$

Both $\mathbf{U}$ and $-\mathbf{U}$ are unit vectors whose representations are perpendicular to the plane.

**30.** If $\mathbf{A}$ and $\mathbf{B}$ are any two vectors in $V_3$, prove that $(\mathbf{A} - \mathbf{B}) \times (\mathbf{A} + \mathbf{B}) = 2(\mathbf{A} \times \mathbf{B})$.

SOLUTION: Applying Theorem 18.6.4, we have

$$(\mathbf{A} - \mathbf{B}) \times (\mathbf{A} + \mathbf{B}) = (\mathbf{A} - \mathbf{B}) \times \mathbf{A} + (\mathbf{A} - \mathbf{B}) \times \mathbf{B} \tag{1}$$

Next, we apply Theorem 18.6.3 to the two cross products on the right side of (1). Thus,

$$(\mathbf{A} - \mathbf{B}) \times (\mathbf{A} + \mathbf{B}) = (-\mathbf{A}) \times (\mathbf{A} - \mathbf{B}) + (-\mathbf{B}) \times (\mathbf{A} - \mathbf{B}) \tag{2}$$

Applying Theorems 18.6.4 and 18.6.5 to the right side of (2), we get

$$(\mathbf{A} - \mathbf{B}) \times (\mathbf{A} + \mathbf{B}) = (-\mathbf{A}) \times \mathbf{A} + \mathbf{A} \times \mathbf{B} + (-\mathbf{B}) \times \mathbf{A} + \mathbf{B} \times \mathbf{B} \tag{3}$$

Applying Theorems 18.6.2 and 18.6.3 to the right side of (3), we obtain

$$(\mathbf{A}-\mathbf{B}) \times (\mathbf{A}+\mathbf{B}) = \mathbf{0} + \mathbf{A}\times\mathbf{B} + \mathbf{A}\times\mathbf{B} + \mathbf{0}$$
$$= 2(\mathbf{A}\times\mathbf{B})$$

**34.** Find the perpendicular distance between the two given skew lines.

$$\frac{x+1}{2} = \frac{y+2}{-4} = \frac{z-1}{-3} \quad \text{and} \quad \frac{x-1}{5} = \frac{y-1}{3} = \frac{z+1}{2}$$

SOLUTION: The vector $\langle 2, -4, -3\rangle$ is a direction vector of the first line, and the vector $\langle 5, 3, 2\rangle$ is a direction vector of the second line. We take

$$\mathbf{N} = \langle 2, -4, -3\rangle \times \langle 5, 3, 2\rangle$$
$$= \langle 1, -19, 26\rangle$$

Because the lines do not intersect, there exist parallel planes each containing one of the lines. Because $\mathbf{N}$ is orthogonal to both of the direction vectors of the lines, then $\mathbf{N}$ is a normal vector of both of the parallel planes. The first line contains the point $P(-1, -2, 1)$ and the second line contains the point $Q(1, 1, -1)$. The number of units in the perpendicular distance between the lines is the absolute value of the scalar projection of $\mathbf{V}(\overrightarrow{PQ})$ on $\mathbf{N}$. Because $\mathbf{V}(\overrightarrow{PQ}) = \langle 2, 3, -2\rangle$, we take

$$\left|\frac{\mathbf{N}\cdot\mathbf{V}(\overrightarrow{PQ})}{|\mathbf{N}|}\right| = \left|\frac{\langle 1, -19, 26\rangle \cdot \langle 2, 3, -2\rangle}{\sqrt{1^2 + (-19)^2 + 26^2}}\right|$$
$$= \frac{107}{\sqrt{1038}}$$

Thus, the perpendicular distance between the lines is $107/\sqrt{1038}$ units.

## 18.7 CYLINDERS AND SURFACES OF REVOLUTION

**18.7.1 Definition** A *cylinder* is a surface that is generated by a line moving along a given plane curve in such a way that it always remains parallel to a fixed line not lying in the plane of the given curve. The moving line is called a *generator* of the cylinder and the given plane curve is called a *directrix* of the cylinder. Any position of a generator is called a *ruling* of the cylinder.

**18.7.2 Theorem** In three-dimensional space, the graph of an equation in two of the three variables $x, y$, and $z$ is a cylinder whose rulings are parallel to the axis associated with the missing variable and whose directrix is a curve in the plane associated with the two variables appearing in the equation.

**18.7.3 Definition** If a plane curve is revolved about a fixed line lying in the plane of the curve, the surface generated is called a *surface of revolution*. The fixed line is called the *axis* of the surface of revolution, and the plane curve is called the *generating curve*.

In Table A we summarize the three types of surfaces of revolution that we consider in this book.

**Table A**

| Equation of generating curve in the coordinate plane | Axis of revolution | Equation of surface of revolution |
|---|---|---|
| $y = f(x)$ in the $xy$ plane, or $z = f(x)$ in the $xz$ plane | $x$-axis | $y^2 + z^2 = [f(x)]^2$ |
| $x = f(y)$ in the $xy$ plane, or $z = f(y)$ in the $yz$ plane | $y$-axis | $x^2 + z^2 = [f(y)]^2$ |
| $x = f(z)$ in the $xz$ plane, or $y = f(z)$ in the $yz$ plane | $z$-axis | $x^2 + y^2 = [f(z)]^2$ |

## *Exercises 18.7*

In Exercises 1-8, draw a sketch of the cylinder having the given equation.

2. $x^2 - z^2 = 4$

SOLUTION: The directrix in the $xz$ plane is the hyperbola whose equation is $x^2 - z^2 = 4$. The rulings of the cylinder are parallel to the $y$-axis. In Fig. 18.7.2 we show a sketch of the cylinder. Note that the $xz$ plane is horizontal.

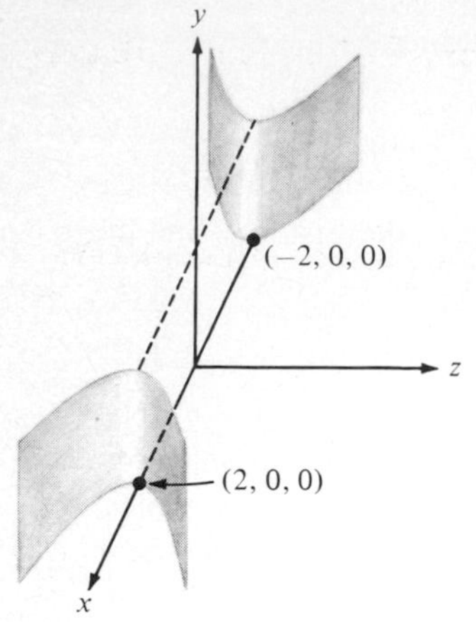


Figure 18.7.2

8. $z^2 = 4y^2$

SOLUTION: The rulings are parallel to the $x$-axis and perpendicular to the $yz$ plane. The directrix in the $yz$ plane is the pair of intersecting lines $z = \pm 2y$. In Fig. 18.7.8 we show a sketch of the cylinder, which is the union of two planes that have the $x$-axis as their line of intersection.

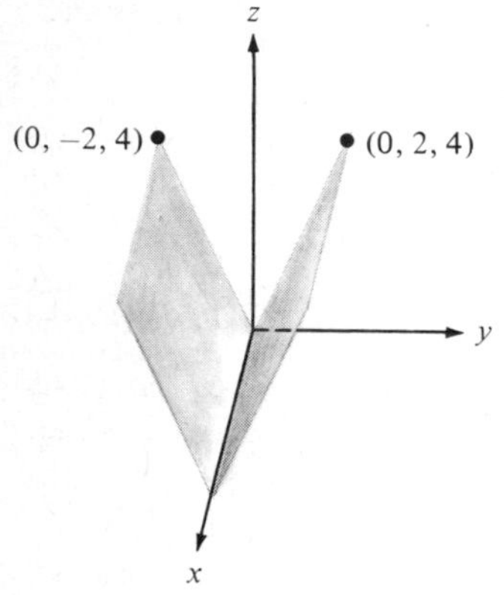


Figure 18.7.8

In Exercises 9-14, find an equation of the surface of revolution generated by revolving the given plane curve about the indicated axis. Draw a sketch of the surface.

**10.** $x^2 = 4y$ in the $xy$ plane about the $x$-axis

SOLUTION: Because we revolve the curve about the $x$-axis, we first express $y$ as a function of $x$. Thus, $y = \frac{1}{4}x^2$. Let $f(x) = \frac{1}{4}x^2$. An equation of the surface of revolution if the curve $y = f(x)$ is revolved about the $x$-axis is given by

$$y^2 + z^2 = [f(x)]^2$$

Thus,

$$y^2 + z^2 = \left(\frac{1}{4}x^2\right)^2$$

$$16y^2 + 16z^2 = x^4$$

is an equation of the surface of revolution. In Fig. 18.7.10 we show a sketch of the surface. Note that we have taken the $xz$ plane as the horizontal plane so that the surface may be more clearly illustrated.

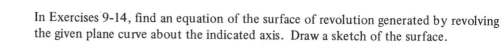

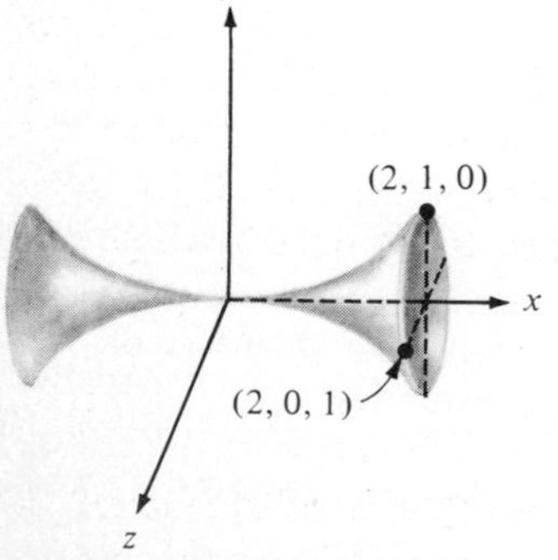


Figure 18.7.10

**14.** $y^2 = z^3$ in the $yz$ plane about the $z$-axis

SOLUTION: Because the $z$-axis is the axis of revolution, we solve for $y$ in terms of $z$. Thus, $y = \pm z^{3/2}$. If $f(z) = z^{3/2}$, an equation of the surface of revolution if the curve $y = \pm f(z)$ is revolved about the $z$-axis is given by

$$x^2 + y^2 = [f(z)]^2$$

Hence,

$$x^2 + y^2 = [z^{3/2}]^2$$
$$x^2 + y^2 = z^3$$

is an equation of the surface of revolution. In Fig. 18.7.14 we show a sketch of the surface.

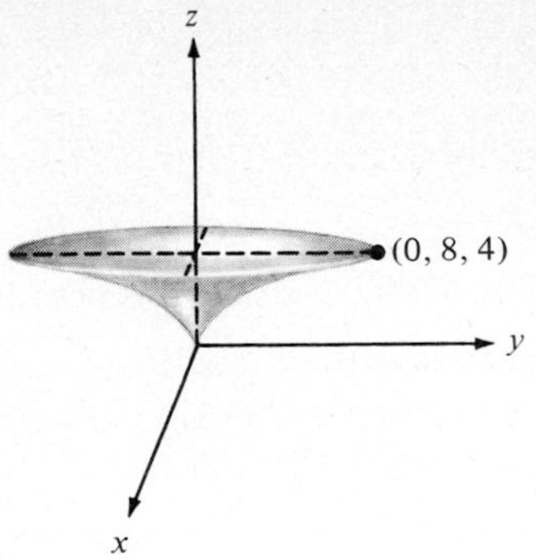

Figure 18.7.14

**18.** Find a generating curve and the axis for the given surface of revolution. Draw a sketch of the surface.

$$4x^2 + 9y^2 + 4z^2 = 36$$

SOLUTION: We write the given equation in a form given in Table A. Thus, we have

$$4x^2 + 4z^2 = 36 - 9y^2$$

$$x^2 + z^2 = \frac{9}{4}(4 - y^2)$$

$$x^2 + z^2 = \left(\frac{3}{2}\sqrt{4 - y^2}\right)^2$$

We take

$$f(y) = \frac{3}{2}\sqrt{4 - y^2}$$

A generating curve for the surface of revolution is $z = \pm f(y)$, with axis the $y$-axis. Thus,

$$z = \pm\frac{3}{2}\sqrt{4 - y^2}$$

or, equivalently,

$$9y^2 + 4z^2 = 36$$

is a generating curve in the $yz$ plane. The $y$-axis is the axis of revolution. A sketch of the surface, which is an ellipsoid of revolution, is shown in Fig. 18.7.18. Note that the horizontal plane is the $xz$ plane.

A generating curve in the $xy$ plane is $x = \pm f(y)$, or, equivalently,

$$4x^2 + 9y^2 = 36$$

with the $y$-axis the axis of revolution.

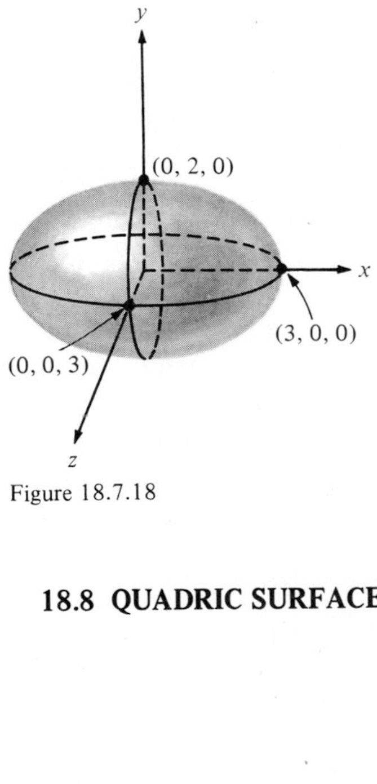

Figure 18.7.18

## 18.8 QUADRIC SURFACES

If $A \neq 0$, $B \neq 0$, and $C \neq 0$, the graph of the equation

$$Ax^2 + By^2 + Cz^2 = 1 \qquad \text{(I)}$$

is symmetric with respect to each of the coordinate planes and is called a *central quadric*. If $A$, $B$, and $C$ are all positive, the graph of Eq. (I) is an *ellipsoid*. If two of the numbers $A$, $B$, and $C$ are positive and one of the numbers $A$, $B$, or $C$ is negative, the graph of (I) is an *elliptic hyperboloid of one sheet* with axis corresponding to the variable whose coefficient is negative. If one of the numbers $A$, $B$, or $C$ is positive and two of the numbers $A$, $B$, and $C$ are negative, the graph of (I) is an *elliptic hyperboloid of two sheets* with axis corresponding to the variable whose coefficient is positive. If $A$, $B$, and $C$ are all negative, the graph of (I) is the empty set.

If $A \neq 0$ and $B \neq 0$, the graphs of the equations

$$z = Ax^2 + By^2 \qquad y = Ax^2 + Bz^2 \qquad x = Ay^2 + Bz^2 \qquad \text{(II)}$$

contain the origin and are symmetric with respect to two of the coordinate planes but are not symmetric with respect to the coordinate plane corresponding to the two squared variables. If $A$ and $B$ are either both positive or both negative, the graphs of Eqs. (II) are *elliptic paraboloids* with axis corresponding to the variable that is not squared. If $A$ and $B$ have opposite signs, the graphs of (II) are *hyperbolic paraboloids* with axis corresponding to the variable that is not squared.

If $A > 0$ and $B > 0$, the graphs of the equations

$$\begin{aligned} Ax^2 + By^2 - z^2 &= 0 \\ Ax^2 - y^2 + Bz^2 &= 0 \\ -x^2 + Ay^2 + Bz^2 &= 0 \end{aligned} \qquad \textbf{(III)}$$

are elliptic cones with axis corresponding to the variable whose coefficient is negative.

## *Exercises 18.8*

In Exercises 1-12, draw a sketch of the graph of the given equation and name the surface.

**6.** $x^2 = y^2 + z^2$

SOLUTION: Because the given equation may be written as

$$-x^2 + y^2 + z^2 = 0$$

the equation is of type (III), and hence, its graph is an elliptic cone with axis the $x$-axis. Each cross section in the plane $x = k$ with $k \neq 0$ is the circle $y^2 + z^2 = k^2$. Thus, the cone is a right circular cone. A sketch of the graph is shown in Fig. 18.8.6. We take the $xz$ plane as the horizontal plane.

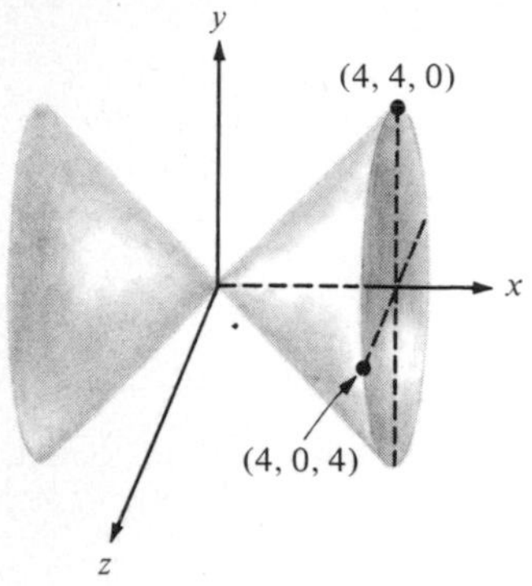

Figure 18.8.6

**8.** $\dfrac{y^2}{25} + \dfrac{x^2}{36} = 4$

SOLUTION: Because there is no term involving $z$ in the equation, the graph is a cylinder with rulings perpendicular to the $xy$ plane. The given equation may be written as

$$\frac{x^2}{12^2} + \frac{y^2}{10^2} = 1$$

Thus, the directrix in the $xy$ plane is an ellipse with vertices at $(\pm 12, 0, 0)$ and ends of the minor axis at $(0, \pm 10, 0)$. A sketch of the graph is shown in Fig. 18.8.8.

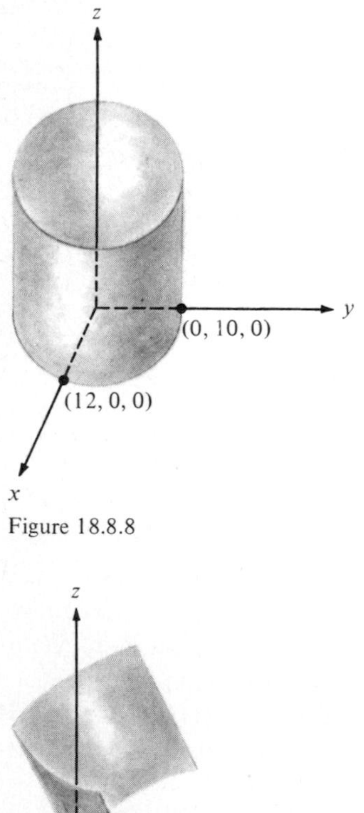

Figure 18.8.8

**10.** $x^2 = 2y + 4z$

SOLUTION: The cross sections in the planes $z = k$ are the parabolas $x^2 = 2y + 4k$. In particular, the cross section in the $xy$ plane is the parabola $x^2 = 2y$. The cross sections in the planes $x = k$ are the lines $2y + 4z = k^2$. In particular, the cross section in the $yz$ plane is the line $2y + 4z = 0$. Therefore, the surface is a parabolic cylinder with rulings parallel to the line $x = 0, 2y + 4z = 0$, and with directrix the parabola $x^2 = 2y$ in the $xy$ plane. (It is possible to eliminate the term involving $z$ by rotating the $z$-axis until it coincides with the line $x = 0, 2y + 4z = 0$.) In Fig. 18.8.10 we show a sketch of the surface.

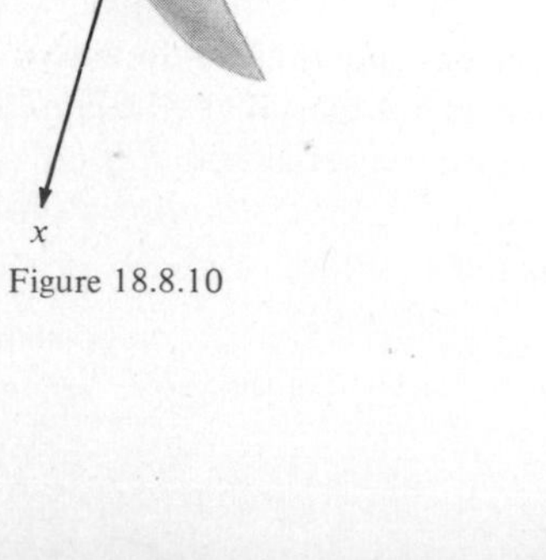

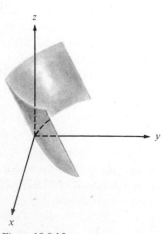

Figure 18.8.10

**14.** Find the vertex and locus of the parabola that is the intersection of the plane $y = 2$ with the hyperbolic paraboloid

$$\frac{y^2}{16} - \frac{x^2}{4} = \frac{z}{9}$$

SOLUTION: We let $y = 2$ in the given equation and write the resulting equation in standard form. Thus, we have

$$\frac{1}{4} - \frac{x^2}{4} = \frac{z}{9}$$

$$x^2 = -\frac{4}{9}\left(z - \frac{9}{4}\right)$$

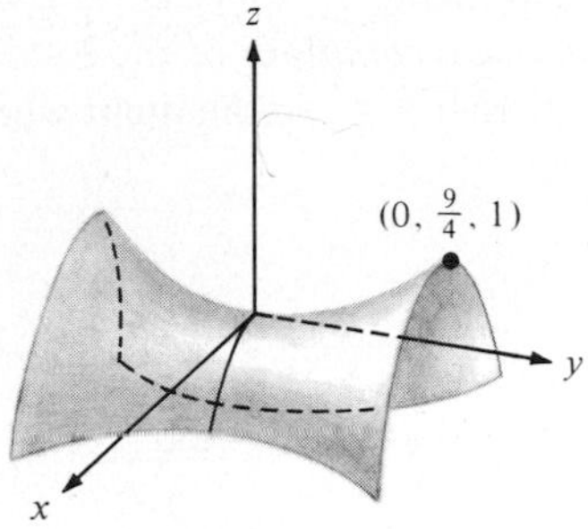

Figure 18.8.14

The cross section of the given surface in the plane $y = 2$ is a parabola with vertex $(0, \frac{9}{4}, 1)$. The parabola has axis parallel to the $z$-axis and opens downward. In Fig. 18.8.14 we show a sketch of the given hyperbolic paraboloid and the parabola that is the cross section in the plane $y = 2$.

**16.** Show that the intersection of the hyperbolic paraboloid $y^2/b^2 - x^2/a^2 = z/c$ and the plane $z = bx + ay$ consists of two intersecting straight lines.

SOLUTION: Substituting $z = bx + ay$ into the equation of the hyperbolic paraboloid, we have

$$\frac{y^2}{b^2} - \frac{x^2}{a^2} = \frac{bx + ay}{c}$$

We write the equation in zero form and factor. Thus,

$$\left(\frac{y}{b} + \frac{x}{a}\right)\left(\frac{y}{b} - \frac{x}{a}\right) - \frac{ab}{c}\left(\frac{x}{a} + \frac{y}{b}\right) = 0$$

$$\left(\frac{y}{b} + \frac{x}{a}\right)\left(\frac{y}{b} - \frac{x}{a} - \frac{ab}{c}\right) = 0$$

Setting each factor equal to zero, we obtain

$$\frac{y}{b} + \frac{x}{a} = 0 \quad \text{and} \quad \frac{y}{b} - \frac{x}{a} = \frac{ab}{c} \tag{1}$$

Thus, the plane $z = bx + ay$ intersects the hyperbolic paraboloid in the two lines

$$z = bx + ay \qquad \frac{y}{b} + \frac{x}{a} = 0 \tag{2}$$

and

$$z = bx + ay \qquad \frac{y}{b} - \frac{x}{a} = \frac{ab}{c} \tag{3}$$

We show that the lines (2) and (3) intersect. Solving Eqs. (1) simultaneously, we get

$$x = -\frac{a^2 b}{2c} \quad \text{and} \quad y = \frac{ab^2}{2c} \tag{4}$$

Substituting from (4) into the equation

$$z = bx + ay$$

we get

$$z = \frac{-a^2 b^2}{2c} + \frac{a^2 b^2}{2c} = 0$$

Therefore, the lines (2) and (3) intersect in the point $(-a^2 b/2c, ab^2/2c, 0)$.

## 18.9 CURVES IN $R^3$

**18.9.1 Definition** Let $f_1, f_2$, and $f_3$ be three real-valued functions of a real variable $t$. Then for every number $t$ in the domain common to $f_1, f_2$, and $f_3$ there is a vector $\mathbf{R}$ defined by

$$\mathbf{R}(t) = f_1(t)\mathbf{i} + f_2(t)\mathbf{j} + f_3(t)\mathbf{k}$$

and $\mathbf{R}$ is called a *vector-valued function.*

The graph of the vector-valued function of Definition 18.9.1 is the curve $C$ that consists of the set of all points $(x, y, z)$ such that

$$x = f_1(t) \qquad y = f_2(t) \qquad z = f_3(t) \tag{1}$$

Eqs. (1) are called *parametric equations* of curve $C$. Eliminating $t$ from Eqs. (1), we obtain two equations in $x, y$, and $z$ that are called *Cartesian equations* of the curve $C$. The length of arc for the curve $C$ from the point where $t = t_1$ to the point where $t = t_2$ is given by

$$L = \int_{t_1}^{t_2} |\mathbf{R}'(t)|\, dt$$

If $|\mathbf{R}'(t)| \neq 0$, we define the *unit tangent vector* by

$$\mathbf{T}(t) = \frac{\mathbf{R}'(t)}{|\mathbf{R}'(t)|}$$

The *curvature vector* is given by

$$\mathbf{K}(t) = \frac{\mathbf{T}'(t)}{|\mathbf{R}'(t)|}$$

The *curvature* is the magnitude of the curvature vector; that is,

$$K(t) = |\mathbf{K}(t)|$$

If the curvature is not zero, we define the *unit normal vector* to the curve $C$ by

$$\mathbf{N}(t) = \frac{\mathbf{K}(t)}{|\mathbf{K}(t)|}$$

and the *unit binormal vector* to the curve $C$ by

$$\mathbf{B}(t) = \mathbf{T}(t) \times \mathbf{N}(t)$$

The three mutually orthogonal vectors, $\mathbf{T}, \mathbf{N}$, and $\mathbf{B}$ of the curve $C$, are called the *moving trihedral* of $C$.

If the parameter $t$ measures time and a particle moves along the curve $C$, then $\mathbf{R}(t)$ is called the *position vector* of the particle; $\mathbf{V}(t)$ is called the *velocity vector* of the particle; and $\mathbf{A}(t)$ is called the *acceleration vector* of the particle, where

$$\mathbf{V}(t) = \mathbf{R}'(t)$$
$$\mathbf{A}(t) = \mathbf{V}'(t)$$

The magnitude of the velocity vector is called the *speed* of the particle. That is, the speed is given by

$$\frac{ds}{dt} = |\mathbf{V}(t)|$$

## *Exercises 18.9*

**4.** Find the unit tangent vector for the curve having the given vector equation

$$\mathbf{R}(t) = t^2\mathbf{i} + \left(t + \frac{1}{3}t^3\right)\mathbf{j} + \left(t - \frac{1}{3}t^3\right)\mathbf{k}$$

SOLUTION:

$$\mathbf{R}'(t) = 2t\,\mathbf{i} + (1 + t^2)\mathbf{j} + (1 - t^2)\mathbf{k}$$

$$|\mathbf{R}'(t)| = \sqrt{(2t)^2 + (1+t^2)^2 + (1-t^2)^2}$$
$$= \sqrt{2(1+t^2)^2}$$
$$= \sqrt{2}(1+t^2)$$

Thus,

$$\mathbf{T}(t) = \frac{\mathbf{R}'(t)}{|\mathbf{R}'(t)|}$$
$$= \frac{1}{\sqrt{2}(1+t^2)}[2t\mathbf{i} + (1+t^2)\mathbf{j} + (1-t^2)\mathbf{k}]$$

**6.** Find the length of arc of the curve having the vector equation

$$\mathbf{R}(t) = \sin 2t\,\mathbf{i} + \cos 2t\,\mathbf{j} + 2t^{3/2}\mathbf{k}$$

from $t_1 = 0$ to $t_2 = 1$.

SOLUTION:

$$\mathbf{R}'(t) = 2\cos 2t\,\mathbf{i} - 2\sin 2t\,\mathbf{j} + 3t^{1/2}\mathbf{k}$$
$$|\mathbf{R}'(t)| = \sqrt{4\cos^2 2t + 4\sin^2 2t + 9t}$$
$$= \sqrt{4+9t}$$

Thus,

$$L = \int_{t_1}^{t_2} |\mathbf{R}'(t)|dt$$
$$= \int_0^1 \sqrt{4+9t}\,dt$$
$$= \frac{1}{9}\cdot\frac{2}{3}(4+9t)^{3/2}\Big]_0^1$$
$$= \frac{2}{27}[13^{3/2} - 8]$$

**12.** Prove that the unit tangent vector of the circular helix of Example 1 makes an angle of constant radian measure with the unit vector $\mathbf{k}$.

SOLUTION: From Example 1 we have

$$\mathbf{R}(t) = a\cos t\,\mathbf{i} + a\sin t\,\mathbf{j} + t\mathbf{k} \quad \text{where } a > 0$$

From Example 2 we have

$$\mathbf{T}(t) = \frac{1}{\sqrt{a^2+1}}(-a\sin t\,\mathbf{i} + a\cos t\,\mathbf{j} + \mathbf{k}) \tag{1}$$

Let $\theta$ be the radian measure of the angle between $\mathbf{T}(t)$ and the unit vector $\mathbf{k}$. From Theorem 18.3.4 we have

$$\cos\theta = \frac{\mathbf{T}(t)\cdot\mathbf{k}}{|\mathbf{T}(t)||\mathbf{k}|} \tag{2}$$

Because both $\mathbf{T}(t)$ and $\mathbf{k}$ are unit vectors, $|\mathbf{T}(t)| = 1$ and $|\mathbf{k}| = 1$. Thus, substituting from (1) into (2), we obtain

$$\cos\theta = \frac{1}{\sqrt{a^2+1}}(-a\sin t\,\mathbf{i} + a\cos t\,\mathbf{j} + \mathbf{k})\cdot\mathbf{k}$$

$$= \frac{1}{\sqrt{a^2 + 1}}$$

Therefore, the angle between $\mathbf{T}(t)$ and $\mathbf{k}$ has constant radian measure.

**18.** Find the moving trihedral and the curvature of the given curve at $t = t_1$ if they exist.

The curve of Exercise 4; $t_1 = 1$

SOLUTION: From Exercise 4, we have

$$\mathbf{R}(t) = t^2\mathbf{i} + \left(t + \frac{1}{3}t^3\right)\mathbf{j} + \left(t - \frac{1}{3}t^3\right)\mathbf{k}$$

$$|\mathbf{R}'(t)| = \sqrt{2}(1 + t^2) \tag{1}$$

$$\mathbf{T}(t) = \frac{1}{\sqrt{2}(1 + t^2)}[2t\mathbf{i} + (1 + t^2)\mathbf{j} + (1 - t^2)\mathbf{k}] \tag{2}$$

Thus, the unit tangent vector at $t_1 = 1$ is

$$\mathbf{T}(1) = \frac{1}{2\sqrt{2}}(2\mathbf{i} + 2\mathbf{j})$$

$$= \frac{1}{2}\sqrt{2}(\mathbf{i} + \mathbf{j})$$

From (2) we have

$$\mathbf{T}(t) = \frac{1}{\sqrt{2}}\left[\frac{2t}{1 + t^2}\mathbf{i} + \mathbf{j} + \frac{1 - t^2}{1 + t^2}\mathbf{k}\right]$$

Thus,

$$\mathbf{T}'(t) = \frac{1}{\sqrt{2}}\left[\frac{2(1 - t^2)}{(1 + t^2)^2}\mathbf{i} - \frac{4t}{(1 + t^2)^2}\mathbf{k}\right]$$

and

$$\mathbf{T}'(1) = -\frac{1}{2}\sqrt{2}\,\mathbf{k} \tag{3}$$

From (1) we obtain

$$|\mathbf{R}'(1)| = 2\sqrt{2} \tag{4}$$

Thus, the curvature vector at $t_1 = 1$ is given by

$$\mathbf{K}(1) = \frac{\mathbf{T}'(1)}{|\mathbf{R}'(1)|}$$

$$= \frac{-\frac{1}{2}\sqrt{2}\,\mathbf{k}}{2\sqrt{2}}$$

$$= -\frac{1}{4}\mathbf{k} \tag{5}$$

Hence, the curvature is given by

$$K(1) = |\mathbf{K}(1)| = \frac{1}{4}$$

We obtain the unit normal vector at $t_1 = 1$ from (5). Thus,

$$N(1) = \frac{K(1)}{|K(1)|}$$

$$= \frac{-\frac{1}{4}k}{\frac{1}{4}}$$

$$= -k$$

The unit binormal vector at $t_1 = 1$ is given by

$$B(1) = T(1) \times N(1)$$

$$= \frac{1}{2}\sqrt{2}(i + j) \times (-k)$$

$$= -\frac{1}{2}\sqrt{2}(i \times k + j \times k)$$

$$= -\frac{1}{2}\sqrt{2}(-j + i)$$

$$= -\frac{1}{2}\sqrt{2}(i - j)$$

**22.** A particle is moving along the given curve. Find the velocity vector, the acceleration vector, and the speed at $t = t_1$. Draw a sketch of a portion of the curve at $t = t_1$ and draw the velocity and acceleration vectors there.

$$x = \frac{1}{2(t^2 + 1)} \qquad y = \ln(1 + t^2) \qquad z = \tan^{-1} t \qquad t_1 = 1$$

SOLUTION: We have

$$R(t) = \frac{1}{2}(t^2 + 1)^{-1}i + \ln(1 + t^2)j + \tan^{-1} t\, k$$

Thus, the velocity vector is given by

$$V(t) = R'(t) = -\frac{t}{(1 + t^2)^2}i + \frac{2t}{1 + t^2}j + \frac{1}{1 + t^2}k$$

And the acceleration vector is given by

$$A(t) = V'(t) = \frac{3t^2 - 1}{(1 + t^2)^3}i + \frac{2(1 - t^2)}{1 + t^2}j - \frac{2t}{(1 + t^2)^2}k$$

At $t = 1$, we have

$$V(1) = -\frac{1}{4}i + j + \frac{1}{2}k$$

$$A(1) = \frac{1}{4}i - \frac{1}{2}k$$

And the speed of the particle at $t = 1$ is given by

$$|V(1)| = \sqrt{\left(-\frac{1}{4}\right)^2 + 1^2 + \left(\frac{1}{2}\right)^2}$$

$$= \frac{1}{4}\sqrt{21}$$

Let $P(x, y, z)$ be the position of the particle when $t = 1$. Because $R(1) = \frac{1}{4}i + \ln 2j + \frac{1}{4}\pi k$, we have $P \approx (0.25, 0.7, 0.8)$. If $\overrightarrow{PQ}$ is a representation of $V(1)$, then

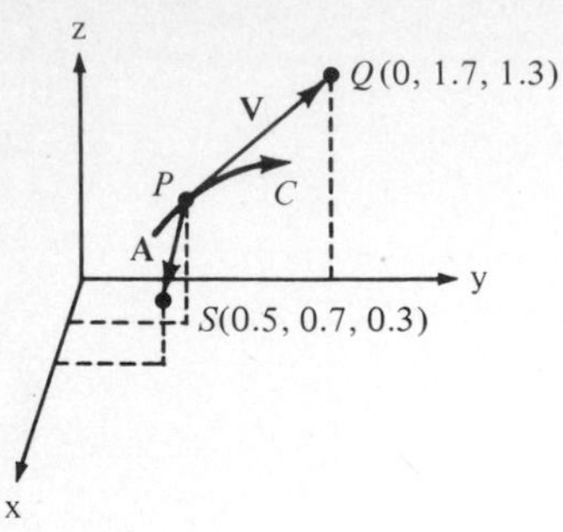

Figure 18.9.22

$Q \approx (0, 1.7, 1.3)$. If $\overrightarrow{PS}$ is a representation of $\mathbf{A}(1)$, then $S \approx (0.5, 0.7, 0.3)$. In Fig. 18.9.22 we show a sketch of the curve and the representations $\overrightarrow{PQ}$ and $\overrightarrow{PS}$.

**24.** Prove that for the twisted cubic of Illustration 2, if $t \neq 0$, no two of the vectors $\mathbf{R}(t)$, $\mathbf{V}(t)$, and $\mathbf{A}(t)$ are orthogonal.

SOLUTION: In Illustration 2, we are given

$$\mathbf{R}(t) = t\mathbf{i} + t^2\mathbf{j} + t^3\mathbf{k}$$

Thus,

$$\begin{aligned} \mathbf{V}(t) &= \mathbf{R}'(t) = \mathbf{i} + 2t\mathbf{j} + 3t^2\mathbf{k} \\ \mathbf{A}(t) &= \mathbf{V}'(t) = \quad 2\mathbf{j} + 6t\mathbf{k} \end{aligned}$$

Two vectors are orthogonal if and only if their dot product is zero. We have

$$\begin{aligned} \mathbf{R}(t) \cdot \mathbf{V}(t) &= (t\mathbf{i} + t^2\mathbf{j} + t^3\mathbf{k}) \cdot (\mathbf{i} + 2t\mathbf{j} + 3t^2\mathbf{k}) \\ &= t + 2t^3 + 3t^5 \\ &= t(3t^4 + 2t^2 + 1) \end{aligned}$$

We are given that $t \neq 0$. If $3t^4 + 2t^2 + 1 = 0$, then by the quadratic formula

$$t^2 = \frac{-2 \pm \sqrt{-8}}{6}$$

which is impossible, since $t^2$ is real. Therefore $\mathbf{R}(t) \cdot \mathbf{V}(t) \neq 0$, and the vectors $\mathbf{R}(t)$ and $\mathbf{V}(t)$ are not orthogonal for any $t \neq 0$. Furthermore,

$$\begin{aligned} \mathbf{R}(t) \cdot \mathbf{A}(t) &= (t\mathbf{i} + t^2\mathbf{j} + t^3\mathbf{k}) \cdot (2\mathbf{j} + 6t\mathbf{k}) \\ &= 2t^2 + 6t^4 \\ &= 2t^2(1 + 3t^2) \end{aligned}$$

Because $t^2 \neq 0$ if $t \neq 0$, and $1 + 3t^2 \neq 0$, then $\mathbf{R}(t) \cdot \mathbf{A}(t) \neq 0$ if $t \neq 0$. Thus, the vectors $\mathbf{R}(t)$ and $\mathbf{A}(t)$ are not orthogonal for any $t \neq 0$. Finally, we have

$$\begin{aligned} \mathbf{V}(t) \cdot \mathbf{A}(t) &= (\mathbf{i} + 2t\mathbf{j} + 3t^2\mathbf{k}) \cdot (2\mathbf{j} + 6t\mathbf{k}) \\ &= 4t + 18t^3 \\ &= t(4 + 18t^2) \end{aligned}$$

Because $4 + 18t^2 \neq 0$, then $\mathbf{V}(t) \cdot \mathbf{A}(t) \neq 0$ for all $t \neq 0$. Thus, the vectors $\mathbf{V}(t)$ and $\mathbf{A}(t)$ are not orthogonal for any $t \neq 0$.

**28.** Find the curvature of the given curve at the indicated point.

$$x = e^t \qquad y = e^{-t} \qquad z = t \qquad t = 0$$

SOLUTION: We use the formula of Exercise 25.

$$K(t) = \frac{|D_t\mathbf{R}(t) \times D_t^2\mathbf{R}(t)|}{|D_t\mathbf{R}(t)|^3} \tag{1}$$

We have

$$\begin{aligned} \mathbf{R}(t) &= e^t\mathbf{i} + e^{-t}\mathbf{j} + t\mathbf{k} \\ D_t\mathbf{R}(t) &= e^t\mathbf{i} - e^{-t}\mathbf{j} + \mathbf{k} \\ D_t^2\mathbf{R}(t) &= e^t\mathbf{i} + e^{-t}\mathbf{j} \end{aligned}$$

At $t = 0$, we get

$$\begin{aligned} D_t\mathbf{R}(0) &= \mathbf{i} - \mathbf{j} + \mathbf{k} \\ D_t^2\mathbf{R}(0) &= \mathbf{i} + \mathbf{j} \end{aligned}$$

Therefore,

$$\begin{aligned} D_t\mathbf{R}(0) \times D_t^2\mathbf{R}(0) &= (\mathbf{i}-\mathbf{j}+\mathbf{k}) \times (\mathbf{i}+\mathbf{j}) \\ &= -\mathbf{i}+\mathbf{j}+2\mathbf{k} \end{aligned}$$

$$\begin{aligned} |D_t\mathbf{R}(0) \times D_t^2\mathbf{R}(0)| &= \sqrt{(-1)^2+1^2+2^2} \\ &= \sqrt{6} \end{aligned} \tag{2}$$

Furthermore,

$$\begin{aligned} |D_t\mathbf{R}(0)|^3 &= (\sqrt{1^2+(-1)^2+1^2})^3 \\ &= 3\sqrt{3} \end{aligned} \tag{3}$$

Substituting from (2) and (3) into (1), we obtain

$$\begin{aligned} K(0) &= \frac{\sqrt{6}}{3\sqrt{3}} \\ &= \frac{1}{3}\sqrt{2} \end{aligned}$$

## 18.10 CYLINDRICAL AND SPHERICAL COORDINATES

We take the polar axis as the positive $x$-axis with the polar plane as the $xy$ plane. If cylindrical coordinates $(r, \theta, z)$ of a point $P$ are given, we use the equations

$$x = r\cos\theta \qquad y = r\sin\theta \qquad z = z \tag{I}$$

to uniquely determine the Cartesian coordinates $(x, y, z)$ of $P$. If the Cartesian coordinates of $P$ are given, then cylindrical coordinates are not unique. Any numbers $r, \theta$, and $z$ that simultaneously satisfy all three equations in (I) may be used as cylindrical coordinates. The following equations hold if $(x, y, z)$ and $(r, \theta, z)$ represent the same point, but the equations are not sufficient conditions for changing from one coordinate system to the other.

$$r^2 = x^2 + y^2 \qquad \tan\theta = \frac{y}{x} \qquad z = z \tag{II}$$

If spherical coordinates $(\rho, \theta, \phi)$ of a point $P$ are given, we use the equations

$$x = \rho\sin\phi\cos\theta \qquad y = \rho\sin\phi\sin\theta \qquad z = \rho\cos\phi \tag{III}$$

to uniquely determine the Cartesian coordinates of $P$. Furthermore, a set of cylindrical coordinates of $P$ are given by

$$r = \rho\sin\phi \qquad \theta = \theta \qquad z = \rho\cos\phi \tag{IV}$$

If the Cartesian coordinates of $P$ are given and $P$ is not the origin, then the spherical coordinates $\rho$ and $\phi$ are uniquely determined by

$$\rho = \sqrt{x^2+y^2+z^2} \qquad \phi = \cos^{-1}\frac{z}{\rho} \tag{V}$$

However, $\theta$ is not unique. Any numbers $\rho, \theta$, and $\phi$ that simultaneously satisfy all three equations in (III) may be used as spherical coordinates of $P$. Spherical coordinates for the origin are $(0, \theta, \phi)$, where $\theta$ is any real number and $0 \leqslant \phi \leqslant \pi$. Thus, for any point we have $\rho \geqslant 0$ and $0 \leqslant \phi \leqslant \pi$.

*Exercises 18.10*

4. Find a set of spherical coordinates of the point having the given Cartesian coordinates:

(a) $(1, -1, -\sqrt{2})$ (b) $(-1, \sqrt{3}, 2)$ (c) $(2, 2, 2)$

SOLUTION:

(a) We use formulas (V). Thus,

$$\rho = \sqrt{x^2 + y^2 + z^2} = \sqrt{1^2 + (-1)^2 + (-\sqrt{2})^2} = 2$$

$$\phi = \cos^{-1}\frac{z}{\rho} = \cos^{-1}\left(\frac{-\sqrt{2}}{2}\right) = \frac{3}{4}\pi$$

From formula (II) we have

$$\tan\theta = \frac{y}{x} = \frac{-1}{1} = -1 \tag{1}$$

Because $x > 0$ and $y < 0$, then $\theta$ is the radian measure of a fourth-quadrant angle that satisfies (1). Therefore, we take $\theta = -\frac{1}{4}\pi$. Spherical coordinates are $(2, -\frac{1}{4}\pi, \frac{3}{4}\pi)$.

(b) As in part (a), we have

$$\rho = \sqrt{x^2 + y^2 + z^2} = \sqrt{(-1)^2 + (\sqrt{3})^2 + 2^2} = 2\sqrt{2}$$

$$\phi = \cos^{-1}\frac{z}{\rho} = \cos^{-1}\frac{2}{2\sqrt{2}} = \frac{1}{4}\pi$$

Furthermore,

$$\tan\theta = \frac{y}{x} = -\sqrt{3}$$

Because $x < 0$ and $y > 0$, then $\theta$ measures a second-quadrant angle. Thus, we take $\theta = \frac{2}{3}\pi$. Spherical coordinates are $(2\sqrt{2}, \frac{2}{3}\pi, \frac{1}{4}\pi)$.

(c) We have

$$\rho = \sqrt{2^2 + 2^2 + 2^2} = 2\sqrt{3}$$

$$\phi = \cos^{-1}\left(\frac{2}{2\sqrt{3}}\right) = \cos^{-1}\frac{1}{3}\sqrt{3}$$

and

$$\tan\theta = 1$$

Because $x > 0$ and $y > 0$, we take $\theta = \frac{1}{4}\pi$. Therefore, spherical coordinates are $(2\sqrt{3}, \frac{1}{4}\pi, \cos^{-1}\frac{1}{3}\sqrt{3})$.

6. Find a set of spherical coordinates of the point having the given cylindrical coordinates:

(a) $\left(3, \frac{1}{6}\pi, 3\right)$ (b) $\left(3, \frac{1}{2}\pi, 2\right)$ (c) $\left(2, \frac{5}{6}\pi, -4\right)$

SOLUTION:

(a) $$\rho = \sqrt{x^2 + y^2 + z^2} = \sqrt{r^2 + z^2} = \sqrt{3^2 + 3^2} = 3\sqrt{2}$$

$$\theta = \frac{1}{6}\pi$$

$$\phi = \cos^{-1}\frac{z}{\rho} = \cos^{-1}\frac{3}{3\sqrt{2}} = \frac{1}{4}\pi$$

Therefore, spherical coordinates are $(3\sqrt{2}, \frac{1}{6}\pi, \frac{1}{4}\pi)$.

(b) As in part (a),

$$\rho = \sqrt{r^2 + z^2} = \sqrt{3^2 + 2^2} = \sqrt{13}$$

$$\theta = \frac{1}{2}\pi$$

$$\phi = \cos^{-1}\frac{z}{\rho} = \cos^{-1}\frac{2}{\sqrt{13}} = \cos^{-1}\frac{2}{13}\sqrt{13}$$

Thus, spherical coordinates are $(\sqrt{13}, \frac{1}{2}\pi, \cos^{-1}\frac{2}{13}\sqrt{13})$.

(c) $\rho = \sqrt{r^2 + z^2} = \sqrt{2^2 + (-4)^2} = 2\sqrt{5}$

$$\theta = \frac{5}{6}\pi$$

$$\phi = \cos^{-1}\frac{z}{\rho} = \cos^{-1}\frac{-4}{2\sqrt{5}} = \cos^{-1} -\frac{2}{5}\sqrt{5}$$

Spherical coordinates are $(2\sqrt{5}, \frac{5}{6}\pi, \cos^{-1} -\frac{2}{5}\sqrt{5})$.

**10.** Find an equation in cylindrical coordinates of the given surface and identify the surface.

$$x^2 + y^2 = z^2$$

SOLUTION: Because the given equation is equivalent to $x^2 + y^2 - z^2 = 0$, the equation is of type III in Section 18.8. Thus, the graph is a right circular cone with axis the $z$-axis. Because $x^2 + y^2 = r^2$, a cylindrical-coordinate equation is $r^2 = z^2$. Because the point $(r, \theta, z)$ is the same as the point $(-r, \theta + \pi, z)$, the graph of the equation $r = z$ is the same as the graph of $r = -z$. Thus, the graph of $r^2 = z^2$ is the same as the graph of $r = z$, which is a cylindrical-coordinate equation of the cone $x^2 + y^2 = z^2$.

**14.** Find an equation in spherical coordinates of the given surface and identify the surface

$$x^2 + y^2 - 2z$$

SOLUTION: From (II) in Section 18.8, the graph of the equation is a paraboloid of revolution. The given equation is equivalent to

$$r^2 = 2z$$

Substituting from (IV), we obtain

$$\rho^2 \sin^2\phi = 2\rho\cos\phi$$

Thus, either $\rho = 0$, or

$$\rho\sin^2\phi = 2\cos\phi \tag{1}$$

Because Eq. (1) contains the origin $(0, 0, \frac{1}{2}\pi)$, we do not need $\rho = 0$. Thus, (1) is a spherical-coordinate equation of the paraboloid.

**18.** Find an equation in Cartesian coordinates for the surface whose equation is given in cylindrical coordinates.

$$z^2\sin^3\theta = r^3$$

SOLUTION: Because the graph contains the origin, we may multiply both sides by $r^3$ without affecting the graph. Thus,

$$z^2(r\sin\theta)^3 = (r^2)^3$$

We let $r\sin\theta = y$ and $r^2 = x^2 + y^2$. Thus,

$$z^2y^3 = (x^2 + y^2)^3$$

22. Find an equation in Cartesian coordinates for the surface whose equation is given in spherical coordinates.

$$\rho = 6 \sin \phi \sin \theta + 3 \cos \phi$$

SOLUTION: Because $(0, 0, \frac{1}{2}\pi)$ satisfies the given equation, the graph contains the origin. Thus, we may multiply on both sides by $\rho$. We have

$$\rho^2 = 6\rho \sin \phi \sin \theta + 3\rho \cos \phi$$

Substituting from (III) and (V) into the above, we get

$$x^2 + y^2 + z^2 = 6y + 3z$$

26. A *conical helix* winds around a cone in a way similar to that in which a circular helix winds around a cylinder. Use the formula of Exercise 24 to find the length of arc from $t = 0$ to $t = 2\pi$ of the conical helix having parametric equations

$$\rho = t \qquad \theta = t \qquad \phi = \frac{1}{4}\pi \tag{1}$$

SOLUTION: The formula of Exercise 24 is

$$L = \int_a^b \sqrt{(D_t\rho)^2 + \rho^2 \sin^2 \phi (D_t\theta)^2 + \rho^2 (D_t\phi)^2}\, dt \tag{2}$$

From (1), we obtain

$$D_t\rho = 1 \qquad D_t\theta = 1 \qquad D_t\phi = 0 \tag{3}$$

Substituting from (3) and (1) into formula (2), we get

$$L = \int_0^{2\pi} \sqrt{1 + \frac{1}{2}t^2}\, dt$$

Let $\bar{\theta} = \tan^{-1} \frac{1}{2}\sqrt{2}t$. Then $\sqrt{1 + \frac{1}{2}t^2} = \sec \bar{\theta}$; $dt = \sqrt{2} \sec^2 \bar{\theta}\, d\bar{\theta}$. Substituting into (4), we get

$$L = \sqrt{2} \int_0^{\tan^{-1} \sqrt{2}\pi} \sec^3 \bar{\theta}\, d\bar{\theta}$$

We use the result of Example 6 in Section 11.2. Thus,

$$L = \frac{1}{2}\sqrt{2}\,[\sec \bar{\theta} \tan \bar{\theta} + \ln|\sec \bar{\theta} + \tan \bar{\theta}|]_0^{\tan^{-1} \sqrt{2}\pi}$$

We have $\tan(\tan^{-1}\sqrt{2}\,\pi) = \sqrt{2}\,\pi$ and $\sec(\tan^{-1}\sqrt{2}\,\pi) = \sqrt{1 + \tan^2(\tan^{-1}\sqrt{2}\,\pi)} = \sqrt{1 + 2\pi^2}$. Thus,

$$L = \frac{1}{2}\sqrt{2}\,[\sqrt{1 + 2\pi^2}\sqrt{2}\,\pi + \ln(\sqrt{1 + 2\pi^2} + \sqrt{2}\,\pi)]$$

## *Review Exercises*

In Exercises 3-11, describe in words the set of points in $R^3$ satisfying the given equation or the given pair of equations. Draw a sketch of the graph.

4. $\begin{cases} x = z \\ y = z \end{cases}$

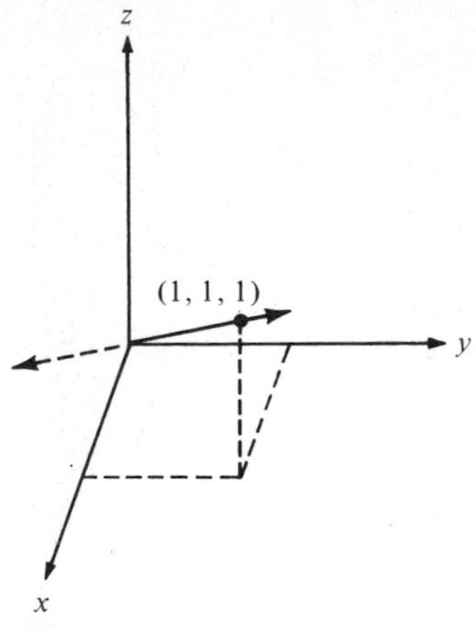

Figure 18.4 R

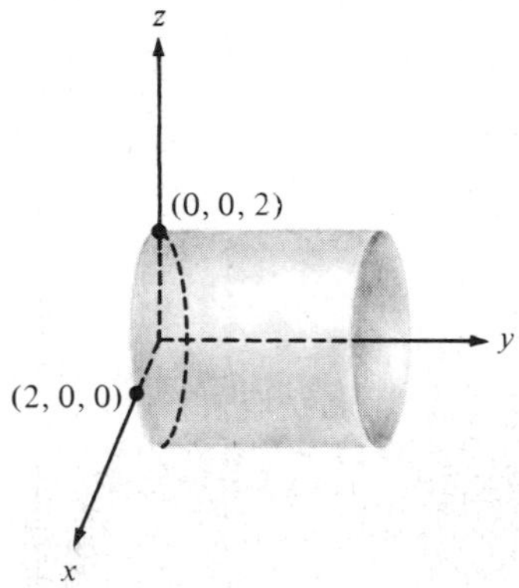

Figure 18.6 R

SOLUTION: Because

$$x = y = z$$

these are symmetric equations of the line that contains the origin and is parallel to representations of the vector $\langle 1, 1, 1\rangle$. In Fig. 18.4R we show a sketch of the line.

6. $x^2 + z^2 = 4$

SOLUTION: Because there is no term containing $y$, the graph is a cylinder with rulings perpendicular to the $xz$ plane. Because the directrix in the $xz$ plane is the circle $x^2 + z^2 = 4$ with center at the origin and radius 2, the cylinder is a right circular cylinder with axis the $y$-axis. A sketch of the cylinder is shown in Fig. 18.6R.

In Exercises 12-17, let $\mathbf{A} = -\mathbf{i} + 3\mathbf{j} + 2\mathbf{k}$, $\mathbf{B} = 2\mathbf{i} + \mathbf{j} - 4\mathbf{k}$, $\mathbf{C} = \mathbf{i} + 2\mathbf{j} - 2\mathbf{k}$, $\mathbf{D} = 3\mathbf{j} - \mathbf{k}$, $\mathbf{E} = 5\mathbf{i} - 2\mathbf{j}$, and find the indicated vector or scalar.

14. $2\mathbf{B} \cdot \mathbf{C} + 3\mathbf{D} \cdot \mathbf{E}$

SOLUTION:

$$\begin{aligned} 2\mathbf{B} \cdot \mathbf{C} &= 2(2\mathbf{i} + \mathbf{j} - 4\mathbf{k}) \cdot (\mathbf{i} + 2\mathbf{j} - 2\mathbf{k}) \\ &= 2[2 \cdot 1 + 1 \cdot 2 + (-4)(-2)] \\ &= 24 \end{aligned}$$

$$\begin{aligned} 3\mathbf{D} \cdot \mathbf{E} &= 3(3\mathbf{j} - \mathbf{k}) \cdot (5\mathbf{i} - 2\mathbf{j}) \\ &= 3[0 \cdot 5 + 3(-2) + (-1)0] \\ &= -18 \end{aligned}$$

Thus,

$$2\mathbf{B} \cdot \mathbf{C} + 3\mathbf{D} \cdot \mathbf{E} = 24 - 18 = 6$$

16. $(\mathbf{A} \times \mathbf{C}) - (\mathbf{D} \times \mathbf{E})$

SOLUTION:

$$\begin{aligned} \mathbf{A} \times \mathbf{C} &= (-\mathbf{i} + 3\mathbf{j} + 2\mathbf{k}) \times (\mathbf{i} + 2\mathbf{j} - 2\mathbf{k}) \\ &= \begin{vmatrix} \mathbf{i} & \mathbf{j} & \mathbf{k} \\ -1 & 3 & 2 \\ 1 & 2 & -2 \end{vmatrix} \\ &= \begin{vmatrix} 3 & 2 \\ 2 & -2 \end{vmatrix}\mathbf{i} - \begin{vmatrix} -1 & 2 \\ 1 & -2 \end{vmatrix}\mathbf{j} + \begin{vmatrix} -1 & 3 \\ 1 & 2 \end{vmatrix}\mathbf{k} \\ &= -10\mathbf{i} - 5\mathbf{k} \end{aligned}$$

$$\begin{aligned} \mathbf{D} \times \mathbf{E} &= (3\mathbf{j} - \mathbf{k}) \times (5\mathbf{i} - 2\mathbf{j}) \\ &= 15(\mathbf{j} \times \mathbf{i}) - 6(\mathbf{j} \times \mathbf{j}) - 5(\mathbf{k} \times \mathbf{i}) + 2(\mathbf{k} \times \mathbf{j}) \\ &= -15\mathbf{k} - 5\mathbf{j} - 2\mathbf{i} \end{aligned}$$

Thus,

$$\begin{aligned} (\mathbf{A} \times \mathbf{C}) - (\mathbf{D} \times \mathbf{E}) &= (-10\mathbf{i} - 5\mathbf{k}) - (-15\mathbf{k} - 5\mathbf{j} - 2\mathbf{i}) \\ &= -8\mathbf{i} + 5\mathbf{j} + 10\mathbf{k} \end{aligned}$$

18. There is only one way that a meaningful expression can be obtained by inserting parentheses. Insert the parentheses and find the indicated vector or scalar if $\mathbf{A} = \langle 3, -2, 4\rangle$, $\mathbf{B} = \langle -5, 7, 2\rangle$, and $\mathbf{C} = \langle 4, 6, -1\rangle$.

$$\mathbf{B} \cdot \mathbf{A} - \mathbf{C}$$

SOLUTION: Because $\mathbf{B}\cdot\mathbf{A}$ is a scalar, $(\mathbf{B}\cdot\mathbf{A})-\mathbf{C}$ is not defined. Thus, we take

$$\begin{aligned}\mathbf{B}\cdot(\mathbf{A}-\mathbf{C}) &= \langle -5,7,2\rangle\cdot(\langle 3,-2,4\rangle-\langle 4,6,-1\rangle)\\ &= \langle -5,7,2\rangle\cdot\langle -1,-8,5\rangle\\ &= (-5)(-1)+7(-8)+2\cdot 5\\ &= -41\end{aligned}$$

24. If $\mathbf{A}$ is any vector, prove that $\mathbf{A}=(\mathbf{A}\cdot\mathbf{i})\mathbf{i}+(\mathbf{A}\cdot\mathbf{j})\mathbf{j}+(\mathbf{A}\cdot\mathbf{k})\mathbf{k}$.

SOLUTION: Let $\mathbf{A}=a\mathbf{i}+b\mathbf{j}+c\mathbf{k}$. Then

$$\begin{aligned}\mathbf{A}\cdot\mathbf{i} &= (a\mathbf{i}+b\mathbf{j}+c\mathbf{k})\cdot\mathbf{i}=a\\ \mathbf{A}\cdot\mathbf{j} &= (a\mathbf{i}+b\mathbf{j}+c\mathbf{k})\cdot\mathbf{j}=b\\ \mathbf{A}\cdot\mathbf{k} &= (a\mathbf{i}+b\mathbf{j}+c\mathbf{k})\cdot\mathbf{k}=c\end{aligned}$$

Thus,

$$\begin{aligned}(\mathbf{A}\cdot\mathbf{i})\mathbf{i}+(\mathbf{A}\cdot\mathbf{j})\mathbf{j}+(\mathbf{A}\cdot\mathbf{k})\mathbf{k} &= a\mathbf{i}+b\mathbf{j}+c\mathbf{k}\\ &=\mathbf{A}\end{aligned}$$

28. Show that there are representations of the three vectors $\mathbf{A}=5\mathbf{i}+\mathbf{j}-3\mathbf{k}$, $\mathbf{B}=\mathbf{i}+3\mathbf{j}-2\mathbf{k}$, and $\mathbf{C}=-4\mathbf{i}+2\mathbf{j}+\mathbf{k}$ that form a triangle.

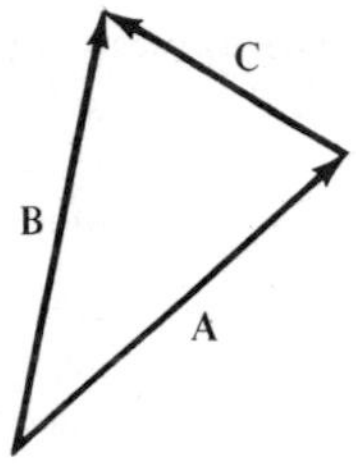

Figure 18.28R

SOLUTION: Representations of the vectors form a triangle if and only if one of the vectors is the sum of the other two. Because

$$\begin{aligned}\mathbf{A}+\mathbf{C} &= (5\mathbf{i}+\mathbf{j}-3\mathbf{k})+(-4\mathbf{i}+2\mathbf{j}+\mathbf{k})\\ &= \mathbf{i}+3\mathbf{j}-2\mathbf{k}\\ &= \mathbf{B}\end{aligned}$$

there are representations of the vectors that form a triangle. See Fig. 18.28.R.

30. Find an equation of the plane containing the points $(1,7,-3)$ and $(3,1,2)$ and which does not intersect the $x$-axis.

SOLUTION: Let $P=(1,7,-3)$ and $Q=(3,1,2)$. We have the vector

$$\mathbf{V}(\overrightarrow{PQ})=2\mathbf{i}-6\mathbf{j}+5\mathbf{k}$$

whose representations are parallel to the plane. Because the plane is parallel to the $x$-axis, representations of the unit vector $\mathbf{i}$ are parallel to the plane. Therefore $\mathbf{i}\times\mathbf{V}(\overrightarrow{PQ})$ is normal to the plane. We take

$$\begin{aligned}\mathbf{N} &= \mathbf{i}\times\mathbf{V}(\overrightarrow{PQ})\\ &= \mathbf{i}\times(2\mathbf{i}-6\mathbf{j}+5\mathbf{k})\\ &= -5\mathbf{j}-6\mathbf{k}\end{aligned}$$

Applying Theorem 18.4.2 with the normal vector $\mathbf{N}$ and the point $P$, we obtain

$$\begin{aligned}0(x-1)-5(y-7)-6(z+3)&=0\\ 5y+6z-17&=0\end{aligned}$$

36. Prove that the lines

$$\frac{x-1}{1}=\frac{y+2}{2}=\frac{z-2}{2}\quad\text{and}\quad\frac{x-2}{2}=\frac{y-5}{3}=\frac{z-5}{1}$$

are skew lines and find the distance between them.

SOLUTION: Let $L_1$ be the first line and $L_2$ be the second line. Because the direction numbers $[1,2,2]$ for line $L_1$ are not proportional to the direction numbers $[2,3,1]$ for line $L_2$, the lines are not parallel. We show that the lines do not intersect. Parametric equations of line $L_1$ are

$$x = t + 1 \qquad y = 2t - 2 \qquad z = 2t + 2 \tag{1}$$

Parametric equations of line $L_2$ are

$$x = 2s + 2 \qquad y = 3s + 5 \qquad z = s + 5 \tag{2}$$

If the lines intersect, there must be a replacement for $t$ and a replacement for $s$ that give the same values for $x, y$, and $z$ in Eq. (1) as in Eq. (2). Thus, there must be replacements for $s$ and $t$ such that the following three equations are all satisfied.

$$t + 1 = 2s + 2 \tag{3}$$

$$2t - 2 = 3s + 5 \tag{4}$$

$$2t + 2 = s + 5 \tag{5}$$

Solving Eqs. (3) and (4), we get $s = 5$ and $t = 11$. Because these values do not satisfy Eq. (5), we conclude that there is no simultaneous solution to Eqs. (3), (4), and (5). Therefore, line $L_1$ does not intersect line $L_2$, and thus $L_1$ and $L_2$ are skew lines.

We have point $P_1(1, -2, 2)$ in line $L_1$ and point $P_2(2, 5, 5)$ in line $L_2$. The distance between the lines is the absolute value of the scalar projection of vector $\mathbf{V}(\overrightarrow{P_1P_2})$ on a vector $\mathbf{N}$, whose representations are perpendicular to both line $L_1$ and line $L_2$. We have

$$\mathbf{V}(\overrightarrow{P_1P_2}) = \mathbf{i} + 7\mathbf{j} + 3\mathbf{k}$$

To find a suitable choice for $\mathbf{N}$, we take the cross product of direction vectors of lines $L_1$ and $L_2$. Thus, we take

$$\begin{aligned} \mathbf{N} &= (\mathbf{i} + 2\mathbf{j} + 2\mathbf{k}) \times (2\mathbf{i} + 3\mathbf{j} + \mathbf{k}) \\ &= -4\mathbf{i} + 3\mathbf{j} - \mathbf{k} \end{aligned}$$

Then the distance between the lines is given by

$$\begin{aligned} \left| \frac{\mathbf{N} \cdot \mathbf{V}(\overrightarrow{P_1P_2})}{|\mathbf{N}|} \right| &= \left| \frac{(-4\mathbf{i} + 3\mathbf{j} - \mathbf{k}) \cdot (\mathbf{i} + 7\mathbf{j} + 3\mathbf{k})}{\sqrt{(-4)^2 + 3^2 + (-1)^2}} \right| \\ &= \frac{14}{\sqrt{26}} \end{aligned}$$

**42.** Find the unit tangent vector and the curvature at any point on the curve having the vector equation

$$\mathbf{R}(t) = e^t\mathbf{i} + e^{-t}\mathbf{j} + 2t\mathbf{k}$$

SOLUTION: We have

$$\begin{aligned} \mathbf{R}'(t) &= e^t\mathbf{i} - e^{-t}\mathbf{j} + 2\mathbf{k} \\ |\mathbf{R}'(t)| &= \sqrt{e^{2t} + e^{-2t} + 4} \end{aligned}$$

Thus, the unit tangent vector is

$$\begin{aligned} \mathbf{T}(t) &= \frac{\mathbf{R}'(t)}{|\mathbf{R}'(t)|} \\ &= (e^{2t} + e^{-2t} + 4)^{-1/2}(e^t\mathbf{i} - e^{-t}\mathbf{j} + 2\mathbf{k}) \end{aligned}$$

Using the formula for the derivative of a scalar times a vector, we obtain

$$\begin{aligned} \mathbf{T}'(t) = {} & (e^{2t} + e^{-2t} + 4)^{-1/2}(e^t\mathbf{i} + e^{-t}\mathbf{j}) \\ & + \left(-\frac{1}{2}\right)(e^{2t} + e^{-2t} + 4)^{-3/2}(2e^{2t} - 2e^{-2t})(e^t\mathbf{i} - e^{-t}\mathbf{j} + 2\mathbf{k}) \end{aligned}$$

$$= (e^{2t} + e^{-2t} + 4)^{-3/2}[(e^{2t} + e^{-2t} + 4)(e^{t}\mathbf{i} + e^{-t}\mathbf{j}) + (-e^{2t} + e^{-2t})(e^{t}\mathbf{i} - e^{-t}\mathbf{j} + 2\mathbf{k})]$$
$$= 2(e^{2t} + e^{-2t} + 4)^{-3/2}[(e^{-t} + 2e^{t})\mathbf{i} + (e^{t} + 2e^{-t})\mathbf{j} + (e^{-2t} - e^{2t})\mathbf{k}]$$

Therefore,

$$\frac{\mathbf{T}'(t)}{|\mathbf{R}'(t)|} = 2(e^{2t} + e^{-2t} + 4)^{-2}[(e^{-t} + 2e^{t})\mathbf{i} + (e^{t} + 2e^{-t})\mathbf{j} + (e^{-2t} - e^{2t})\mathbf{k}]$$

Because the curvature is given by

$$K(t) = \left|\frac{\mathbf{T}'(t)}{|\mathbf{R}'(t)|}\right|$$

we have

$$K(t) = 2(e^{2t} + e^{-2t} + 4)^{-2}\sqrt{(e^{-t} + 2e^{t})^2 + (e^{t} + 2e^{-t})^2 + (e^{-2t} - e^{2t})^2}$$
$$= 2(e^{2t} + e^{-2t} + 4)^{-2}\sqrt{e^{4t} + 5e^{2t} + 5e^{-2t} + e^{-4t} + 6}$$

**44.** Find an equation in spherical coordinates of the graph of each of the equations:

**(a)** $x^2 + y^2 + 4z^2 = 4$ **(b)** $4x^2 - 4y^2 + 9z^2 = 36$

SOLUTION:

**(a)** We have

$$(x^2 + y^2 + z^2) + 3z^2 = 4$$

Because $x^2 + y^2 + z^2 = \rho^2$ and $z = \rho \cos \phi$, we obtain

$$\rho^2 + 3\rho^2 \cos^2 \phi = 4$$
$$\rho^2(1 + 3 \cos^2 \phi) = 4$$

**(b)** Substituting from Eqs. (III) of Section 18.10 into the given equation, we obtain

$$4\rho^2 \sin^2 \phi \cos^2 \theta - 4\rho^2 \sin^2 \phi \sin^2 \theta + 9\rho^2 \cos^2 \phi = 36$$
$$4\rho^2 \sin^2 \phi(\cos^2 \theta - \sin^2 \theta) + 9\rho^2 \cos^2 \phi = 36$$
$$4\rho^2 \sin^2 \phi \cos 2\theta + 9\rho^2 \cos^2 \phi = 36$$
$$\rho^2(4 \sin^2 \phi \cos 2\theta + 9 \cos^2 \phi) = 36$$

# 19

# Differential calculus of functions of several variables

**19.1 FUNCTIONS OF MORE THAN ONE VARIABLE** Let $P$ be an ordered pair of real numbers $(x, y)$, and let $z$ be a real number. A *function of two variables* is a set of ordered pairs of the form $(P, z)$ in which no two distinct pairs have the same first element. The set of all possible replacements of $P$ is called the *domain* of the function, and the set of all possible values of $z$ is called the *range* of the function. The variables $x$ and $y$ are called the *independent variables,* and $z$ is called the *dependent variable.* If $f$ is a function of two variables, then we denote the function value of $f$ at $P$ by either $f(P)$ or $f(x, y)$.

**19.1.3 Definition** If $f$ is a function of a single variable and $g$ is a function of two variables, then the *composite function* $f \circ g$ is the function of two variables defined by

$$(f \circ g)(x, y) = f(g(x, y))$$

and the domain of $f \circ g$ is the set of all points $(x, y)$ in the domain of $g$ such that $g(x, y)$ is in the domain of $f$.

**19.1.5 Definition** If $f$ is a function of two variables, then the *graph* of $f$ is the set of all points $(x, y, z)$ in $R^3$ for which $(x, y)$ is a point in the domain of $f$ and $z = f(x, y)$.

Thus, the graph of a function of two variables is a surface in $R^3$. Because there is only one value of $z$ for each replacement of $(x, y)$ from the domain of the function, a line perpendicular to the $xy$ plane intersects the surface in no more than one point.

Let $C$ be the curve that is the cross section of the surface $z = f(x, y)$ and the plane $z = k$. The projection of $C$ on the $xy$ plane, that is, the curve $f(x, y) = k$ in the $xy$ plane is called the *contour curve* (or *level curve*) of the function $f$ at $k$. The set of all contour curves of $f$ at $k$ if $k = k_1, k_2, \ldots, k_n$ is called a *contour map* of the function $f$. If the numbers $k_1, k_2, \ldots, k_n$ are chosen so that the difference between successive numbers is constant, then the resulting contour map of the function $f$ indicates the steepness of the surface that is the graph of $f$. Closely spaced contour curves indicate that the surface is steep; contour curves that are widely spaced indicate that the surface is not steep.

The definitions for functions of one and two variables are extended to include functions of three or more variables as follows.

**19.1.1 Definition** The set of all ordered $n$-tuples of real numbers is called the *n-dimensional number space* and is denoted by $R^n$. Each ordered $n$-tuple $(x_1, x_2, \ldots, x_n)$ is called a *point* in the $n$-dimensional number space.

**19.1.2 Definition** A *function of n variables* is a set of ordered pairs of the form $(P, w)$ in which no two distinct ordered pairs have the same first element. $P$ is a point in $n$-dimensional number space and $w$ is a real number. The set of all possible values of $P$ is called the *domain* of the function, and the set of all possible values of $w$ is called the *range* of the function.

**19.1.4 Definition** If $f$ is a function of a single variable and $g$ is a function of $n$ variables, then the *composite function* $f \circ g$ is the function of $n$ variables defined by

$$(f \circ g)(x_1, x_2, \ldots, x_n) = f(g(x_1, x_2, \ldots, x_n)).$$

and the domain of $f \circ g$ is the set of all points $(x_1, x_2, \ldots, x_n)$ in the domain of $g$ such that $g(x_1, x_2, \ldots, x_n)$ is in the domain of $f$.

**19.1.6 Definition** If $f$ is a function of $n$ variables, then the *graph* of $f$ is the set of all points $(x_1, x_2, \ldots, x_n, w)$ in $R^{n+1}$ for which $(x_1, x_2, \ldots, x_n)$ is a point in the domain of $f$ and $w = f(x_1, x_2, \ldots, x_n)$.

## *Exercises 19.1*

2. Let the function $g$ of three variables, $x$, $y$, and $z$, be the set of all ordered pairs of the form $(P, w)$ such that $w = \sqrt{4 - x^2 - y^2 - z^2}$. Find: **(a)** $g(1, -1, -1)$; **(b)** $g(-a, 2b, \frac{1}{2}c)$; **(c)** $g(y, -x, -y)$; **(d)** the domain of $g$; **(e)** the range of $g$; **(f)** $[g(x, y, z)]^2 - [g(x + 2, y + 2, z)]^2$. Draw a sketch showing as a shaded solid in $R^3$ the set of points in the domain of $g$.

SOLUTION: Because $w = g(x, y, z)$, we have

$$g(x, y, z) = \sqrt{4 - x^2 - y^2 - z^2}$$

Thus,

**(a)** $g(1, -1, -1) = \sqrt{4 - 1^2 - (-1)^2 - (-1)^2} = 1$

**(b)** $$g\left(-a, 2b, \frac{1}{2}c\right) = \sqrt{4 - (-a)^2 - (2b)^2 - \left(\frac{1}{2}c\right)^2}$$
$$= \sqrt{4 - a^2 - 4b^2 - \frac{1}{4}c^2}$$

**(c)** $$g(y, -x, -y) = \sqrt{4 - y^2 - (-x)^2 - (-y)^2}$$
$$= \sqrt{4 - x^2 - 2y^2}$$

**(d)** The domain of $g$ is $\{(x, y, z) | 4 - x^2 - y^2 - z^2 \geqslant 0\} = \{(x, y, z) | x^2 + y^2 + z^2 \leqslant 4\}$

(e) Because $w = \sqrt{4 - (x^2 + y^2 + z^2)}$ and $0 \leqslant x^2 + y^2 + z^2 \leqslant 4$, then $0 \leqslant w \leqslant 2$. Thus, the range of $g$ is the closed interval $[0, 2]$.

(f) $[g(x, y, z)]^2 = 4 - x^2 - y^2 - z^2$

$$[g(x+2, y+2, z)]^2 = 4 - (x+2)^2 - (y+2)^2 - z^2$$
$$= -x^2 - y^2 - z^2 - 4x - 4y - 4$$

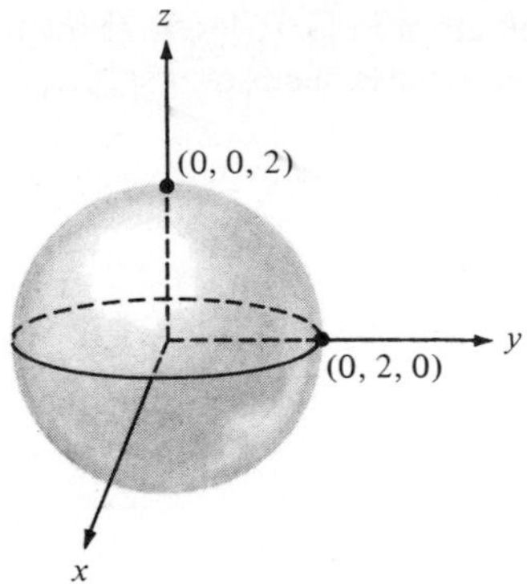

Figure 19.1.2

Thus,

$$[g(x, y, z)]^2 - [g(x+2, y+2, z)]^2 = (4 - x^2 - y^2 - z^2) - (-x^2 - y^2 - z^2 - 4x - 4y - 4) = 4x + 4y + 8$$

The set of all points in the domain of $g$ is the set of all points that are either on the sphere $x^2 + y^2 + z^2 = 4$ or inside the sphere, as illustrated in Fig. 19.1.2.

In Exercises 3-11, find the domain and range of the function $f$ and draw a sketch showing as a shaded region in $R^2$ the set of points in the domain of $f$.

6. $f(x, y) = \sqrt{\dfrac{x - y}{x + y}}$

SOLUTION: The domain of $f$ is

$$\left\{(x, y) \,\middle|\, \frac{x - y}{x + y} \geqslant 0\right\}$$

We consider two cases:

*Case 1:* $x + y > 0$, or, equivalently $y > -x$. Then if we multiply on both sides of the inequality

$$\frac{x - y}{x + y} \geqslant 0$$

by the positive number $x + y$, the result is

$$x - y \geqslant 0$$
$$y \leqslant x$$

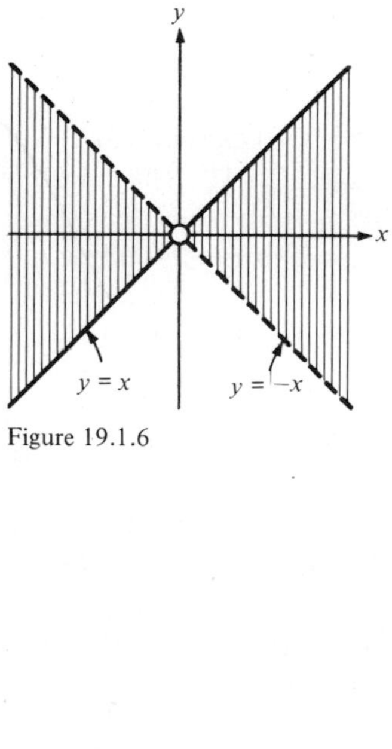

Figure 19.1.6

Thus, we have $\{(x, y) | y > -x$ and $y \leqslant x\}$. The graph of this set is the shaded region in the first and fourth quadrants, shown in Fig. 19.1.6.

*Case 2:* $x + y < 0$, or, equivalently, $y < -x$. Multiplying, on both sides of the inequality

$$\frac{x - y}{x + y} \geqslant 0$$

by the negative number $x + y$, we obtain

$$x - y \leqslant 0$$
$$y \geqslant x$$

Thus, we have $\{(x, y) | y < -x$ and $y \geqslant x\}$ whose graph is the shaded region in the second and third quadrants of Fig. 19.1.6.

The domain of $f$ is the union of the two sets found in Case 1 and Case 2 and is the entire shaded region in Fig. 19.1.6. Note that the boundary line $y = -x$ is not included in the set, and we indicate this by using a broken line. The solid line $y = x$ indicates that this boundary line is included in the set (except for the origin).

The range of $f$ is $[0, +\infty)$ because $f(x, y) = 0$ when $x = y$ and $f(x, y)$ can be made larger than any number by choosing $x + y$ close to zero.

**10.** $f(x, y) = \sin^{-1}(x + y)$

SOLUTION: Because the domain of the inverse sine function is $[-1, 1]$, the domain of $f$ is $\{(x, y) | -1 \leqslant x + y \leqslant 1\}$. Thus, the domain is the intersection of sets $A$ and $B$, where

$$A = \{(x, y) | x + y \leqslant 1\}$$

and

$$B = \{(x, y) | x + y \geqslant -1\}$$

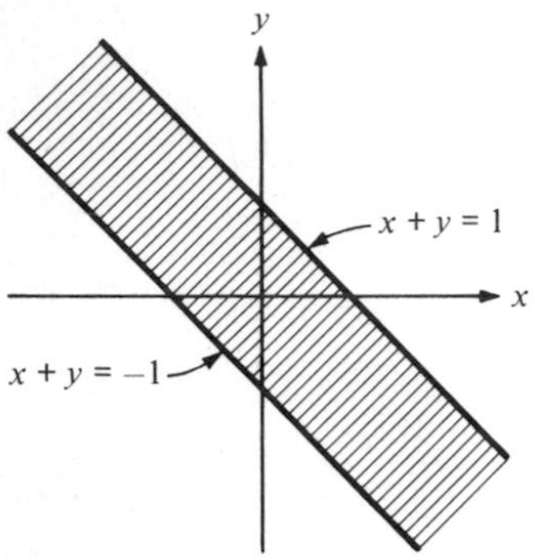

Figure 19.1.10

The graph of set $A$ is the set of all points that are either on or below the line $x + y = 1$. The graph of set $B$ is the set of all points that are either on or above the line $x + y = -1$. Thus, the graph of the domain of $f$ is the set of all points that lie on either of the two lines or between the two lines, which are parallel. See Fig. 19.1.10.

The range of $f$ is $[-\frac{1}{2}\pi, \frac{1}{2}\pi]$ because this is the range of the inverse sine function.

**14.** Find the domain and range of the function $f$.

$$f(x, y, z) = |x| e^{y/z}$$

SOLUTION: The exponential function and the absolute value function are defined at all real numbers. Thus, $f$ is defined at all $x$, $y$, and $z$ for which $z \neq 0$. The domain of $f$ is $\{(x, y, z) | z \neq 0\}$. The range of $f$ is $[0, +\infty)$, because $|x| \geqslant 0$ for all $x$ and the exponential function always has a positive value.

**18.** Find the domain and range of the function $f$ and draw a sketch of the graph.

$$f(x, y) = \sqrt{100 - 25x^2 - 4y^2}$$

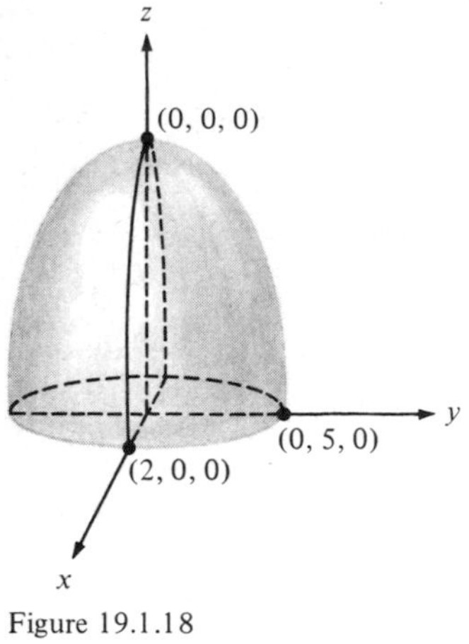

Figure 19.1.18

SOLUTION: Let $z = f(x, y)$. Then

$$z^2 = 100 - 25x^2 - 4y^2$$
$$25x^2 + 4y^2 + z^2 = 100$$

Because $z \geqslant 0$, the graph of the function $f$ is that part of the ellipsoid $25x^2 + 4y^2 + z^2 = 100$ that is not below the $xy$ plane, as shown in Fig. 19.1.18. The domain of $f$ is $\{(x, y) | 100 - 25x^2 - 4y^2 \geqslant 0\} = \{(x, y) | 25x^2 + 4y^2 \leqslant 100\}$. Thus, the domain of $f$ is the set of all points in the $xy$ plane that are either on or inside the ellipse $25x^2 + 4y^2 = 100$. Note that the domain of $f$ is the projection on the $xy$ plane of the surface that is the graph of $f$. The range of $f$ is $\{z | 0 \leqslant z \leqslant 10\} = [0, 10]$, which is the projection of the graph of $f$ on the $z$-axis.

In Exercises 21-26, draw a sketch of the contour map of the function showing the level curves of $f$ at the given numbers.

**22.** The function of Exercise 16 at 5, 4, 3, 2, 1, and 0.

SOLUTION: The function of Exercise 16 is the function $f$ defined by

$$f(x, y) = \sqrt{x + y}$$

The level curve of $f$ at the number 5 is the graph of $f(x, y) = 5$ in the $xy$ plane. If $f(x, y) = 5$, we have

$$\sqrt{x + y} = 5$$
$$x + y = 25$$

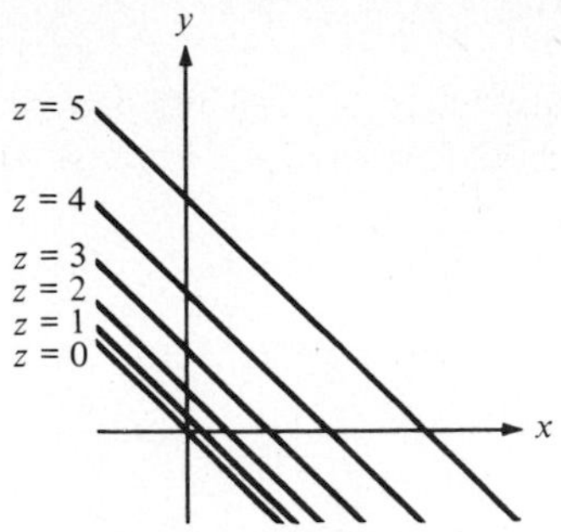

Figure 19.1.22

In a similar manner, we find an equation of the level curve for each of the numbers 5, 4, 3, 2, 1, and 0. The results are summarized in Table 22. We draw a sketch of each level curve on the same coordinate system. This set of curves is the required contour map, as shown in Fig. 19.1.22.

**Table 22**

| $z$ | $f(x, y) = z$ |
|---|---|
| 5 | $x + y = 25$ |
| 4 | $x + y = 16$ |
| 3 | $x + y = 9$ |
| 2 | $x + y = 4$ |
| 1 | $x + y = 1$ |
| 0 | $x + y = 0$ |

**26.** The function $f$ for which $f(x, y) = (x - 3)/(y + 2)$ at $4, 2, 1, \frac{1}{2}, \frac{1}{4}, 0, -\frac{1}{4}, -\frac{1}{2}, -1, -2$, and $-4$.

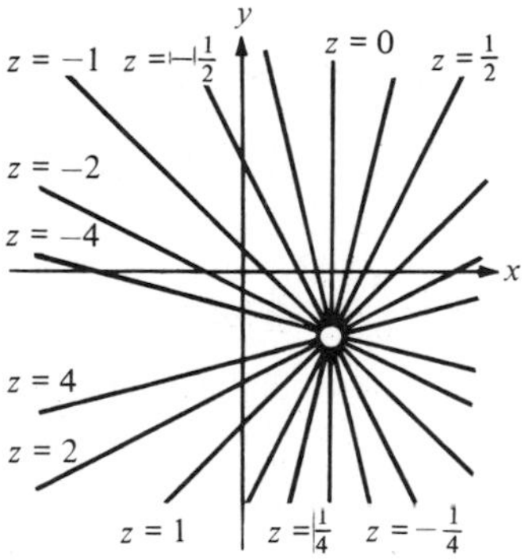

Figure 19.1.26

SOLUTION: If $f(x, y) = k$, then

$$\frac{x-3}{y+2} = k \tag{1}$$

If $k \neq 0$, we may solve for $y$, obtaining

$$y = \frac{1}{k}x - 2 - \frac{3}{k}$$

Thus, if $k \neq 0$, the graph in the $xy$ plane of (1) is a line with the point $(3, -2)$ deleted. The line has slope $1/k$. If $k = 0$, the graph of (1) is the line $x = 3$ with the point $(3, -2)$ deleted. We show the contour map, obtained by letting $k$ take on each of the given values, in Fig. 19.1.26.

**30.** Given $f(x, y) = x/y^2$, $g(x) = x^2$, $h(x) = \sqrt{x}$. Find: **(a)** $(h \circ f)(2, 1)$; **(b)** $f(g(2), h(4))$; **(c)** $f(g(\sqrt{x}), h(x^2))$; **(d)** $h((g \circ f)(x, y))$; **(e)** $(h \circ g)(f(x, y))$.

SOLUTION:

**(a)** $(h \circ f)(2, 1) = h(f(2, 1))$. Because

$$f(2, 1) = \frac{2}{1^2} = 2$$

then

$$(h \circ f)(2, 1) = h(2) = \sqrt{2}$$

**(b)** $g(2) = 2^2 = 4$ and $h(4) = \sqrt{4} = 2$. Thus,

$$\begin{aligned} f(g(2), h(4)) &= f(4, 2) \\ &= \frac{4}{2^2} \\ &= 1 \end{aligned}$$

**(c)** $g(\sqrt{x}) = (\sqrt{x})^2 = x$ and $h(x^2) = \sqrt{x^2} = |x|$. Thus,

$$f(g(\sqrt{x}), h(x^2)) = f(x, |x|)$$

$$= \frac{x}{|x|^2}$$

$$= \frac{1}{x}$$

(d) $(g \circ f)(x, y) = g(f(x, y))$
$= [f(x, y)]^2$
$= \dfrac{x^2}{y^4}$

Thus,

$$h((g \circ f)(x, y)) = h\left(\frac{x^2}{y^4}\right) = \sqrt{\frac{x^2}{y^4}} = \frac{|x|}{y^2}$$

(e) $(h \circ g)(f(x, y)) = h(g(f(x, y))$
$= h((g \circ f)(x, y))$

Because this is the same function as in part (d), we have

$$(h \circ g)(f(x, y)) = \frac{|x|}{y^2}$$

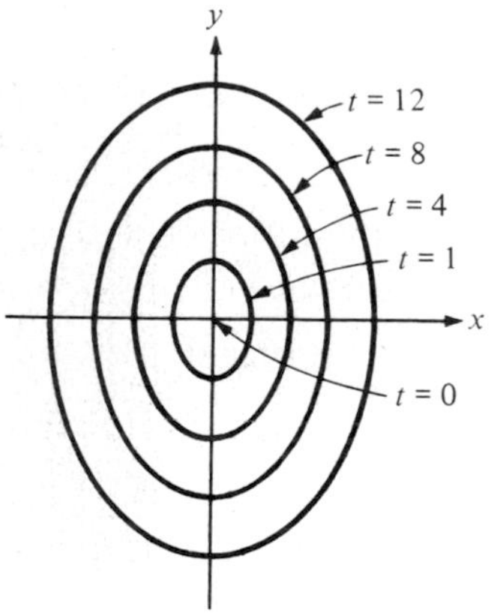

Figure 19.1.32

32. The temperature at a point $(x, y)$ of a flat metal plate is $t$ degrees and $t = 4x^2 + 2y^2$. Draw the isothermals for $t = 12, 8, 4, 1, 0$.

SOLUTION: Let $f(x, y) = 4x^2 + 2y^2$. The isothermals are the level curves of $f$. When $t = 12$, we have the ellipse

$$4x^2 + 2y^2 = 12$$
$$2x^2 + y^2 = 6$$

which has vertices at $(0, \pm\sqrt{6})$ and ends of the minor axis at $(\pm\sqrt{3}, 0)$. For each $t > 0$, we have an ellipse with vertices at $(0, \pm\sqrt{\frac{1}{2}t})$ and ends of the minor axis at $(\pm\frac{1}{2}\sqrt{t}, 0)$. For $t = 0$, we have the single point $(0, 0)$. The isothermals for $t = 12, 8, 4, 1, 0$ are shown in Fig. 19.1.32.

## 19.2 LIMITS OF FUNCTIONS OF MORE THAN ONE VARIABLE

We consider functions of two variables. The *open disk* $B((x_0, y_0); r)$ is the set of all points in the $xy$ plane that are inside the circle with center at the point $(x_0, y_0)$ and radius $r$. Thus, a point $(x, y)$ is in $B((x_0, y_0); r)$ if and only if $\sqrt{(x - x_0)^2 + (y - y_0)^2} < r$.

**19.2.5 Definition** Let $f$ be a function of two variables that is defined on some open disk $B((x_0, y_0); r)$, except possibly at the point $(x_0, y_0)$ itself. Then

$$\lim_{(x, y) \to (x_0, y_0)} f(x, y) = L$$

if for any $\epsilon > 0$, however small, there exists a $\delta > 0$ such that

$$|f(x, y) - L| < \epsilon \quad \text{whenever } 0 < \sqrt{(x - x_0)^2 + (y - y_0)^2} < \delta$$

Let $S$ be a set of points in the $xy$ plane. A point $(x_0, y_0)$ is said to be an *accumulation* point of $S$ if and only if every open disk $B((x_0, y_0); r)$ contains infinitely many points of $S$.

The next definition and the following two theorems extend to functions of two variables the concept of one-sided limits for functions of one variable.

**19.2.7 Definition** Let $f$ be a function defined on a set of points $S$ in $R^2$, and let $(x_0, y_0)$ be an accumulation point of $S$. Then the *limit of* $f(x, y)$ *as* $(x, y)$ *approaches* $(x_0, y_0)$ *in* $S$ *is* $L$, written as

$$\lim_{\substack{(x, y) \to (x_0, y_0) \\ (P \text{ in } S)}} f(x, y) = L$$

if for any $\epsilon > 0$, however small, there exists a $\delta > 0$ such that

$$|f(x, y) - L| < \epsilon \quad \text{whenever } 0 < \sqrt{(x - x_0)^2 + (y - y_0)^2} < \delta$$

and $(x, y)$ is in $S$.

**19.2.8 Theorem** Suppose that the function $f$ is defined for all points on an open disk having its center at $(x_0, y_0)$, except possibly at $(x_0, y_0)$ itself, and

$$\lim_{(x, y) \to (x_0, y_0)} f(x, y) = L$$

Then if $S$ is any set of points in $R^2$ having $(x_0, y_0)$ as an accumulation point,

$$\lim_{\substack{(x, y) \to (x_0, y_0) \\ (P \text{ in } S)}} f(x, y)$$

exists and always has the value $L$.

**19.2.9 Theorem** If the function $f$ has different limits as $(x, y)$ approaches $(x_0, y_0)$ through two distinct sets of points having $(x_0, y_0)$ as an accumulation point, then $\lim_{(x, y) \to (x_0, y_0)} f(x, y)$ does not exist.

The limit theorems of Chapter 2, with minor modifications, apply to functions of two variables. We also have the following theorem regarding the limit of a composite function of two variables.

**19.2.10 Theorem** If $g$ is a function of two variables and $\lim_{(x, y) \to (x_0, y_0)} g(x, y) = b$, and $f$ is a function of a single variable continuous at $b$, then

$$\lim_{(x, y) \to (x_0, y_0)} (f \circ g)(x, y) = f(b)$$

or, equivalently,

$$\lim_{(x, y) \to (x_0, y_0)} f(g(x, y)) = f\left(\lim_{(x, y) \to (x_0, y_0)} g(x, y)\right)$$

We may be able to find the limit of a function of two variables by applying the limit theorems of Chapter 2 together with Theorem 19.2.10. If these theorems fail, as they do, for example, when both the numerator and denominator of a fraction have limit zero, then the following steps may sometimes be used to determine whether or not $\lim_{(x, y) \to (x_0, y_0)} f(x, y)$ exists.

1. Let $S_1$ be a set of points on a curve containing $(x_0, y_0)$ and find $L_1$, if it exists, where

$$L_1 = \lim_{\substack{(x, y) \to (x_0, y_0) \\ (P \text{ in } S_1)}} f(x, y)$$

2. If $L_1$ does not exist, then $\lim_{(x, y) \to (x_0, y_0)} f(x, y)$ does not exist.

3. If $L_1$ does exist, then let $S_2$ be a set of points on a curve containing $(x_0, y_0)$ with $S_2 \neq S_1$, and find $L_2$, if it exists, where

$$L_2 = \lim_{\substack{(x,y)\to(x_0,y_0) \\ (P \text{ in } S_2)}} f(x,y)$$

4. If $L_2$ does not exist, or if $L_1 \neq L_2$, then $\lim_{(x,y)\to(x_0,y_0)} f(x,y)$ does not exist.

5. If $L_1 = L_2$, try to use Definition 19.2.5 with $L = L_1 = L_2$.

It is possible that $L_1 = L_2$ and yet the limit does not exist. (Refer to Example 4 in Section 19.2 of the text.)

The following are generalizations of the corresponding definitions and theorems for functions of two variables. They extend these definitions and theorems to functions of more than two variables.

**19.2.1 Definition** If $P(x_1, x_2, \ldots, x_n)$ and $A(a_1, a_2, \ldots, a_n)$ are two points in $R$, then the distance between $P$ and $A$, denoted by $\|P - A\|$, is given by

$$\|P - A\| = \sqrt{(x_1 - a_1)^2 + (x_2 - a_2)^2 + \cdots + (x_n - a_n)^2}$$

**19.2.2 Definition** If $A$ is a point in $R^n$ and $r$ is a positive number, then the *open ball* $B(A; r)$ is defined to be the set of all points $P$ in $R^n$ such that $\|P - A\| < r$.

**19.2.4 Definition** Let $f$ be a function of $n$ variables that is defined on some open ball $B(A; r)$, except possibly at the point $A$ itself. Then the *limit of $f(P)$ as $P$ approaches $A$ is $L$*, written as

$$\lim_{P\to A} f(P) = L$$

if for any $\epsilon > 0$, however small, there exists a $\delta > 0$ such that

$$|f(P) - L| < \epsilon \quad \text{whenever } 0 < \|P - A\| < \delta$$

**19.2.6 Definition** A point $P_0$ is said to be an *accumulation point* of a set $S$ of points in $R^n$ if every open ball $B(P_0; r)$ contains infinitely many points of $S$.

## *Exercises 19.2*

**4.** Establish the limit by finding a $\delta > 0$ for any $\epsilon > 0$ so that Definition 19.2.5 holds.

$$\lim_{(x,y)\to(2,3)} (2x^2 - y^2) = -1$$

SOLUTION: For any $\epsilon > 0$ we must find a $\delta > 0$ such that

$$|2x^2 - y^2 + 1| < \epsilon \quad \text{whenever } 0 < \sqrt{(x-2)^2 + (y-3)^2} < \delta \tag{1}$$

Applying the corollary of the triangle inequality (1.2.9), we have

$$\begin{aligned} |2x^2 - y^2 + 1| &= |2(x^2 - 4) - (y^2 - 9)| \\ &\leq 2|x-2||x+2| + |y-3||y+3| \end{aligned} \tag{2}$$

If $\delta \leq 1$ and $0 < \sqrt{(x-2)^2 + (y-3)^2} < \delta$, then

$$|x - 2| \leq \sqrt{(x-2)^2 + (y-3)^2} < \delta \leq 1 \tag{3}$$

and

$$|y - 3| \leq \sqrt{(x-2)^2 + (y-3)^2} < \delta \leq 1 \tag{4}$$

Furthermore, from (3) we have $|x-2|<1$, and thus

$$\begin{aligned} -1 &< x-2 < 1 \\ 3 &< x+2 < 5 \\ |x+2| &< 5 \end{aligned} \tag{5}$$

And, from (4) we have $|y-3|<1$. Thus,

$$\begin{aligned} -1 &< y-3 < 1 \\ 5 &< y+3 < 7 \\ |y+3| &< 7 \end{aligned} \tag{6}$$

From (3), (4), (5), and (6), we obtain

$$2|x-2||x+2|+|y-3||y+3|<2\delta\cdot 5+\delta\cdot 7=17\delta \tag{7}$$

Therefore, for any $\epsilon>0$ we take $\delta=\min(1, \frac{1}{17}\epsilon)$. Whenever $0<\sqrt{(x-2)^2+(y-3)^2}<\delta\leqslant 1$, inequality (7) holds. Hence, from (2) and (7), we have

$$\begin{aligned} |2x^2-y^2+1| &\leqslant 2|x-2||x+2|+|y-3||y+3| \\ &\leqslant 17\delta \\ &\leqslant \epsilon \quad \text{whenever } 0<\sqrt{(x-2)^2+(y-3)^2}<\delta \end{aligned}$$

Therefore,

$$\lim_{(x,y)\to(2,3)} (2x^2-y^2)=-1$$

**10.** Prove that for the given function $f$, $\displaystyle\lim_{(x,y)\to(0,0)} f(x,y)$ does not exist.

$$f(x,y)=\frac{x^4+3x^2y^2+2xy^3}{(x^2+y^2)^2}$$

SOLUTION: We use Theorem 19.2.9. Let $S_1$ be the set of all points on the $y$-axis. Because $x=0$ for all points $P$ in $S_1$, we have

$$\lim_{\substack{(x,y)\to(0,0)\\(P \text{ in } S_1)}} f(x,y)=\lim_{y\to 0} f(0,y)=\lim_{y\to 0}\frac{0}{y^4}=0$$

Next, let $S_2$ be the set of all points on the line $y=x$. Then

$$\lim_{\substack{(x,y)\to(0,0)\\(P \text{ in } S_2)}} f(x,y)=\lim_{x\to 0} f(x,x)=\lim_{x\to 0}\frac{x^4+3x^4+2x^4}{(x^2+x^2)^2}=\frac{3}{2}$$

Because

$$\lim_{\substack{(x,y)\to(0,0)\\(P \text{ in } S_1)}} f(x,y)\neq\lim_{\substack{(x,y)\to(0,0)\\(P \text{ in } S_2)}} f(x,y)$$

then $\displaystyle\lim_{(x,y)\to(0,0)} f(x,y)$ does not exist.

**14.** Prove that $\displaystyle\lim_{(x,y)\to(0,0)} f(x,y)$ exists.

$$f(x,y)=\frac{x^3+y^3}{x^2+y^2}$$

SOLUTION: Let $S$ be the set of points on the $x$-axis. Thus $y=0$ in $S$, and

$$\lim_{\substack{(x,y)\to(0,0)\\(P \text{ in } S)}} f(x,y) = \lim_{x\to 0} f(x,0) = \lim_{x\to 0}\frac{x^3}{x^2} = \lim_{x\to 0} x = 0$$

Therefore, if $\lim_{(x,y)\to(0,0)} f(x,y)$ exists, then $\lim_{(x,y)\to(0,0)} f(x,y) = 0$. We use Definition 19.2.5 to prove that the limit is zero. For any $\epsilon > 0$, we must find some $\delta > 0$ such that

$$\left|\frac{x^3+y^3}{x^2+y^2}\right| < \epsilon \quad \text{whenever } 0 < \sqrt{x^2+y^2} < \delta$$

Because $x^2 \leqslant x^2+y^2$ and $y^2 \leqslant x^2+y^2$, then $|x| \leqslant \sqrt{x^2+y^2}$ and $|y| \leqslant \sqrt{x^2+y^2}$. Whenever $0 < \sqrt{x^2+y^2} < \delta$, we have by the triangle inequality

$$\begin{aligned}\left|\frac{x^3+y^3}{x^2+y^2}\right| &\leqslant \left|\frac{x^3}{x^2+y^2}\right| + \left|\frac{y^3}{x^2+y^2}\right| \\ &= \frac{|x|x^2}{x^2+y^2} + \frac{|y|y^2}{x^2+y^2} \\ &\leqslant \frac{|x|(x^2+y^2)}{x^2+y^2} + \frac{|y|(x^2+y^2)}{x^2+y^2} \\ &\leqslant |x| + |y| \\ &\leqslant \sqrt{x^2+y^2} + \sqrt{x^2+y^2} \\ &< \delta + \delta \\ &= 2\delta \end{aligned} \tag{1}$$

Thus, if we take $\delta = \frac{1}{2}\epsilon$, then whenever $0 < \sqrt{x^2+y^2} < \delta$, (1) follows. Thus,

$$\left|\frac{x^3+y^3}{x^2+y^2}\right| < 2\delta = \epsilon \quad \text{whenever } 0 < \sqrt{x^2+y^2} < \delta$$

Therefore

$$\lim_{(x,y)\to(0,0)} f(x,y) = 0$$

**18.** Evaluate the given limit by the use of limit theorems.

$$\lim_{(x,y)\to(-2,4)} y\sqrt[3]{x^3+2y}$$

SOLUTION:

$$\begin{aligned}\lim_{(x,y)\to(-2,4)} y\sqrt[3]{x^3+2y} &= \left(\lim_{y\to 4} y\right)\cdot \sqrt[3]{\lim_{(x,y)\to(-2,4)} (x^3+2y)} \\ &= 4\sqrt[3]{0} \\ &= 0\end{aligned}$$

**22.** Show the application of Theorem 19.2.10 to find the indicated limit.

$$\lim_{(x,y)\to(-2,3)} \left[\!\left[5x + \frac{1}{2}y^2\right]\!\right]$$

SOLUTION: Let $g$ be the function such that $g(x,y) = 5x + \frac{1}{2}y^2$, and let $f$ be the function defined by $f(t) = [\![t]\!]$. Then

$$\lim_{(x,y)\to(-2,3)} g(x,y) = \lim_{(x,y)\to(-2,3)} \left(5x + \frac{1}{2}y^2\right) = -\frac{11}{2}$$

Because $f$ is continuous at all $t$ if $t$ is not an integer, then $f$ is continuous at $-\frac{11}{2}$. Thus, by Theorem 19.2.10, we have

$$\lim_{(x,y)\to(-2,3)} \left[\!\left[5x + \frac{1}{2}y^2\right]\!\right] = \left[\!\left[\lim_{(x,y)\to(-2,3)} \left(5x + \frac{1}{2}y^2\right)\right]\!\right]$$
$$= \left[\!\left[-\frac{11}{2}\right]\!\right]$$
$$= -6$$

**26.** Determine if the indicated limit exists.

$$\lim_{(x,y)\to(2,-2)} \frac{\sin(x+y)}{x+y}$$

SOLUTION: Let $S$ be the set of points on the line $y = -x$. Note that $S$ has $(2, -2)$ as an accumulation point. Because,

$$\lim_{\substack{(x,y)\to(2,-2)\\(P \text{ in } S)}} \frac{\sin(x+y)}{x+y} = \lim_{x\to 2} \frac{\sin(x-x)}{x-x} = \lim_{x\to 2} \frac{\sin 0}{0}$$

which does not exist, since division by zero is not defined, then by Theorem 19.2.8,

$$\lim_{(x,y)\to(2,-2)} \frac{\sin(x+y)}{x+y}$$

does not exist.

**30.** **(a)** Give a definition, similar to Definition 19.2.5, of the limit of a function of three variables as a point $(x, y, z)$ approaches a point $(x_0, y_0, z_0)$.
**(b)** Give a definition, similar to Definition 19.2.7, of the limit of a function of three variables as a point $(x, y, z)$ approaches a point $(x_0, y_0, z_0)$ in a specific set of points $S$ in $R^3$.

SOLUTION:

**(a)** Let $f$ be a function of three variables that is defined on some open ball $B((x_0, y_0, z_0); r)$, except possibly at the point $(x_0, y_0, z_0)$ itself. Then

$$\lim_{(x,y,z)\to(x_0,y_0,z_0)} f(x, y, z) = L$$

If for any $\epsilon > 0$, however small, there exists a $\delta > 0$ such that

$$|f(x, y, z) - L| < \epsilon \quad \text{whenever } 0 < \sqrt{(x-x_0)^2 + (y-y_0)^2 + (z-z_0)^2} < \delta$$

**(b)** Let $f$ be a function defined on a set of points $S$ in $R^3$, and let $(x_0, y_0, z_0)$ be an accumulation point of $S$. Then

$$\lim_{\substack{(x,y,z)\to(x_0,y_0,z_0)\\(P \text{ in } S)}} f(x, y, z) = L$$

if for any $\epsilon > 0$, however small, there exists a $\delta > 0$ such that

$$|f(x, y, z) - L| < \epsilon \quad \text{whenever } 0 < \sqrt{(x-x_0)^2 + (y-y_0)^2 + (z-z_0)^2} < \delta$$

and $(x, y, z)$ is in $S$.

**32.** Use the definitions and theorems of Exercises 30 and 31 to prove that $\lim_{(x,y,z)\to(0,0,0)} f(x, y, z)$ does not exist.

$$f(x, y, z) = \frac{x^4 + yx^3 + z^2x^2}{x^4 + y^4 + z^4}$$

SOLUTION: In Exercise 31 we extend Theorem 19.2.9 to functions of three variables. We apply the extended theorem. Let $S_1$ be the set of all points in the $yz$ plane. Then $(0, 0, 0)$ is an accumulation point of $S_1$ and $x = 0$ for every point in $S$. Thus,

$$\lim_{\substack{(x, y, z) \to (0, 0, 0) \\ (P \text{ in } S_1)}} f(x, y, z) = \lim_{(y, z) \to (0, 0)} f(0, y, z)$$

$$= \lim_{(y, z) \to (0, 0)} \frac{0}{y^4 + z^4} = 0$$

Let $S_2$ be the set of all points on the line $x = y = z$. Then $(0, 0, 0)$ is an accumulation point of $S_2$ and

$$\lim_{\substack{(x, y, z) \to (0, 0, 0) \\ (P \text{ in } S_2)}} f(x, y, z) = \lim_{x \to 0} f(x, x, x) = \lim_{x \to 0} \frac{3x^4}{3x^4} = 1$$

Because

$$\lim_{\substack{(x, y, z) \to (0, 0, 0) \\ (P \text{ in } S_1)}} f(x, y, z) \neq \lim_{\substack{(x, y, z) \to (0, 0, 0) \\ (P \text{ in } S_2)}} f(x, y, z)$$

then $\lim_{(x, y, z) \to (0, 0, 0)} f(x, y, z)$ does not exist.

**36.** Use the definition in Exercise 30(a) to prove that $\lim_{(x, y, z) \to (0, 0, 0)}$ exists.

$$f(x, y, z) = \frac{y^3 + xz^2}{x^2 + y^2 + z^2}$$

SOLUTION: Let $S$ be the line $x = y = z$. Then

$$\lim_{\substack{(x, y, z) \to (0, 0, 0) \\ (P \text{ in } S)}} f(x, y, z) = \lim_{x \to 0} \frac{x^3 + x \cdot x^2}{x^2 + x^2 + x^2} = \lim_{x \to 0} \frac{2}{3}x = 0$$

Thus, we take $L = 0$ as the possible limit. For any $\epsilon > 0$ we must find a $\delta > 0$ such that

$$\left| \frac{y^3 + xz^2}{x^2 + y^2 + z^2} \right| < \epsilon \quad \text{whenever } 0 < \sqrt{x^2 + y^2 + z^2} < \delta$$

Now whenever $0 < \sqrt{x^2 + y^2 + z^2} < \delta$, we have

$$|x| \leqslant \sqrt{x^2 + y^2 + z^2} < \delta \qquad |y| \leqslant \sqrt{x^2 + y^2 + z^2} < \delta \qquad |z| \leqslant \sqrt{x^2 + y^2 + z^2} < \delta$$

Therefore,

$$\left| \frac{y^3 + xz^2}{x^2 + y^2 + z^2} \right| \leqslant \left| \frac{y^3}{x^2 + y^2 + z^2} \right| + \left| \frac{xz^2}{x^2 + y^2 + z^2} \right|$$

$$= \frac{|y|y^2}{x^2 + y^2 + z^2} + \frac{|x|z^2}{x^2 + y^2 + z^2}$$

$$\leqslant \frac{|y|(x^2 + y^2 + z^2)}{x^2 + y^2 + z^2} + \frac{|x|(x^2 + y^2 + z^2)}{x^2 + y^2 + z^2}$$

$$= |y| + |x|$$

$$\leqslant \sqrt{x^2 + y^2 + z^2} + \sqrt{x^2 + y^2 + z^2}$$

$$< \delta + \delta$$
$$= 2\delta \qquad (1)$$

We take $\delta = \frac{1}{2}\epsilon$. Hence, whenever $0 < \sqrt{x^2 + y^2 + z^2} < \delta$, from inequality (1) we have

$$\left| \frac{y^3 + xz^2}{x^2 + y^2 + z^2} \right| < 2\delta = \epsilon$$

Therefore, $\lim_{(x, y, z) \to (0, 0, 0)} f(x, y, z) = 0$.

## 19.3 CONTINUITY OF FUNCTIONS OF MORE THAN ONE VARIABLE

The definitions and theorems of this section are extensions of the corresponding definitions and theorems for continuity of functions of one variable. First, we extend to functions of two variables.

**19.3.2 Definition** The function $f$ of two variables $x$ and $y$ is said to be *continuous* at the point $(x_0, y_0)$ if and only if the following three conditions are satisfied:

(i) $f(x_0, y_0)$ exists;
(ii) $\lim_{(x, y) \to (x_0, y_0)} f(x, y)$ exists;
(iii) $\lim_{(x, y) \to (x_0, y_0)} f(x, y) = f(x_0, y_0)$.

**19.3.3 Theorem** If $f$ and $g$ are two functions that are continuous at the point $(x_0, y_0)$, then

(i) $f + g$ is continuous at $(x_0, y_0)$;
(ii) $f - g$ is continuous at $(x_0, y_0)$;
(iii) $fg$ is continuous at $(x_0, y_0)$;
(iv) $f/g$ is continuous at $(x_0, y_0)$, provided that $g(x_0, y_0) \neq 0$.

**19.3.4 Theorem** A polynomial function of two variables is continuous at every point in $R^2$.

**19.3.5 Theorem** A rational function of two variables is continuous at every point in its domain.

**19.3.7 Theorem** Suppose that $f$ is a function of a single variable and $g$ is a function of two variables. Suppose further that $g$ is continuous at $(x_0, y_0)$ and $f$ is continuous at $g(x_0, y_0)$. Then the composite function $f \circ g$ is continuous at $(x_0, y_0)$.

The following two definitions are extensions of the preceding definitions to functions of more than two variables.

**19.3.1 Definition** Suppose that $f$ is a function of $n$ variables and $A$ is a point in $R^n$. Then $f$ is said to be *continuous* at the point $A$ if and only if the following three conditions are satisfied:

(i) $f(A)$ exists;
(ii) $\lim_{P \to A} f(P)$ exists;
(iii) $\lim_{P \to A} f(P) = f(A)$.

If one or more of these three conditions fail to hold at the point $A$, then $f$ is said to be *discontinuous* at $A$.

**19.3.6 Definition** The function $f$ of $n$ variables is said to be *continuous on an open ball* if it is continuous at every point of the open ball.

*Exercises 19.3*

4. Discuss the continuity of $f$.

$$f(x, y) = \begin{cases} \dfrac{x^3 + y^3}{x^2 + y^2} & \text{if } (x, y) \neq (0, 0) \\ 0 & \text{if } (x, y) = (0, 0) \end{cases}$$

SOLUTION: In Exercise 14 of Exercises 19.2 we proved that

$$\lim_{(x, y) \to (0, 0)} \frac{x^3 + y^3}{x^2 + y^2} = 0$$

Because $f(0, 0) = 0$, then the three conditions of Definition 19.3.2 are satisfied, and hence $f$ is continuous at $(0, 0)$. Let $(x_0, y_0)$ be any point such that $(x_0, y_0) \neq (0, 0)$. Then

$$\lim_{(x, y) \to (x_0, y_0)} f(x, y) = \frac{x_0^3 + y_0^3}{x_0^2 + y_0^2} = f(x_0, y_0)$$

Thus, by Definition 19.3.2, $f$ is continuous at $(x_0, y_0)$. We conclude, therefore, that $f$ is continuous everywhere.

In Exercises 8-17, determine the region of continuity of $f$ and draw a sketch showing as a shaded region in $R^2$ the region of continuity of $f$.

8. $f(x, y) = \dfrac{y}{\sqrt{x^2 - y^2 - 4}}$

SOLUTION: Because the function $g$ defined by $g(x, y) = y$ is continuous at all points in $R^2$ and the function $h$ defined by $h(x, y) = \sqrt{x^2 - y^2 - 4}$ is continuous at all points $(x, y)$ for which

$$x^2 - y^2 - 4 > 0$$
$$x^2 - y^2 > 4 \qquad (1)$$

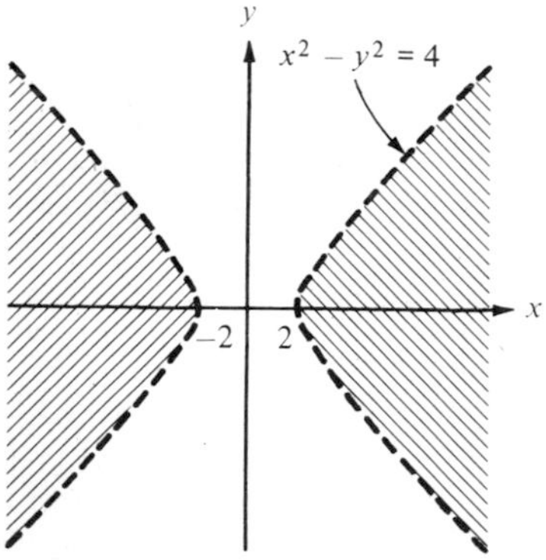

Figure 19.3.8

then, by Theorem 19.3.3(iv), $f$ is continuous at all points $(x, y)$, for which inequality (1) holds. This is the region that is "outside" of the two branches of the hyperbola, as indicated by the shading in Fig. 19.3.8.

If $x^2 - y^2 = 4$, then $f(x, y)$ is not defined because the denominator of the fraction has value zero. Thus, $f$ is discontinuous at every point on the hyperbola $x^2 - y^2 = 4$. If $x^2 - y^2 < 4$ and $y \neq 0$, then $f(x, y)$ does not have a real value, and hence $f$ is discontinuous at all points $(x, y)$ that are between the two branches of the hyperbola and not on the $x$-axis. If $x^2 - y^2 < 4$ and $y = 0$, then $f(x, y) = 0$. Thus, the points on the $x$-axis between the branches of the hyperbola $x^2 - y^2 = 4$ are in the domain of $f$. Let $(a, 0)$ be such a point, and let $S$ be the set of points on the line $x = a$. We have

$$\lim_{\substack{(x, y) \to (a, 0) \\ (P \text{ in } S)}} f(x, y) = \lim_{y \to 0} \frac{y}{\sqrt{a^2 - y^2 - 4}} \qquad (2)$$

Because $(a, 0)$ is between the branches of the hyperbola, then $-2 < a < 2$. Thus,

$$a^2 - 4 < 0$$

and hence

$$a^2 - 4 - y^2 < 0$$

for all $y \neq 0$. Therefore, $\sqrt{a^2 - y^2 - 4}$ is not real for all $y \neq 0$, and hence the limit in (2) does not exist. We conclude, therefore, that $f$ is discontinuous at all points on the $x$-axis that are between the branches of the hyperbola $x^2 - y^2 = 4$. Thus, the only points at which $f$ is continuous are those which satisfy inequality (1).

**12.** $f(x, y) = \ln(x^2 + y^2 - 9) - \ln(1 - x^2 - y^2)$

SOLUTION: Because the natural logarithm function is defined for positive numbers only, $f(x, y)$ is defined only when both

$$x^2 + y^2 - 9 > 0 \quad \text{and} \quad 1 - x^2 - y^2 > 0$$

or, equivalently, only when both

$$x^2 + y^2 > 9 \quad \text{and} \quad x^2 + y^2 < 1 \tag{1}$$

Because there are no points $(x, y)$ that simultaneously satisfy both conditions in (1), we conclude that $f(x, y)$ is not defined for any points in $R^2$. Thus, $f$ is discontinuous at every point in $R^2$.

**16.** $f(x, y) = \begin{cases} \dfrac{x^2 - y^2}{x - y} & \text{if } x \neq y \\ x - y & \text{if } x = y \end{cases}$

SOLUTION: If $(x_0, y_0)$ is any point for which $x_0 - y_0 \neq 0$, then

$$\lim_{(x, y) \to (x_0, y_0)} f(x, y) = \frac{x_0^2 - y_0^2}{x_0 - y_0} = f(x_0, y_0)$$

Thus, $f$ is continuous at all points not on the line $y = x$. Let $(x_0, y_0)$ be a point such that $x_0 = y_0$ and $(x_0, y_0) \neq (0, 0)$. Then $(x_0, y_0)$ is a point on the line $y = x$ that is not the origin. We have

$$f(x_0, y_0) = x_0 - y_0 = 0$$

Let $S$ be the line $y = y_0$. Then $(x_0, y_0)$ is an accumulation point of $S$, and

$$\begin{aligned}
\lim_{\substack{(x, y) \to (x_0, y_0) \\ (P \text{ in } S)}} f(x, y) &= \lim_{x \to x_0} f(x, y_0) \\
&= \lim_{x \to x_0} \frac{x^2 - y_0^2}{x - y_0} \\
&= \lim_{x \to x_0} (x + y_0) \\
&= x_0 + y_0 \\
&\neq 0
\end{aligned}$$

Because

$$\lim_{(x, y) \to (x_0, y_0)} f(x, y) \neq f(x_0, y_0)$$

then $f$ is not continuous at $(x_0, y_0)$.

Finally, we consider the origin. Let $S_1$ be the line $y = x$. Then

$$\lim_{\substack{(x, y) \to (0, 0) \\ (P \text{ in } S_1)}} f(x, y) = \lim_{(x, y) \to (0, 0)} (x - x) = 0$$

Let $S_2$ be the set of all points in $R^2$ for which $y \neq x$. Then

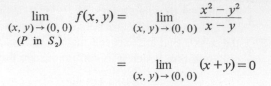

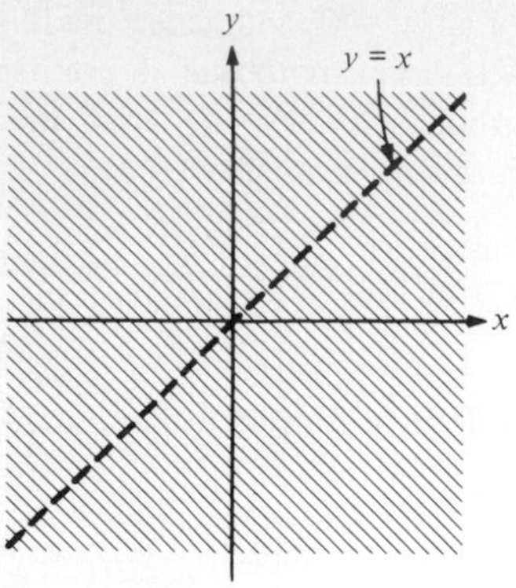

Figure 19.3.16

$$\lim_{\substack{(x, y) \to (0, 0) \\ (P \text{ in } S_2)}} f(x, y) = \lim_{(x, y) \to (0, 0)} \frac{x^2 - y^2}{x - y}$$

$$= \lim_{(x, y) \to (0, 0)} (x + y) = 0$$

Because $S_1 \cup S_2 = R^2$, we conclude that

$$\lim_{(x, y) \to (0, 0)} f(x, y) = 0$$

Furthermore, $f(0, 0) = 0$. Thus, $f$ is continuous at the origin. We have shown, therefore, that $f$ is continuous at all points except those on the line $y = x$, excluding the origin. We show the region of continuity by the shading in Fig. 19.3.16. The broken line indicates that the points on the line are not included in the region.

**20.** Prove that the function is discontinuous at the origin. Then determine if the discontinuity is removable or essential. If the discontinuity is removable, define $f(0, 0)$ so that the discontinuity is removed.

$$f(x, y) = \frac{x^2 y^2}{x^2 + y^2}$$

SOLUTION: Because $f(0, 0)$ is not defined, $f$ is discontinuous at the origin. Let $S$ be the $x$-axis. Then

$$\lim_{\substack{(x, y) \to (0, 0) \\ (P \text{ in } S)}} f(x, y) = \lim_{x \to 0} f(x, 0) = \lim_{x \to 0} \frac{0}{x^2} = 0$$

We show that $\lim_{(x, y) \to (0, 0)} f(x, y) = 0$ by using the Definition 19.2.5. For any $\epsilon > 0$ we must find some $\delta > 0$ such that

$$\left| \frac{x^2 y^2}{x^2 + y^2} \right| < \epsilon \quad \text{whenever } 0 < \sqrt{x^2 + x^2} < \delta$$

Now, whenever $0 < \sqrt{x^2 + y^2} < \delta$, we have $x^2 + y^2 < \delta^2$. Thus, whenever $0 < \sqrt{x^2 + y^2} < \delta$, we have

$$\left| \frac{x^2 y^2}{x^2 + y^2} \right| \leqslant \frac{x^2 (y^2 + x^2)}{x^2 + y^2} = x^2 \leqslant x^2 + y^2 < \delta^2 \tag{1}$$

Thus, if we take $\delta = \sqrt{\epsilon}$ for any $\epsilon > 0$, then from inequality (1)

$$\left| \frac{x^2 y^2}{x^2 + y^2} \right| < \epsilon \quad \text{whenever } 0 < \sqrt{x^2 + y^2} < \delta$$

Therefore,

$$\lim_{(x, y) \to (0, 0)} f(x, y) = 0$$

Hence, the discontinuity at the origin is removable, and we may remove the discontinuity if we define $f(0, 0) = 0$.

**24.** Use the definitions and theorems of Exercise 22 to discuss the continuity of the given function.

$$f(x, y, z) = \ln(36 - 4x^2 - y^2 - 9z^2)$$

SOLUTION: In Exercise 22 we extend Theorem 19.3.7 to include functions of three variables. Let $g$ be the function such that

$$g(x, y, z) = 36 - 4x^2 - y^2 - 9z^2$$

and let $h$ be the function defined by

$$h(t) = \ln t$$

Then $f$ is the composite function $h \circ g$. Because $h$ is continuous for all $t > 0$ and $g$ is continuous at all points in $R^3$, by Theorem 19.3.7 $h \circ g$ is continuous at all points $(x, y, z)$ such that

$$36 - 4x^2 - y^2 - 9z^2 > 0$$
$$4x^2 + y^2 + 9z^2 < 36$$

Thus, $f$ is continuous at all points inside the ellipsoid $4x^2 + y^2 + 9z^2 = 36$, and $f$ is discontinuous at all points that are either on the ellipsoid or outside the ellipsoid.

## 19.4 PARTIAL DERIVATIVES

**19.4.1 Definition** Let $f$ be a function of two variables, $x$ and $y$. The *partial derivative of $f$ with respect to $x$* is that function, denoted by $D_1f$, such that its function value at any point $(x, y)$ in the domain of $f$ is given by

$$D_1f(x, y) = \lim_{\Delta x \to 0} \frac{f(x + \Delta x, y) - f(x, y)}{\Delta x}$$

if this limit exists. Similarly, the *partial derivative of $f$ with respect to $y$* is that function, denoted by $D_2f$, such that its function value at any point $(x, y)$ in the domain of $f$ is given by

$$D_2f(x, y) = \lim_{\Delta y \to 0} \frac{f(x, y + \Delta y) - f(x, y)}{\Delta y}$$

if this limit exists.

There are various notations for partial derivative function values. We have

$$D_1f(x, y) = f_1(x, y) = f_x(x, y) = \frac{\partial f}{\partial x}$$

and

$$D_2f(x, y) = f_2(x, y) = f_y(x, y) = \frac{\partial f}{\partial x}$$

Moreover, if $z = f(x, y)$, then

$$\frac{\partial z}{\partial x} = D_1f(x, y) \quad \text{and} \quad \frac{\partial z}{\partial y} = D_2f(x, y)$$

We may use the following formulas to find the value of the partial derivatives at a particular point $(x_0, y_0)$.

$$D_1f(x_0, y_0) = \lim_{x \to x_0} \frac{f(x, y_0) - f(x_0, y_0)}{x - x_0}$$

if this limit exists, and

$$D_2f(x_0, y_0) = \lim_{y \to y_0} \frac{f(x_0, y) - f(x_0, y_0)}{y - y_0}$$

if this limit exists.

We do not usually have to apply Definition 19.4.1 to find the partial derivatives of a function. To find $D_1f(x, y)$, differentiate with respect to $x$, with $y$ regarded as a constant, and use the differentiation formulas for ordinary derivatives. To find $D_2f(x, y)$, differentiate with respect to $y$, with $x$ regarded as a constant, and use the differentiation formulas for ordinary derivatives.

The line tangent to the curve that is the intersection of the surface $z = f(x, y)$ and the plane $x = x_0$ at the point $(x_0, y_0, f(x_0, y_0))$ has $[0, 1, D_2f(x_0, y_0)]$ for a set of direction numbers. Thus, equations of the line are

$$x = x_0 \qquad z - z_0 = D_2f(x_0, y_0)(y - y_0)$$

The line tangent to the curve that is the intersection of the surface $z = f(x, y)$ and the plane $y = y_0$ at the point $(x_0, y_0, f(x_0, y_0))$ has $[1, 0, D_1f(x_0, y_0)]$ for a set of direction numbers. Thus, equations of the line are

$$y = y_0 \qquad z - z_0 = D_1f(x_0, y_0)(x - x_0)$$

Partial derivatives for functions of three or more variables are defined as follows.

**19.4.2 Definition** Let $P(x_1, x_2, \ldots, x_n)$ be a point in $R^n$, and let $f$ be a function of the $n$ variables $x_1, x_2, \ldots, x_n$. Then the partial derivative of $f$ with respect to $x_k$ is that function, denoted by $D_kf$, such that its function value at any point $P$ in the domain of $f$ is given by

$$D_kf(x_1, x_2, \ldots, x_n) =$$

$$\lim_{\Delta x_k \to 0} \frac{f(x_1, x_2, \ldots, x_{k-1}, x_k + \Delta x_k, x_{k+1}, \ldots, x_n) - f(x_1, x_2, \ldots, x_n)}{\Delta x_k}$$

if this limit exists.

We may find $D_kf(x_1, x_2, \ldots, x_k, \ldots, x_n)$ by using the formulas for ordinary differentiation if we regard all variables except $x_k$ as constants and differentiate with respect to $x_k$.

## *Exercises 19.4*

**4.** Apply Definition 19.4.1 to find the partial derivative.

$$f(x, y) = xy^2 - 5y + 6; \qquad D_2f(x, y)$$

SOLUTION:

$$\begin{aligned} D_2f(x, y) &= \lim_{\Delta y \to 0} \frac{f(x, y + \Delta y) - f(x, y)}{\Delta y} \\ &= \lim_{\Delta y \to 0} \frac{[x(y + \Delta y)^2 - 5(y + \Delta y) + 6] - [xy^2 - 5y + 6]}{\Delta y} \\ &= \lim_{\Delta y \to 0} \frac{xy^2 + 2xy\Delta y + x(\Delta y)^2 - 5y - 5\Delta y + 6 - xy^2 + 5y - 6}{\Delta y} \\ &= \lim_{\Delta y \to 0} \frac{\Delta y(2xy + x\Delta y - 5)}{\Delta y} \\ &= \lim_{\Delta y \to 0} (2xy + x\Delta y - 5) \\ &= 2xy - 5 \end{aligned}$$

**8.** Apply Definition 19.4.2 to find the partial derivative.

$$f(x, y, z) = x^2 + 4y^2 + 9z^2; \qquad D_1f(x, y, z)$$

SOLUTION:

$$D_1f(x, y, z) = \lim_{\Delta x \to 0} \frac{f(x + \Delta x, y, z) - f(x, y, z)}{\Delta x}$$

$$= \lim_{\Delta x \to 0} \frac{[(x+\Delta x)^2 + 4y^2 + 9z^2] - [x^2 + 4y^2 + 9z^2]}{\Delta x}$$

$$= \lim_{\Delta x \to 0} \frac{x^2 + 2x\Delta x + (\Delta x)^2 - x^2}{\Delta x}$$

$$= \lim_{\Delta x \to 0} \frac{\Delta x(2x + \Delta x)}{\Delta x}$$

$$= \lim_{\Delta x \to 0} (2x + \Delta x)$$

$$= 2x$$

**12.** For the function $f$ defined by $f(x, y) = x^2 - 9y^2$, find $D_2f(2, 1)$ by: **(a)** applying formula (4), **(b)** applying formula (5), and **(c)** applying formula (2) and then replacing $x$ and $y$ by 2 and 1, respectively.

SOLUTION:

(a) $$D_2f(2, 1) = \lim_{\Delta y \to 0} \frac{f(2, 1 + \Delta y) - f(2, 1)}{\Delta y}$$

$$= \lim_{\Delta y \to 0} \frac{[4 - 9(1 + \Delta y)^2] - [4 - 9]}{\Delta y}$$

$$= \lim_{\Delta y \to 0} \frac{4 - 9 - 18\Delta y - 9(\Delta y)^2 + 5}{\Delta y}$$

$$= \lim_{\Delta y \to 0} (-18 - 9\Delta y)$$

$$= -18$$

(b) $$D_2f(2, 1) = \lim_{y \to 1} \frac{f(2, y) - f(2, 1)}{y - 1}$$

$$= \lim_{y \to 1} \frac{(4 - 9y^2) - (-5)}{y - 1}$$

$$= \lim_{y \to 1} \frac{-9(y + 1)(y - 1)}{y - 1}$$

$$= \lim_{y \to 1} [-9(y + 1)]$$

$$= -18$$

(c) $$D_2f(x, y) = \lim_{\Delta y \to 0} \frac{f(x, y + \Delta y) - f(x, y)}{\Delta y}$$

$$= \lim_{\Delta y \to 0} \frac{[x^2 - 9(y + \Delta y)^2 - [x^2 - 9y^2]}{\Delta y}$$

$$= \lim_{\Delta y \to 0} \frac{x^2 - 9y^2 - 18y\Delta y - 9(\Delta y)^2 - x^2 + 9y^2}{\Delta y}$$

$$= \lim_{\Delta y \to 0} \frac{\Delta y(-18y - 9\Delta y)}{\Delta y}$$

$$= \lim_{\Delta y \to 0} (-18y - 9\Delta y)$$

$$= -18y$$

Thus,

$$D_2f(2, 1) = -18(1) = -18$$

In Exercises 13-24, find the indicated partial derivatives by holding all but one of the variables constant and applying theorems for ordinary differentiation.

**16.** $f(r, \theta) = r^2 \cos\theta - 2r \tan\theta;\ D_2f(r, \theta)$

SOLUTION: We differentiate with respect to $\theta$ with $r$ regarded as a constant. Thus,

$$D_2f(r, \theta) = -r^2 \sin\theta - 2r \sec^2\theta$$

**20.** $u = \tan^{-1}(xyzw);\ \dfrac{\partial u}{\partial w}$

SOLUTION: We differentiate with respect to $w$ with $x, y$, and $z$ held constant. Thus,

$$\frac{\partial u}{\partial w} = \frac{1}{1 + (xyzw)^2} \cdot xyz$$

$$= \frac{xyz}{1 + x^2y^2z^2w^2}$$

**24.** $f(r, \theta, \phi) = 4r^2 \sin\theta + 5e^r \cos\theta \sin\phi - 2\cos\phi;\ f_2(r, \theta, \phi)$

SOLUTION: We differentiate with respect to $\theta$ with $r$ and $\phi$ held constant. Thus,

$$f_2(r, \theta, \phi) = 4r^2 \cos\theta - 5e^r \sin\theta \sin\phi$$

**26.** Find $f_x(x, y)$ and $f_y(x, y)$.

$$f(x, y) = \int_x^y e^{\cos t}\, dt$$

SOLUTION: We use Theorem 7.6.1. If $x$ is held constant and we differentiate $f$ with respect to $y$, we obtain

$$f_y(x, y) = e^{\cos y}$$

If we hold $y$ constant and differentiate with respect to $x$, then in order to apply Theorem 7.6.1, we must first interchange the limits of integration so that the upper limit is $x$. Thus,

$$f(x, y) = \int_y^x -e^{\cos t}\, dt$$

Hence,

$$f_x(x, y) = -e^{\cos x}$$

**30.** Given

$$f(x, y) = \begin{cases} \dfrac{x^2 - xy}{x + y} & \text{if } (x, y) \neq (0, 0) \\ 0 & \text{if } (x, y) = (0, 0) \end{cases}$$

Find: **(a)** $f_1(0, y)$ if $y \neq 0$ and **(b)** $f_1(0, 0)$.

SOLUTION:

(a) If $y \neq 0$, then

$$f_1(0, y) = \lim_{x \to 0} \frac{f(x, y) - f(0, y)}{x - 0}$$

$$= \lim_{x \to 0} \frac{\dfrac{x^2 - xy}{x + y} - \dfrac{0}{y}}{x}$$

$$= \lim_{x \to 0} \frac{x - y}{x + y}$$

$$= -1$$

(b) $$f_1(0, 0) = \lim_{x \to 0} \frac{f(x, y) - f(0, 0)}{x - 0}$$

$$= \lim_{x \to 0} \frac{\dfrac{x^2 - xy}{x + y} - 0}{x}$$

$$= \lim_{x \to 0} \frac{x - y}{x + y}$$

$$= -1$$

**34.** Find equations of the tangent line to the curve of intersection of the surface $x^2 + y^2 + z^2 = 9$ with the plane $y = 2$ at the point $(1, 2, 2)$.

SOLUTION: We hold $y$ constant and differentiate implicitly with respect to $x$ on both sides of the equation $x^2 + y^2 + z^2 = 9$. Thus,

$$2x + 2z \frac{\partial z}{\partial x} = 0$$

$$\frac{\partial z}{\partial x} = -\frac{x}{z}$$

At the point $(1, 2, 2)$, we have

$$\frac{\partial z}{\partial x} = -\frac{1}{2}$$

The projecting plane of the tangent line to the curve of intersection of the surface and the plane has the equation

$$z - 2 = -\frac{1}{2}(x - 1)$$

Thus, equations of the tangent line are

$$z - 2 = -\frac{1}{2}(x - 1) \quad \text{and} \quad y = 2$$

## 19.5 DIFFERENTIABILITY AND THE TOTAL DIFFERENTIAL

**19.5.1 Definition** If $f$ is a function of two variables $x$ and $y$, then the *increment of* $f$ at the point $(x_0, y_0)$, denoted by $\Delta f(x_0, y_0)$, is given by

$$\Delta f(x_0, y_0) = f(x_0 + \Delta x, y_0 + \Delta y) - f(x_0, y_0)$$

**19.5.2 Definition** If $f$ is a function of two variables $x$ and $y$ and the increment of $f$ at $(x_0, y_0)$ can be written as

$$\Delta f(x_0, y_0) = D_1 f(x_0, y_0)\Delta x + D_2 f(x_0, y_0)\Delta y + \epsilon_1 \Delta x + \epsilon_2 \Delta y$$

where $\epsilon_1$ and $\epsilon_2$ are functions of $\Delta x$ and $\Delta y$ such that $\epsilon_1 \to 0$ and $\epsilon_2 \to 0$ as $(\Delta x, \Delta y) \to (0, 0)$, then $f$ is said to be *differentiable* at $(x_0, y_0)$.

**19.5.3 Theorem** If a function $f$ of two variables is differentiable at a point, it is continuous at that point.

**19.5.5 Theorem** Let $f$ be a function of two variables $x$ and $y$. Suppose that $D_1 f$ and $D_2 f$ exist on an open disk $B(P_0; r)$, where $P_0$ is the point $(x_0, y_0)$. Then if $D_1 f$ and $D_2 f$ are continuous at $P_0$, $f$ is differentiable at $P_0$.

**19.5.6 Definition** If $f$ is a function of two variables $x$ and $y$, and $f$ is differentiable at $(x, y)$, then the *total differential* of $f$ is the function $df$ having function values given by

$$df(x, y, \Delta x, \Delta y) = D_1 f(x, y)\Delta x + D_2 f(x, y)\Delta y$$

If $z = f(x, y)$, then the equation in Definition 19.5.6 may be written as follows.

$$dz = \frac{\partial z}{\partial x} dx + \frac{\partial z}{\partial y} dy$$

If $\Delta x$ and $\Delta y$ are "small," then $dz$ is approximately equal to $\Delta z$.

The following definitions extend the preceding concepts to functions of $n$ variables.

**19.5.7 Definition** If $f$ is a function of the $n$ variables $x_1, x_2, \ldots, x_n$ and $\bar{P}$ is the point $(\bar{x}_1, \bar{x}_2, \ldots, \bar{x}_n)$, then the *increment of* $f$ at $\bar{P}$ is given by

$$\Delta f(\bar{P}) = f(\bar{x}_1 + \Delta x_1, \bar{x}_2 + \Delta x_2, \ldots, \bar{x}_n + \Delta x_n) - f(\bar{P})$$

**19.5.8 Definition** If $f$ is a function of the $n$ variables $x_1, x_2, \ldots, x_n$ and the increment of $f$ at the point $\bar{P}$ can be written as

$$\Delta f(\bar{P}) = D_1 f(\bar{P})\Delta x_1 + D_2 f(\bar{P})\Delta x_2 + \cdots + D_n f(\bar{P})\Delta x_n + \epsilon_1 \Delta x_1 + \epsilon_2 \Delta x_2 + \cdots + \epsilon_n \Delta x_n$$

where $\epsilon_1 \to 0, \epsilon_2 \to 0, \ldots, \epsilon_n \to 0$, as

$$(\Delta x_1, \Delta x_2, \ldots, \Delta x_n) \to (0, 0, \ldots, 0)$$

then $f$ is said to be *differentiable* at $\bar{P}$.

**19.5.9 Definition** If $f$ is a function of the $n$ variables $x_1, x_2, \ldots, x_n$ and $f$ is differentiable at $P$, then the *total differential* of $f$ is the function $df$ having function values given by

$$df(P, \Delta x_1, \Delta x_2, \ldots, \Delta x_n) = D_1 f(P)\Delta x_1 + D_2 f(P)\Delta x_2 + \cdots + D_n f(P)\Delta x_n$$

The following is the mean-value theorem extended to functions of two variables.

**19.5.4 Theorem** Let $f$ be a function of two variables defined for all $x$ in the closed interval $[a, b]$ and all $y$ in the closed interval $[c, d]$.

(i) If $D_1 f(x, y_0)$ exists for some $y_0$ in $[c, d]$ and for all $x$ in $[a, b]$, then there is a number $\xi_1$ in the open interval $(a, b)$ such that

$$f(b, y_0) - f(a, y_0) = (b - a)D_1 f(\xi_1, y_0)$$

(ii) If $D_2 f(x_0, y)$ exists for some $x_0$ in $[a, b]$ and for all $y$ in $[c, d]$, then there is a number $\xi_2$ in the open interval $(c, d)$ such that

$$f(x_0, d) - f(x_0, c) = (d - c)D_2 f(x_0, \xi_2)$$

*Exercises 19.5*

**4.** If $f(x, y, z) = x^2y + 2xyz - z^3$, $\Delta x = 0.01$, $\Delta y = 0.03$, and $\Delta z = -0.01$, find: **(a)** the increment of $f$ at $(-3, 0, 2)$ and **(b)** the total differential of $f$ at $(-3, 0, 2)$.

SOLUTION:

**(a)** We use Definition 19.5.7. The increment of $f$ at $(x_0, y_0, z_0)$ is given by

$$\Delta f(x_0, y_0, z_0) = f(x_0 + \Delta x, y_0 + \Delta y, z_0 + \Delta z) - f(x_0, y_0, z_0)$$

We take $(x_0, y_0, z_0) = (-3, 0, 2)$ and substitute the given values for $\Delta x$, $\Delta y$, and $\Delta z$. Thus,

$$\begin{aligned}\Delta f(-3, 0, 2) &= f(-3 + 0.01, 0 + 0.03, 2 - 0.01) - f(-3, 0, 2)\\ &= f(-2.99, 0.03, 1.99) - f(-3, 0, 2)\\ &= [(-2.99)^2(0.03) + 2(-2.99)(0.03)(1.99) - (1.99)^3]\\ &\quad - [(-3)^2 \cdot 0 + 2(-3)(0)(2) - 2^3]\\ &= (-7.969402) - (-8)\\ &= 0.030598\end{aligned}$$

**(b)** We use Definition 19.5.9. The total differential of $f$ is given by

$$\begin{aligned}df(x, y, z, \Delta x, \Delta y, \Delta z) &= D_1f(x, y, z)\Delta x + D_2f(x, y, z)\Delta y\\ &\quad + D_3f(x, y, z)\Delta z\\ &= (2xy + 2yz)\Delta x + (x^2 + 2xz)\Delta y\\ &\quad + (2xy - 3z^2)\Delta z\end{aligned}$$

Substituting the given values for $x, y, z, \Delta x, \Delta y$, and $\Delta z$, we obtain

$$\begin{aligned}df(-3, 0, 2, 0.01, 0.03, -0.01) &= (0)(0.01) + (-3)(0.03) + (-12)(-0.01)\\ &= 0.03\end{aligned}$$

**8.** Prove that $f$ is differentiable at all points in its domain by doing each of the following:

**(a)** Find $\Delta f(x_0, y_0)$ for the given function.
**(b)** Find an $\epsilon_1$ and an $\epsilon_2$ so that Eq. (3) holds.
**(c)** Show that the $\epsilon_1$ and the $\epsilon_2$ found in part (b) both approach zero as $(\Delta x, \Delta y) \to (0, 0)$.

$$f(x, y) = \frac{y}{x}$$

SOLUTION:

**(a)**
$$\begin{aligned}\Delta f(x_0, y_0) &= f(x_0 + \Delta x, y_0 + \Delta y) - f(x_0, y_0)\\ &= \frac{y_0 + \Delta y}{x_0 + \Delta x} - \frac{y_0}{x_0}\\ &= \frac{x_0\Delta y - y_0\Delta x}{x_0(x_0 + \Delta x)} \qquad (1)\end{aligned}$$

**(b)** Eq. (3) is as follows:

$$\Delta f(x_0, y_0) = D_1f(x_0, y_0)\Delta x + D_2f(x_0, y_0)\Delta y + \epsilon_1\Delta x + \epsilon_2\Delta y$$

or, equivalently,

$$\epsilon_1\Delta x + \epsilon_2\Delta y = \Delta f(x_0, y_0) - D_1f(x_0, y_0)\Delta x - D_2f(x_0, y_0)\Delta y \qquad (2)$$

Taking the partial derivatives of $f$, we obtain

$$D_1 f(x_0, y_0)\Delta x = -\frac{y_0}{x_0^2}\Delta x \tag{3}$$

and

$$D_2 f(x_0, y_0)\Delta y = \frac{1}{x_0}\Delta y \tag{4}$$

Substituting from (1), (3), and (4) into (2), we get

$$\begin{aligned}
\epsilon_1 \Delta x + \epsilon_2 \Delta y &= \frac{x_0 \Delta y - y_0 \Delta x}{x_0(x_0 + \Delta x)} + \frac{y_0 \Delta x}{x_0^2} - \frac{\Delta y}{x_0} \\
&= \frac{x_0(x_0 \Delta y - y_0 \Delta x) + y_0 \Delta x (x_0 + \Delta x) - x_0 \Delta y (x_0 + \Delta x)}{x_0^2(x_0 + \Delta x)} \\
&= \frac{y_0(\Delta x)^2 - x_0 \Delta x \, \Delta y}{x_0^2(x_0 + \Delta x)} \\
&= \frac{y_0 \Delta x}{x_0^2(x_0 + \Delta x)}\Delta x + \frac{-\Delta x}{x_0(x_0 + \Delta x)}\Delta y \qquad (5)
\end{aligned}$$

Equating the coefficients of $\Delta x$ on the left and right sides of (5) and equating the coefficients of $\Delta y$ on the left and right sides of (5), we obtain

$$\epsilon_1 = \frac{y_0 \Delta x}{x_0^2(x_0 + \Delta x)} \quad \text{and} \quad \epsilon_2 = \frac{-\Delta x}{x_0(x_0 + \Delta x)} \tag{6}$$

**(c)** If $(\Delta x, \Delta y) \to (0, 0)$, then $\Delta x \to 0$. Furthermore, if $x_0 \neq 0$, we have from (6)

$$\lim_{\Delta x \to 0} \epsilon_1 = \lim_{\Delta x \to 0} \frac{y_0 \Delta x}{x_0^2(x_0 + \Delta x)} = 0$$

$$\lim_{\Delta x \to 0} \epsilon_2 = \lim_{\Delta x \to 0} \frac{-\Delta x}{x_0(x_0 + \Delta x)} = 0$$

Therefore, $f$ is differentiable at all points $(x_0, y_0)$ for which $x_0 \neq 0$.

**10.** Given

$$f(x, y) = \begin{cases} \dfrac{3x^2y^2}{x^4 + y^4} & \text{if } (x, y) \neq (0, 0) \\ 0 & \text{if } (x, y) = (0, 0) \end{cases}$$

Prove that $D_1 f(0, 0)$ and $D_2 f(0, 0)$ exist but that $f$ is not differentiable at $(0, 0)$.

SOLUTION: We use limits to find the partial derivatives. We have

$$D_1 f(x_0, y_0) = \lim_{x \to x_0} \frac{f(x, y_0) - f(x_0, y_0)}{x - x_0}$$

Thus,

$$\begin{aligned}
D_1 f(0, 0) &= \lim_{x \to 0} \frac{f(x, 0) - f(0, 0)}{x - 0} \\
&= \lim_{x \to 0} \frac{\dfrac{3x^2 \cdot 0}{x^4 + 0} - 0}{x} \\
&= 0
\end{aligned}$$

Similarly,

$$D_2f(x_0, y_0) = \lim_{y \to y_0} \frac{f(x_0, y) - f(x_0, y_0)}{y - y_0}$$

Thus,

$$\begin{aligned} D_2f(0, 0) &= \lim_{y \to 0} \frac{f(0, y) - f(0, 0)}{y - 0} \\ &= \lim_{y \to 0} \frac{\dfrac{3 \cdot 0 \cdot y^2}{0 + y^4} - 0}{y} \\ &= 0 \end{aligned}$$

To show that $f$ is not differentiable at $(0, 0)$, we show that $f$ is discontinuous at $(0, 0)$. Let $S$ be the set of all points on the line $y = x$. Thus, $(0, 0)$ is an accumulation point of $S$. Moreover,

$$\lim_{\substack{(x, y) \to (0, 0) \\ (P \text{ in } S)}} f(x, y) = \lim_{x \to 0} f(x, x) = \lim_{x \to 0} \frac{3x^4}{2x^4} = \frac{3}{2} \tag{1}$$

Because $f(0, 0) = 0$, by (1) and Theorem 19.2.8, we have

$$\lim_{(x, y) \to (0, 0)} f(x, y) \neq f(0, 0)$$

Therefore, $f$ is discontinuous at $(0, 0)$. By Theorem 19.5.3, this proves that $f$ is not differentiable at $(0, 0)$.

**16.** Use Theorem 19.5.4 to find either a $\xi_1$ or a $\xi_2$, whichever applies.

$$f(x, y) = \frac{4x}{x + y}; \ y \text{ is in } [-2, 2]; \ x = 4$$

SOLUTION: Because $x$ is constant, we want to find a $\xi_2$ in $(c, d)$ such that

$$f(x_0, d) - f(x_0, c) = (d - c)D_2f(x_0, \xi_2) \tag{1}$$

We are given $x_0 = 4$ and $[c, d] = [-2, 2]$. Thus,

$$\begin{aligned} f(x_0, d) - f(x_0, c) &= f(4, 2) - f(4, -2) \\ &= -\frac{16}{3} \end{aligned} \tag{2}$$

Because

$$D_2f(x, y) = \frac{-4x}{(x + y)^2}$$

we have

$$(d - c)D_2f(x_0, \xi_2) = -\frac{64}{(4 + \xi_2)^2} \tag{3}$$

Substituting from (2) and (3) into (1), we obtain

$$-\frac{16}{3} = -\frac{64}{(4 + \xi_2)^2}$$

$$\begin{aligned} (4 + \xi_2)^2 &= 12 \\ 4 + \xi_2 &= \pm\sqrt{12} \\ \xi_2 &= -4 \pm 2\sqrt{3} \end{aligned}$$

Because $\xi_2$ is in $(-2, 2)$, we disregard the negative square root. Thus, $\xi_2 = -4 + 2\sqrt{3} \approx -0.54$.

**20.** Prove that $f$ is differentiable at all points in $R^3$ by doing each of the following: **(a)** find $\Delta f(x_0, y_0, z_0)$; **(b)** find an $\epsilon_1$, $\epsilon_2$, and $\epsilon_3$, so that Eq. (30) holds; and **(c)** show that the $\epsilon_1$, $\epsilon_2$, and $\epsilon_3$ found in (b) all approach zero as $(\Delta x, \Delta y, \Delta z)$ approaches $(0, 0, 0)$.

$$f(x, y, z) = 2x^2z - 3yz^2$$

SOLUTION:

**(a)**
$$\begin{aligned}\Delta f(x_0, y_0, z_0) &= f(x_0 + \Delta x, y_0 + \Delta y, z_0 + \Delta z) - f(x_0, y_0, z_0)\\ &= [2(x_0 + \Delta x)^2(z_0 + \Delta z) - 3(y_0 + \Delta y)(z_0 + \Delta z)^2]\\ &\quad - [2x_0^2z_0 - 3y_0z_0^2]\\ &= 2x_0^2\Delta z + 4x_0z_0\Delta x + 4x_0\Delta x\,\Delta z + 2z_0(\Delta x)^2\\ &\quad + 2(\Delta x)^2\Delta z - 6y_0z_0\Delta z - 3y_0(\Delta z)^2 - 3z_0^2\Delta y\\ &\quad - 6z_0\Delta y\,\Delta z - 3\Delta y(\Delta z)^2 \end{aligned} \qquad (1)$$

**(b)** From Eq. (3), we have

$$\begin{aligned}\epsilon_1\Delta x + \epsilon_2\Delta y + \epsilon_3\Delta z &= \Delta f(x_0, y_0, z_0) - D_1f(x_0, y_0, z_0)\Delta x\\ &\quad - D_2f(x_0, y_0, z_0)\Delta y - D_3f(x_0, y_0, z_0)\Delta z\end{aligned} \qquad (2)$$

Taking partial derivatives of the given function $f$, we obtain

$$D_1f(x_0, y_0, z_0)\Delta x = 4x_0z_0\Delta x \qquad (3)$$

$$D_2f(x_0, y_0, z_0)\Delta y = -3z_0^2\Delta y \qquad (4)$$

$$D_3f(x_0, y_0, z_0)\Delta z = (2x_0^2 - 6y_0z_0)\Delta z \qquad (5)$$

Substituting from (1), (3), (4), and (5) into (2), we get

$$\begin{aligned}\epsilon_1\Delta x + \epsilon_2\Delta y + \epsilon_3\Delta z &= 4x_0\Delta x\,\Delta z + 2z_0(\Delta x)^2 + 2(\Delta x)^2\Delta z\\ &\quad - 3y_0(\Delta z)^2 - 6z_0\Delta y\,\Delta z - 3\Delta y(\Delta z)^2\\ &= (4x_0\Delta z + 2z_0\Delta x + 2\Delta x\,\Delta z)\Delta x\\ &\quad + (-6z_0\Delta z - 3(\Delta z)^2)\Delta y + (-3y_0\Delta z)\Delta z\end{aligned}$$

Equating corresponding coefficients of $\Delta x$, $\Delta y$, and $\Delta z$ in the above, we take

$$\epsilon_1 = 4x_0\Delta z + 2z_0\Delta x + 2\Delta x\,\Delta z$$

$$\epsilon_2 = -6z_0\Delta z - 3(\Delta z)^2$$

$$\epsilon_3 = -3y_0\Delta z$$

**(c)**
$$\lim_{(\Delta x, \Delta y, \Delta z)\to(0,0,0)} \epsilon_1 = 4x_0 \cdot 0 + 2z_0 \cdot 0 + 2 \cdot 0 \cdot 0 = 0$$

$$\lim_{(\Delta x, \Delta y, \Delta z)\to(0,0,0)} \epsilon_2 = -6z_0 \cdot 0 - 3 \cdot 0^2 = 0$$

$$\lim_{(\Delta x, \Delta y, \Delta z)\to(0,0,0)} \epsilon_3 = -3y_0 \cdot 0 = 0$$

**24.** Use the total differential to find approximately the greatest error in calculating the area of a right triangle from the lengths of the legs if they are measured to be 6 in. and 8 in., respectively, with a possible error of 0.1 in. for each measurement. Also find the approximate percent error.

SOLUTION: Let $x$ inches and $y$ inches be the lengths of the legs of a right triangle, and let $A$ square inches be the area. Then

$$A = \frac{1}{2}xy$$

We take the total differential of $A$. Thus,

$$dA = \frac{\partial A}{\partial x}dx + \frac{\partial A}{\partial y}dy$$

$$= \frac{1}{2}y\,dx + \frac{1}{2}x\,dy \qquad (1)$$

We are given that $x = 6, y = 8$, $|dx| \leqslant 0.1$, and $|dy| \leqslant 0.1$. Using the triangle inequality with (1), we have

$$|dA| \leqslant \left|\frac{1}{2}y\right| |dx| + \left|\frac{1}{2}x\right| |dy|$$

$$\leqslant 4(0.1) + 3(0.1)$$

$$= 0.7$$

Thus, the greatest error in calculating the error is approximately 0.7 square inches. Because, $A = \frac{1}{2}(6)(8) = 24$, the percent error is found by

$$\frac{|dA|}{A} \leqslant \frac{0.7}{24} = 0.029$$

Thus, the error is approximately 2.9 percent of the calculated error.

**28.** A wooden box is to be made of lumber that is $\frac{2}{3}$ in. thick. The inside length is to be 6 ft.; the inside width is to be 3 ft.; the inside depth is to be 4 ft., and the box is to have no top. Use the total differential to find the approximate amount of lumber to be used in the box.

SOLUTION: Let $x$ ft. be the length, $y$ ft. be the width, $z$ ft. be the depth, and $V$ cu. ft. be the volume of a rectangular box. Then

$$V = xyz$$

We have $x = 6, y = 3$, and $z = 4$. The thickness of the box is $\frac{2}{3}$ in. $= \frac{1}{18}$ ft. Thus, take $dx = 2 \cdot \frac{1}{18} = \frac{1}{9}$ and $dy = \frac{1}{9}$. Because there is no top for the box, we take $dz = \frac{1}{18}$. Hence,

$$dV = \frac{\partial V}{\partial x}dx + \frac{\partial V}{\partial y}dy + \frac{\partial V}{\partial z}dz$$

$$= yz\,dx + xz\,dy + xy\,dz$$

$$= 3 \cdot 4 \cdot \frac{1}{9} + 6 \cdot 4 \cdot \frac{1}{9} + 6 \cdot 3 \cdot \frac{1}{18}$$

$$= 5$$

Therefore, we need approximately 5 cubic feet of lumber.

## 19.6 THE CHAIN RULE

**19.6.1 Theorem** (*The Chain Rule*) If $u$ is a differentiable function of $x$ and $y$, defined by $u = f(x, y)$, and $x = F(r, s)$, $y = G(r, s)$, and $\partial x/\partial r$, $\partial x/\partial s$, $\partial y/\partial r$, and $\partial y/\partial s$ all exist, then $u$ is a function of $r$ and $s$ and

$$\frac{\partial u}{\partial r} = \left(\frac{\partial u}{\partial x}\right)\left(\frac{\partial x}{\partial r}\right) + \left(\frac{\partial u}{\partial y}\right)\left(\frac{\partial y}{\partial r}\right)$$

$$\frac{\partial u}{\partial s} = \left(\frac{\partial u}{\partial x}\right)\left(\frac{\partial x}{\partial s}\right) + \left(\frac{\partial u}{\partial y}\right)\left(\frac{\partial y}{\partial s}\right)$$

If in Theorem 19.6.1 both $x$ and $y$ are differentiable functions of the single variable $t$, then $u$ is a function of $t$, and the *total derivative* of $u$ with respect to $t$ is given by

$$\frac{du}{dt} = \left(\frac{\partial u}{\partial x}\right)\left(\frac{dx}{dt}\right) + \left(\frac{\partial u}{\partial y}\right)\left(\frac{dy}{dt}\right)$$

The chain rule may be extended to functions of three or more variables as follows.

**19.6.2 Theorem** (*The General Chain Rule*) Suppose that $u$ is a differentiable function of the $n$ variables $x_1, x_2, \ldots, x_n$ and each of these variables is in turn a function of the $m$ variables $y_1, y_2, \ldots, y_m$. Suppose further that each of the partial derivatives $\partial x_i/\partial y_j$ $(i = 1, 2, \ldots, n;\ j = 1, 2, \ldots, m)$ exists. Then $u$ is a function of $y_1, y_2, \ldots, y_m$, and

$$\frac{\partial u}{\partial y_1} = \left(\frac{\partial u}{\partial x_1}\right)\left(\frac{\partial x_1}{\partial y_1}\right) + \left(\frac{\partial u}{\partial x_2}\right)\left(\frac{\partial x_2}{\partial y_1}\right) + \cdots + \left(\frac{\partial u}{\partial x_n}\right)\left(\frac{\partial x_n}{\partial y_1}\right)$$

$$\frac{\partial u}{\partial y_2} = \left(\frac{\partial u}{\partial x_1}\right)\left(\frac{\partial x_1}{\partial y_2}\right) + \left(\frac{\partial u}{\partial x_2}\right)\left(\frac{\partial x_2}{\partial y_2}\right) + \cdots + \left(\frac{\partial u}{\partial x_n}\right)\left(\frac{\partial x_n}{\partial y_2}\right)$$

$$\vdots$$

$$\frac{\partial u}{\partial y_m} = \left(\frac{\partial u}{\partial x_1}\right)\left(\frac{\partial x_1}{\partial y_m}\right) + \left(\frac{\partial u}{\partial x_2}\right)\left(\frac{\partial x_2}{\partial y_m}\right) + \cdots + \left(\frac{\partial u}{\partial x_n}\right)\left(\frac{\partial x_n}{\partial y_m}\right)$$

If in Theorem 19.6.2 each $x_i$ is a differentiable function of $t$, then $u$ is a function of $t$, and the total derivative of $u$ with respect to $t$ is given by the following.

$$\frac{du}{dt} = \left(\frac{\partial u}{\partial x_1}\right)\left(\frac{dx_1}{dt}\right) + \left(\frac{\partial u}{\partial x_2}\right)\left(\frac{dx_2}{dt}\right) + \cdots + \left(\frac{\partial u}{\partial x_n}\right)\left(\frac{dx_n}{dt}\right)$$

## *Exercises 19.6*

**4.** Find the indicated partial derivative by two methods: **(a)** use the chain rule and **(b)** make the substitutions for $x$ and $y$ before differentiating.

$$u = x^2 + y^2;\ x = \cosh r \cos t;\ y = \sinh r \sin t;\ \frac{\partial u}{\partial r};\ \frac{\partial u}{\partial t}$$

SOLUTION:

**(a)**
$$\frac{\partial u}{\partial r} = \left(\frac{\partial u}{\partial x}\right)\left(\frac{\partial x}{\partial r}\right) + \left(\frac{\partial u}{\partial y}\right)\left(\frac{\partial y}{\partial r}\right)$$

$$= 2x \sinh r \cos t + 2y \cosh r \sin t \qquad \textbf{(1)}$$

and

$$\frac{\partial u}{\partial t} = \left(\frac{\partial u}{\partial x}\right)\left(\frac{\partial x}{\partial t}\right) + \left(\frac{\partial u}{\partial y}\right)\left(\frac{\partial y}{\partial t}\right)$$

$$= -2x \cosh r \sin t + 2y \sinh r \cos t \qquad \textbf{(2)}$$

**(b)** Eliminating $x$ and $y$ from the equation for $u$, we have

$$u = x^2 + y^2$$
$$= \cosh^2 r \cos^2 t + \sinh^2 r \sin^2 t$$

Thus,

$$\frac{\partial u}{\partial r} = 2 \cosh r \sinh r \cos^2 t + 2 \sinh r \cosh r \sin^2 t \tag{3}$$

$$= 2 \sinh r \cosh r (\cos^2 t + \sin^2 t)$$
$$= 2 \sinh r \cosh r$$

and

$$\frac{\partial u}{\partial t} = -2 \cosh^2 r \sin t \cos t + 2 \sinh^2 r \sin t \cos t \tag{4}$$

$$= -2 \sin t \cos t (\cosh^2 r - \sinh^2 r)$$
$$= -2 \sin t \cos t$$

We show that the partial derivatives found in part (a) agree with those found in part (b). Eliminating $x$ and $y$ from Eq. (1) we obtain

$$\frac{\partial u}{\partial r} = 2(\cosh r \cos t)(\sinh r \cos t) + 2(\sinh r \sin t)(\cosh r \sin t)$$

$$= 2 \sinh r \cosh r \cos^2 t + 2 \sinh r \cosh r \sin^2 t$$

which agrees with Eq. (3). Eliminating $x$ and $y$ from Eq. (2), we obtain

$$\frac{\partial u}{\partial t} = -2(\cosh r \cos t)(\cosh r \sin t) + 2(\sinh r \sin t)(\sinh r \cos t)$$

$$= -2 \cosh^2 r \sin t \cos t + 2 \sinh^2 r \sin t \cos t$$

which agrees with Eq. (4).

8. Find the indicated partial derivatives by using the chain rule.

$$u = xy + xz + yz; \; x = rs; \; y = r^2 - s^2; \; z = (r - s)^2; \; \frac{\partial u}{\partial r}; \frac{\partial u}{\partial s}$$

SOLUTION:

$$\frac{\partial u}{\partial r} = \left(\frac{\partial u}{\partial x}\right)\left(\frac{\partial x}{\partial r}\right) + \left(\frac{\partial u}{\partial y}\right)\left(\frac{\partial y}{\partial r}\right) + \left(\frac{\partial u}{\partial z}\right)\left(\frac{\partial z}{\partial r}\right)$$
$$= (y + z)s + (x + z)(2r) + (x + y)2(r - s)$$
$$= 4xr - 2xs + 2yr - ys + 2zr + zs$$

and

$$\frac{\partial u}{\partial s} = \left(\frac{\partial u}{\partial x}\right)\left(\frac{\partial x}{\partial s}\right) + \left(\frac{\partial u}{\partial y}\right)\left(\frac{\partial y}{\partial s}\right) + \left(\frac{\partial u}{\partial z}\right)\left(\frac{\partial z}{\partial s}\right)$$
$$= (y + z)r + (x + z)(-2s) + (x + y)(-2)(r - s)$$
$$= -2xr - yr + 2ys + zr - 2zs$$

12. Find the total derivative $du/dt$ by two methods: **(a)** use the chain rule and **(b)** make the substitutions for $x$ and $y$ before differentiating.

$$u = \ln xy + y^2 \qquad x = e^t \qquad y = e^{-t}$$

SOLUTION: We have

$$u = \ln x + \ln y + y^2$$

Thus,

**(a)** $$\frac{du}{dt} = \left(\frac{\partial u}{\partial x}\right)\left(\frac{dx}{dt}\right) + \left(\frac{\partial u}{\partial y}\right)\left(\frac{dy}{dt}\right)$$

$$= \frac{1}{x} e^t + \left(\frac{1}{y} + 2y\right)(-e^{-t})$$

$$= \frac{e^t}{x} - \frac{1 + 2y^2}{y} e^{-t} \tag{1}$$

**(b)** Eliminating $x$ and $y$ from the equation for $u$, we have

$$\begin{aligned} u &= \ln x + \ln y + y^2 \\ &= \ln e^t + \ln e^{-t} + (e^{-t})^2 \\ &= t - t + e^{-2t} \\ &= e^{-2t} \end{aligned}$$

Thus,

$$\frac{du}{dt} = -2e^{-2t} \tag{2}$$

We show that the result in part (a) agrees with the result in part (b). Eliminating $x$ and $y$ from Eq. (1), we obtain

$$\begin{aligned} \frac{du}{dt} &= \frac{e^t}{e^t} - \frac{1 + 2e^{-2t}}{e^{-t}} e^{-t} \\ &= 1 - 1 - 2e^{-2t} \\ &= -2e^{-2t} \end{aligned}$$

which agrees with Eq. (2).

**16.** Find $du/dt$ by using the chain rule; do not express $u$ as a function of $t$ before differentiating.

$$u = xy + xz + yz \qquad x = t\cos t \qquad y = t \sin t \qquad z = t$$

SOLUTION:

$$\begin{aligned} \frac{du}{dt} &= \left(\frac{\partial u}{\partial x}\right)\left(\frac{dx}{dt}\right) + \left(\frac{\partial u}{\partial y}\right)\left(\frac{dy}{dt}\right) + \left(\frac{\partial u}{\partial z}\right)\left(\frac{dz}{dt}\right) \\ &= (y + z)(\cos t - t \sin t) + (x + z)(\sin t + t\cos t) + (x + y) \end{aligned}$$

**20.** Assume that the given equation defines $z$ as a function of $x$ and $y$. Differentiate implicitly to find $\partial z/\partial x$ and $\partial z/\partial y$.

$$z = (x^2 + y^2)\sin xz$$

SOLUTION: For $\partial z/\partial x$, we regard $y$ as a constant and differentiate on both sides with respect to $x$. Thus,

$$\begin{aligned} \frac{\partial z}{\partial x} &= (x^2 + y^2)\left[\frac{\partial(\sin xz)}{\partial x}\right] + \sin xz \left[\frac{\partial(x^2 + y^2)}{\partial x}\right] \\ &= (x^2 + y^2)(\cos xz)\left[\frac{\partial(xz)}{\partial x}\right] + (\sin xz)2x \\ &= (x^2 + y^2)(\cos xz)\left[x\left(\frac{\partial z}{\partial x}\right) + z\right] + 2x \sin xz \end{aligned}$$

Solving for $\partial z/\partial x$, we have

$$\frac{\partial z}{\partial x} = [x(x^2 + y^2)\cos xz]\left(\frac{\partial z}{\partial x}\right) + z(x^2 + y^2)\cos xz + 2x \sin xz$$

$$\frac{\partial z}{\partial x}[1 - x(x^2 + y^2)\cos xz] = z(x^2 + y^2)\cos xz + 2x \sin xz$$

$$\frac{\partial z}{\partial x} = \frac{z(x^2 + y^2)\cos xz + 2x \sin xz}{1 - x(x^2 + y^2)\cos xz}$$

For $\partial z/\partial y$, we regard $x$ as a constant and differentiate on both sides of the given equation with respect to $y$. Thus,

$$\frac{\partial z}{\partial y} = (x^2 + y^2)(\cos xz)(x)\left(\frac{\partial z}{\partial y}\right) + (\sin xz)(2y)$$

$$\frac{\partial z}{\partial y}[1 - x(x^2 + y^2)\cos xz] = 2y \sin xz$$

$$\frac{\partial z}{\partial y} = \frac{2y \sin xz}{1 - x(x^2 + y^2)\cos xz}$$

**24.** If $f$ is a differentiable function of two variables $u$ and $v$, let $u = x - y$ and $v = y - x$ and prove that $z = f(x - y, y - x)$ satisfies the equation $\partial z/\partial x + \partial z/\partial y = 0$.

SOLUTION: We have $z = f(u, v)$. By the chain rule

$$\begin{aligned}\frac{\partial z}{\partial x} &= f_1(u, v)\left(\frac{\partial u}{\partial x}\right) + f_2(u, v)\left(\frac{\partial v}{\partial x}\right)\\ &= f_1(u, v) - f_2(u, v) \qquad (1)\end{aligned}$$

and

$$\begin{aligned}\frac{\partial z}{\partial y} &= f_1(u, v)\left(\frac{\partial u}{\partial y}\right) + f_2(u, v)\left(\frac{\partial v}{\partial y}\right)\\ &= -f_1(u, v) + f_2(u, v) \qquad (2)\end{aligned}$$

Adding the members of (1) and (2), we obtain

$$\frac{\partial z}{\partial x} + \frac{\partial z}{\partial y} = 0$$

which is the desired result.

**28.** Suppose $f$ is a differentiable function of $x, y$, and $z$ and $u = f(x, y, z)$. Then if $x = r \sin \phi \cos \theta$, $y = r \sin \phi \sin \theta$, and $z = r \cos \phi$, express $\partial u/\partial r$, $\partial u/\partial \phi$, and $\partial u/\partial \theta$ in terms of $\partial u/\partial x$, $\partial u/\partial y$, and $\partial u/\partial z$.

SOLUTION:

$$\begin{aligned}\frac{\partial u}{\partial r} &= \left(\frac{\partial u}{\partial x}\right)\left(\frac{\partial x}{\partial r}\right) + \left(\frac{\partial u}{\partial y}\right)\left(\frac{\partial y}{\partial r}\right) + \left(\frac{\partial u}{\partial z}\right)\left(\frac{\partial z}{\partial r}\right)\\ &= \left(\frac{\partial u}{\partial x}\right)\sin \phi \cos \theta + \left(\frac{\partial u}{\partial y}\right)\sin \phi \sin \theta + \left(\frac{\partial u}{\partial z}\right)\cos \phi\end{aligned}$$

and

$$\begin{aligned}\frac{\partial u}{\partial \phi} &= \left(\frac{\partial u}{\partial x}\right)\left(\frac{\partial x}{\partial \phi}\right) + \left(\frac{\partial u}{\partial y}\right)\left(\frac{\partial y}{\partial \phi}\right) + \left(\frac{\partial u}{\partial z}\right)\left(\frac{\partial z}{\partial \phi}\right)\\ &= \left(\frac{\partial u}{\partial x}\right) r \cos \phi \cos \theta + \left(\frac{\partial u}{\partial y}\right) r \cos \phi \sin \theta - \left(\frac{\partial u}{\partial z}\right) r \sin \phi\end{aligned}$$

and

$$\frac{\partial u}{\partial \theta} = \left(\frac{\partial u}{\partial x}\right)\left(\frac{\partial x}{\partial \theta}\right) + \left(\frac{\partial u}{\partial y}\right)\left(\frac{\partial y}{\partial \theta}\right) + \left(\frac{\partial u}{\partial z}\right)\left(\frac{\partial z}{\partial \theta}\right)$$

$$= -\left(\frac{\partial u}{\partial x}\right) r \sin\phi \sin\theta + \left(\frac{\partial u}{\partial y}\right) r \sin\phi \cos\theta$$

**32.** Water is flowing into a tank in the form of a right circular cylinder at the rate of $\frac{4}{5}\pi$ ft$^3$/min. The tank is stretching in such a way that even though it remains cylindrical, its radius is increasing at the rate of 0.002 ft/min. How fast is the surface of the water rising when the radius is 2 ft and the volume of water in the tank is $20\pi$ ft$^3$?

SOLUTION:

$t =$ the number of minutes in the time that has elapsed since the water began to flow into the tank
$r =$ the number of feet in the radius of the tank at $t$ min
$h =$ the number of feet in the depth of the water at $t$ min
$V =$ the number of cubic feet in the volume of the water in the tank at $t$ min

We have the formula

$$V = \pi r^2 h \tag{1}$$

We are given that

$$\frac{dV}{dt} = \frac{4}{5}\pi \quad \text{and} \quad \frac{dr}{dt} = 0.002$$

We want to find the value of $dh/dt$ at the moment when $r = 2$ and $V = 20\pi$. Substituting these values into (1), we obtain $h = 5$. We use the chain rule to differentiate on both sides of (1). Thus,

$$\frac{dV}{dt} = \left(\frac{\partial V}{\partial r}\right)\left(\frac{dr}{dt}\right) + \left(\frac{\partial V}{\partial h}\right)\left(\frac{dh}{dt}\right)$$

$$= 2\pi r h \frac{dr}{dt} + \pi r^2 \frac{dh}{dt} \tag{2}$$

Substituting the given values into (2), we obtain

$$\frac{4}{5}\pi = 2\pi \cdot 2 \cdot 5(0.002) + \pi \cdot 2^2 \frac{dh}{dt}$$

Solving for $dh/dt$, we obtain

$$\frac{dh}{dt} = 0.19$$

Thus, the surface of the water is rising at the rate of 0.19 ft/min.

## 19.7 HIGHER-ORDER PARTIAL DERIVATIVES

If $f$ is a function of $x$ and $y$, then the second partial derivative of $f$ obtained by first partial-differentiating $f$ with respect to $x$ and then partial-differentiating the result with respect to $y$ is represented by each of the following notations.

$$D_2(D_1 f) \qquad D_{12} f \qquad f_{12} \qquad f_{xy} \qquad \frac{\partial^2 f}{\partial y \partial x}$$

The second partial derivative of $f$ obtained by partial-differentiating twice with respect to $x$ is represented by each of the following notations.

$$D_1(D_1 f) \qquad D_{11} f \qquad f_{11} \qquad f_{xx} \qquad \frac{\partial^2 f}{\partial x^2}$$

The second partial derivative of $f$ obtained by partial-differentiating twice with respect to $y$ is represented by each of the following notations.

$$D_2(D_2f) \quad D_{22}f \quad f_{22} \quad f_{yy} \quad \frac{\partial^2 f}{\partial y^2}$$

Actually, there is another possible mixed partial derivative, represented by $f_{21}$, $f_{yx}$, etc., but as the following theorem states, it is often true (but not always) that $f_{21} = f_{12}, f_{yx} = f_{xy}$, etc.

**19.7.1 Theorem** Suppose that $f$ is a function of two variables $x$ and $y$ defined on an open disk $B((x_0, y_0); r)$ and $f_x, f_y, f_{xy}$, and $f_{yx}$ also are defined on $B$. Furthermore, suppose that $f_{xy}$ and $f_{yx}$ are continuous on $B$. Then

$$f_{xy}(x_0, y_0) = f_{yx}(x_0, y_0)$$

## *Exercises 19.7*

In Exercises 1-8, do each of the following: **(a)** find $D_{11}f(x, y)$, **(b)** find $D_{22}f(x, y)$, and **(c)** show that $D_{12}f(x, y) = D_{21}f(x, y)$.

4. $f(x, y) = e^{-x/y} + \ln \dfrac{y}{x}$

SOLUTION: Simplifying $f(x, y)$, we have

$$f(x, y) = e^{-x/y} + \ln y - \ln x$$

Hence,

$$D_1f(x, y) = -\frac{1}{y}e^{-x/y} - \frac{1}{x} \tag{1}$$

and

$$D_2f(x, y) = \frac{x}{y^2}e^{-x/y} + \frac{1}{y} \tag{2}$$

**(a)** We partial-differentiate on both sides of (1) with respect to $x$.

$$D_{11}f(x, y) = \frac{1}{y^2}e^{-x/y} + \frac{1}{x^2}$$

**(b)** We partial-differentiate on both sides of (2) with respect to $y$.

$$D_{22}f(x, y) = \frac{x}{y^2} \cdot \frac{x}{y^2}e^{-x/y} + e^{-x/y}\left(\frac{-2x}{y^3}\right) - \frac{1}{y^2}$$

$$= \frac{x^2}{y^4}e^{-x/y} - \frac{2x}{y^3}e^{-x/y} - \frac{1}{y^2}$$

**(c)** We partial-differentiate on both sides of (1) with respect to $y$.

$$D_{12}f(x, y) = \left(-\frac{1}{y}\right)\left(\frac{x}{y^2}\right)e^{-x/y} + e^{-x/y}\left(\frac{1}{y^2}\right)$$

$$= e^{-x/y}\left[\frac{1}{y^2} - \frac{x}{y^3}\right] \tag{3}$$

We partial-differentiate on both sides of (2) with respect to $x$.

$$D_{21}f(x, y) = \left(\frac{x}{y^2}\right)\left(-\frac{1}{y}\right)e^{-x/y} + e^{-x/y}\left(\frac{1}{y^2}\right)$$

$$= e^{-x/y}\left[\frac{1}{y^2} - \frac{x}{y^3}\right] \tag{4}$$

From (3) and (4) we show that $D_{12}f(x, y) = D_{21}f(x, y)$.

8. $f(x, y) = x \cos y - ye^x$

SOLUTION:

$$D_1 f(x, y) = \cos y - ye^x \tag{1}$$

$$D_2 f(x, y) = -x \sin y - e^x \tag{2}$$

(a) From (1), we obtain

$$D_{11}f(x, y) = -ye^x$$

(b) From (2), we obtain

$$D_{22}f(x, y) = -x \cos y$$

(c) From (1), we obtain

$$D_{12}f(x, y) = -\sin y - e^x$$

From (2), we obtain

$$D_{21}f(x, y) = -\sin y - e^x$$

Thus, $D_{12}f(x, y) = D_{21}f(x, y)$.

12. Find the indicated partial derivatives.

$f(u, v) = \ln \cos(u - v)$; (a) $f_{uuv}(u, v)$; (b) $f_{vuv}(u, v)$

SOLUTION:

(a) $$f_u(u, v) = -\frac{\sin(u - v)}{\cos(u - v)} = -\tan(u - v)$$

$$f_{uu}(u, v) = -\sec^2(u - v)$$

$$f_{uuv}(u, v) = 2\sec^2(u - v)\tan(u - v)$$

(b) $$f_v(u, v) = \frac{\sin(u - v)}{\cos(u - v)} = \tan(u - v)$$

$$f_{vu}(u, v) = \sec^2(u - v)$$

$$f_{vuv}(u, v) = -2\sec^2(u - v)\tan(u - v)$$

16. Show that $u(x, y)$ satisfies the equation

$$\frac{\partial^2 u}{\partial x^2} + \frac{\partial^2 u}{\partial y^2} = 0$$

which is known as *Laplace's equation* in $R^2$.

$u(x, y) = e^x \sin y + e^y \cos x$

SOLUTION:

$$\frac{\partial u}{\partial x} = e^x \sin y - e^y \sin x$$

$$\frac{\partial^2 u}{\partial x^2} = e^x \sin y - e^y \cos x \tag{1}$$

and

$$\frac{\partial u}{\partial y} = e^x \cos y + e^y \cos x$$

$$\frac{\partial^2 u}{\partial y^2} = -e^x \sin y + e^y \cos x \qquad (2)$$

Adding the members of (1) and (2), we obtain

$$\frac{\partial^2 u}{\partial x^2} + \frac{\partial^2 u}{\partial y^2} = 0$$

**20.** For the function of Example 4, show that $f_{12}$ is discontinuous at $(0, 0)$ and hence that the hypothesis of Theorem 19.7.1 is not satisfied if $(x_0, y_0) = (0, 0)$.

SOLUTION: The function of Example 4 is defined by

$$f(x, y) = \begin{cases} (xy)\dfrac{x^2 - y^2}{x^2 + y^2} & \text{if } (x, y) \neq (0, 0) \\ 0 & \text{if } (x, y) = (0, 0) \end{cases}$$

We show that $\lim\limits_{(x, y) \to (0, 0)} f_{12}(x, y)$ does not exist. If $(x, y) \neq (0, 0)$, then

$$f(x, y) = \frac{x^3 y - xy^3}{x^2 + y^2}$$

Thus, if $(x, y) \neq (0, 0)$, then

$$f_1(x, y) = \frac{(x^2 + y^2)(3x^2 y - y^3) - (x^3 y - xy^3)(2x)}{(x^2 + y^2)^2}$$

$$= \frac{x^4 y + 4x^2 y^3 - y^5}{(x^2 + y^2)^2}$$

and

$$f_{12}(x, y) = \frac{(x^2 + y^2)^2 (x^4 + 12x^2 y^2 - 5y^4) - (x^4 y + 4x^2 y^3 - y^5)(4y)(x^2 + y^2)}{(x^2 + y^2)^4}$$

$$= \frac{x^6 + 9x^4 y^2 - 9x^2 y^4 - y^6}{(x^2 + y^2)^3}$$

If $S_1$ is the set of points on the line $y = 0$, then

$$\lim_{\substack{(x, y) \to (0, 0) \\ (P \text{ in } S_1)}} f_{12}(x, y) = \lim_{x \to 0} f_{12}(x, 0) = \lim_{x \to 0} \frac{x^6}{x^6} = 1$$

If $S_2$ is the set of points on the line $x = 0$, then

$$\lim_{\substack{(x, y) \to (0, 0) \\ (P \text{ in } S_2)}} f_{12}(x, y) = \lim_{y \to 0} f_{12}(0, y) = \lim_{y \to 0} \frac{y^6}{y^6} = -1$$

Because

$$\lim_{\substack{(x, y) \to (0, 0) \\ (P \text{ in } S_1)}} f_{12}(x, y) \neq \lim_{\substack{(x, y) \to (0, 0) \\ (P \text{ in } S_2)}} f_{12}(x, y)$$

then $\lim\limits_{(x, y) \to (0, 0)} f_{12}(x, y)$ does not exist. Therefore $f_{12}$ is discontinuous at $(0, 0)$.

**24.** Given that $u = f(x, y)$, $x = F(t)$, and $y = G(t)$, and assuming that $f_{xy} = f_{yx}$, prove by using the chain rule that

$$\frac{d^2u}{dt^2} = f_{xx}(x, y)[F'(t)]^2 + 2f_{xy}(x, y)F'(t)G'(t) + f_{yy}(x, y)[G'(t)]^2 + f_x(x, y)F''(t) + F_y(x, y)G''(t)$$

SOLUTION: By the chain rule, we have

$$\frac{du}{dt} = \frac{\partial u}{\partial x}\frac{dx}{dt} + \frac{\partial u}{\partial y}\frac{dy}{dt}$$

which may be expressed as

$$\frac{du}{dt} = f_x(x, y)F'(t) + f_y(x, y)G'(t) \tag{1}$$

We differentiate with respect to $t$ on both sides of (1), using the derivative of a sum formula and the derivative of a product formula. Thus,

$$\frac{d^2u}{dt^2} = f_x(x, y)F''(t) + F'(t)\frac{d(f_x(x, y))}{dt} + f_y(x, y)G''(t) + G'(t)\frac{d(f_y(x, y))}{dt} \tag{2}$$

We use the chain rule to find $d(f_x(x, y))/dt$. Thus,

$$\begin{aligned}\frac{d(f_x(x, y))}{dt} &= \frac{\partial f_x(x, y)}{\partial x}\frac{dx}{dt} + \frac{\partial f_x(x, y)}{\partial y}\frac{dy}{dt}\\ &= f_{xx}(x, y)F'(t) + f_{xy}(x, y)G'(t)\end{aligned} \tag{3}$$

Similarly, we obtain

$$\frac{d(f_y(x, y))}{dt} = f_{yx}(x, y)F'(t) + f_{yy}(x, y)G'(t) \tag{4}$$

Substituting from (3) and (4) into (2) and replacing $f_{yx}(x, y)$ with $f_{xy}(x, y)$, we obtain

$$\begin{aligned}\frac{d^2u}{dt^2} &= f_x(x, y)F''(t) + F'(t)[f_{xx}(x, y)F'(t) + f_{xy}(x, y)G'(t)] + f_y(x, y)G''(t)\\ &\quad + G'(t)[f_{xy}(x, y)F'(t) + f_{yy}(x, y)G'(t)]\\ &= f_{xx}(x, y)[F'(t)]^2 + 2f_{xy}(x, y)F'(t)G'(t) + f_{yy}(x, y)[G'(t)]^2\\ &\quad + f_x(x, y)F''(t) + f_y(x, y)G''(t)\end{aligned}$$

**28.** Given $u = 3xy - 4y^2$, $x = 2se^r$, $y = re^{-s}$. Find $\partial^2u/\partial s\partial r$ in three ways; **(a)** by first expressing $u$ in terms of $r$ and $s$, **(b)** by using the formula of Exercise 25, and **(c)** by using the chain rule.

SOLUTION:

**(a)**

$$\begin{aligned}u &= 3xy - 4y^2\\ &= 3(2se^r)(re^{-s}) - 4(re^{-s})^2\\ &= 6rse^{r-s} - 4r^2e^{-2s}\end{aligned}$$

Thus,

$$\frac{\partial u}{\partial r} = 6rse^{r-s} + 6se^{r-s} - 8re^{-2s}$$

Hence,

$$\frac{\partial^2 u}{\partial s\partial r} = -6rse^{r-s} + 6re^{r-s} - 6se^{r-s} + 6e^{r-s} + 16re^{-2s}$$

**(b)** The formula of Exercise 25 is as follows: Given $u = f(x, y)$, $x = F(r, s)$, and $y = G(r, s)$, and assuming that $f_{xy} = f_{yx}$, then

$$\frac{\partial^2 u}{\partial r \partial s} = f_{xx}(x, y)F_r(r, s)F_s(r, s) + f_{xy}(x, y)[F_s(r, s)G_r(r, s) + F_r(r, s)G_s(r, s)] + f_{yy}(x, y)G_r(r, s)G_s(r, s) + f_x(x, y)F_{sr}(r, s) + f_y(x, y)G_{sr}(r, s) \tag{1}$$

Because $\partial^2 u/\partial s \partial r = \partial^2 u/\partial r \partial s$, we apply the formula. We take

$$f(x, y) = 3xy - 4y^2$$

Thus,

$$f_x(x, y) = 3y \qquad f_y(x, y) = 3x - 8y \tag{2}$$

$$f_{xx}(x, y) = 0 \qquad f_{xy}(x, y) = 3 \quad \text{and} \quad f_{yy}(x, y) = -8 \tag{3}$$

Furthermore, we have

$$F(r, s) = 2se^r$$

Thus,

$$F_r(r, s) = 2se^r \qquad F_s(r, s) = 2e^r \quad \text{and} \quad F_{sr}(r, s) = 2e^r \tag{4}$$

and

$$G(r, s) = re^{-s}$$

Thus,

$$G_r(r, s) = e^{-s} \qquad G_s(r, s) = -re^{-s} \quad \text{and} \quad G_{sr}(r, s) = -e^{-s} \tag{5}$$

Substituting from (2), (3), (4), and (5) into (1), we obtain

$$\begin{aligned}\frac{\partial^2 u}{\partial s \partial r} &= 0 + 3[2e^r e^{-s} + 2se^r(-re^{-s})] + (-8)e^{-s}(-re^{-s}) \\ &\quad + 3y(2e^r) + (3x - 8y)(-e^{-s}) \\ &= 6e^{r-s} - 6rse^{r-s} + 8re^{-2s} + 6ye^r - 3xe^{-s} + 8ye^{-s}\end{aligned}$$

**(c)** By using the chain rule, we have

$$\begin{aligned}\frac{\partial u}{\partial r} &= \frac{\partial u}{\partial x}\frac{\partial x}{\partial r} + \frac{\partial u}{\partial y}\frac{\partial y}{\partial r} \\ &= 3y \cdot 2se^r + (3x - 8y)e^{-s} \\ &= 6yse^r + (3x - 8y)e^{-s}\end{aligned}$$

Partial-differentiating with respect to $s$ and using the formulas for the derivative of a sum and the derivative of a product, we obtain

$$\begin{aligned}\frac{\partial^2 u}{\partial s \partial r} &= 6y\frac{\partial(se^r)}{\partial s} + se^r\frac{\partial(6y)}{\partial s} + (3x - 8y)\frac{\partial(e^{-s})}{\partial s} + e^{-s}\frac{\partial(3x - 8y)}{\partial s} \\ &= 6ye^r + se^r(-6re^{-s}) + (3x - 8y)(-e^{-s}) + e^{-s}[6e^r + 8re^{-s}] \\ &= 6ye^r - 6rse^{r-s} - 3xe^{-s} + 8ye^{-s} + 6e^{r-s} + 8re^{-2s}\end{aligned}$$

**32.** If $u = f(x, y)$ and $v = g(x, y)$, then the equations

$$\frac{\partial u}{\partial x} = \frac{\partial v}{\partial y} \quad \text{and} \quad \frac{\partial v}{\partial x} = -\frac{\partial u}{\partial y} \tag{1}$$

are called the *Cauchy-Riemann equations.* If $f$ and $g$ and their first and second partial derivatives are continuous, prove that if $u$ and $v$ satisfy the Cauchy-Riemann equations, they also satisfy Laplace's equation (see Exercises 15-18).

SOLUTION: Laplace's equation is

$$\frac{\partial^2 u}{\partial x^2} + \frac{\partial^2 u}{\partial y^2} = 0$$

Partial-differentiating with respect to $x$ on both sides of the first equation in (1), we obtain

$$\frac{\partial^2 u}{\partial x^2} = \frac{\partial^2 v}{\partial x \partial y} \tag{2}$$

Partial-differentiating with respect to $y$ on both sides of the second equation in (1), we obtain

$$\frac{\partial^2 v}{\partial y \partial x} = -\frac{\partial^2 u}{\partial y^2} \tag{3}$$

Because $g$ and its first and second partial derivatives are continuous, then $\partial^2 v/\partial y \partial x = \partial^2 v/\partial x \partial y$. Thus, Eq. (3) is equivalent to

$$\frac{\partial^2 u}{\partial y^2} = -\frac{\partial^2 v}{\partial x \partial y} \tag{4}$$

Adding the members of (2) and (4), we obtain

$$\frac{\partial^2 u}{\partial x^2} + \frac{\partial^2 u}{\partial y^2} = 0$$

which is Laplace's equation. Similarly, we may partial-differentiate with respect to $y$ on both sides of the first equation in (1) and partial-differentiate with respect to $x$ on both sides of the second equation in (1). We obtain

$$\frac{\partial^2 u}{\partial y \partial x} = \frac{\partial^2 v}{\partial y^2} \quad \text{and} \quad \frac{\partial^2 v}{\partial x^2} = -\frac{\partial^2 u}{\partial x \partial y} \tag{5}$$

Because $f$ and its first and second partial derivatives are continuous, then $\partial^2 u/\partial y \partial x = \partial^2 u/\partial x \partial y$. Thus, if we add the members of Eq. (5), the result is

$$\frac{\partial^2 v}{\partial x^2} + \frac{\partial^2 v}{\partial y^2} = 0$$

which is Laplace's equation.

## *Review Exercises*

**4.** Find the indicated partial derivatives.

$$h(x, y) = \tan^{-1}\frac{x^3}{y^2};\ D_1h(x, y),\ D_2h(x, y),\ D_{11}h(x, y)$$

SOLUTION:

$$D_1h(x, y) = \frac{1}{1 + \left(\frac{x^3}{y^2}\right)^2} \cdot \frac{3x^2}{y^2}$$

$$= \frac{3x^2y^2}{x^6 + y^4}$$

$$D_2h(x, y) = \frac{1}{1 + \left(\frac{x^3}{y^2}\right)^2}\left(-\frac{2x^3}{y^3}\right)$$

$$= \frac{-2x^3y}{x^6+y^4}$$

$$D_{11}h(x,y) = \frac{(x^6+y^4)6xy^2 - 3x^2y^2 \cdot 6x^5}{(x^6+y^4)^2}$$

$$= \frac{-6xy^2(2x^6-y^4)}{(x^6+y^4)^2}$$

8. Find $\partial u/\partial t$ and $\partial u/\partial s$ by two methods.

$$u = e^{2x+y}\cos(2y-x) \qquad x = 2s^2 - t^2 \qquad y = s^2 + 2t^2$$

SOLUTION: We use the chain rule. Thus,

$$\frac{\partial u}{\partial t} = \frac{\partial u}{\partial x}\frac{\partial x}{\partial t} + \frac{\partial u}{\partial y}\frac{\partial y}{\partial t}$$

$$= [e^{2x+y}\sin(2y-x) + 2\cos(2y-x)e^{2x+y}](-2t)$$

$$+ [-2e^{2x+y}\sin(2y-x) + e^{2x+y}\cos(2y-x)](4t)$$

$$= -10te^{2x+y}\sin(2y-x)$$

and

$$\frac{\partial u}{\partial s} = \frac{\partial u}{\partial x}\frac{\partial x}{\partial s} + \frac{\partial u}{\partial y}\frac{\partial y}{\partial s}$$

$$= [e^{2x+y}\sin(2y-x) + 2e^{2x+y}\cos(2y-x)]4s$$

$$+ [-2e^{2x+y}\sin(2y-x) + e^{2x+y}\cos(2y-x)]2s$$

$$= 10se^{2x+y}\cos(2y-x)$$

For the second method, we eliminate $x$ and $y$ before partial-differentiating. Because

$$2x + y = 5s^2 \quad \text{and} \quad 2y - x = 5t^2$$

we have

$$u = e^{5s^2}\cos 5t^2$$

Thus,

$$\frac{\partial u}{\partial t} = -10te^{5s^2}\sin 5t^2$$

and

$$\frac{\partial u}{\partial s} = 10se^{5s^2}\cos 5t^2$$

12. Find the domain and range of the function $f$ and draw a sketch showing as a shaded region in $R^2$ the set of points in the domain of $f$.

$$f(x,y) = \sin^{-1}\sqrt{1-x^2-y^2}$$

SOLUTION: The domain of $f$ is the set of all points $(x, y)$ such that

$$1 - x^2 - y^2 \geqslant 0$$
$$x^2 + y^2 \leqslant 1$$

Thus, the domain is the set of all points that lie either on the circle $x^2 + y^2 = 1$ or in the interior of the circle, as shown in Fig. 19.12R. Because $0 \leqslant \sqrt{1-x^2-y^2} \leqslant 1$ for all $(x, y)$ in the domain, then $0 \leqslant \sin^{-1}\sqrt{1-x^2-y^2} \leqslant \frac{1}{2}\pi$. Thus, the range of $f$ is the closed interval $[0, \frac{1}{2}\pi]$.

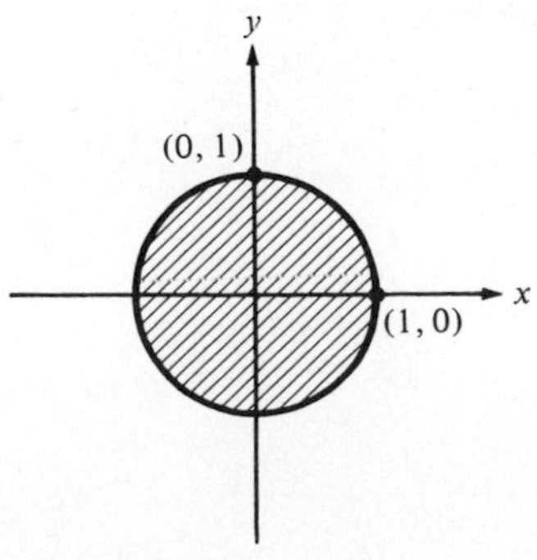

Figure 19.12R

**16.** Find the domain and range of the function $f$.

$$f(x, y, z) = \frac{xy + xz + yz}{xyz}$$

SOLUTION: If $xyz = 0$, then either $x = 0$, $y = 0$, or $z = 0$. Thus, the domain of $f$ is the set of all points in $R^3$ that do not lie on either the $xy$ plane, the $xz$ plane, or the $yz$ plane. The range of $f$ is the set of all real numbers.

**20.** Establish the limit by finding a $\delta > 0$ for any $\epsilon > 0$ so that Definition 19.2.5 holds.

$$\lim_{(x, y) \to (3, 1)} (x^2 - y^2 + 2x - 4y) = 10$$

SOLUTION: For any $\epsilon > 0$, we must find a $\delta > 0$ such that $|x^2 - y^2 + 2x - 4y - 10| < \epsilon$ whenever $0 < \sqrt{(x-3)^2 + (y-1)^2} < \delta$. By the corollary to the triangle inequality, $|a - b| \leqslant |a| + |b|$, we have

$$\begin{aligned} |x^2 - y^2 + 2x - 4y - 10| &= |(x-3)(x+5) - (y-1)(y+5)| \\ &\leqslant |x-3||x+5| + |y-1||y+5| \end{aligned} \tag{1}$$

If $0 < \sqrt{(x-3)^2 + (y-1)^2} < \delta \leqslant 1$, then

$$|x - 3| \leqslant \sqrt{(x-3)^2 + (y-1)^2} < \delta \leqslant 1 \tag{2}$$

and

$$|y - 1| \leqslant \sqrt{(x-3)^2 + (y-1)^2} < \delta \leqslant 1 \tag{3}$$

From (2), we have whenever $0 < \sqrt{(x-3)^2 + (y-1)^2} < \delta \leqslant 1$

$$\begin{aligned} |x - 3| &< 1 \\ -1 < x - 3 &< 1 \\ 7 < x + 5 &< 9 \\ |x + 5| &< 9 \end{aligned} \tag{4}$$

From (3), we have whenever $0 < \sqrt{(x-3)^2 + (y-1)^2} < \delta \leqslant 1$

$$\begin{aligned} |y - 1| &< 1 \\ -1 < y - 1 &< 1 \\ 5 < y + 5 &< 7 \\ |y + 5| &< 7 \end{aligned} \tag{5}$$

Thus, whenever $0 < \sqrt{(x-3)^2 + (y-1)^2} < \delta \leqslant 1$, by (1), (2), (3), (4), and (5) we have

$$|x^2 - y^2 + 2x - 4y - 10| < 9\delta + 7\delta = 16\delta \tag{6}$$

Therefore, we take $\delta$ to be the minimum of 1 and $\frac{1}{16}\epsilon$. Whenever $0 < \sqrt{(x-3)^2 + (y-1)^2} < \delta$, from inequality (6) we have

$$|x^2 - y^2 + 2x - 4y - 10| < \epsilon$$

Thus,

$$\lim_{(x, y) \to (3, 1)} (x^2 - y^2 + 2x - 4y) = 10$$

**24.** Discuss the continuity of $f$.

$$f(x, y) = \begin{cases} \dfrac{x^4 - y^4}{x^4 + y^4} & \text{if } (x, y) \neq (0, 0) \\ 0 & \text{if } (x, y) = (0, 0) \end{cases}$$

SOLUTION: $f$ is continuous at all points $(x, y) \neq (0, 0)$ because $x^4 + y^4 \neq 0$ if $(x, y) \neq (0, 0)$. We are given that $f(0, 0) = 0$. We wish to determine if $\lim_{(x, y) \to (0, 0)} f(x, y)$ exists. Let $S_1$ be the $x$-axis. Then

$$\lim_{\substack{(x, y) \to (0, 0) \\ (P \text{ in } S_1)}} f(x, y) = \lim_{x \to 0} f(x, 0) = \lim_{x \to 0} \frac{x^4}{x^4} = 1$$

Let $S_2$ be the $y$-axis. Then

$$\lim_{\substack{(x, y) \to (0, 0) \\ (P \text{ in } S_2)}} f(x, y) = \lim_{y \to 0} f(0, y) = \lim_{y \to 0} \frac{-y^4}{y^4} = -1$$

Because

$$\lim_{\substack{(x, y) \to (0, 0) \\ (P \text{ in } S_1)}} f(x, y) \neq \lim_{\substack{(x, y) \to (0, 0) \\ (P \text{ in } S_2)}} f(x, y)$$

then

$$\lim_{(x, y) \to (0, 0)} f(x, y)$$

does not exist. Therefore, $f$ is discontinuous at $(0, 0)$.

28. Prove that the function $f$ is differentiable at all points in its domain by showing that Definition 19.5.2 holds.

$$f(x, y) = 3xy^2 - 4x^2 + y^2$$

SOLUTION: By Definition 19.5.2, we must find $\epsilon_1$ and $\epsilon_2$ such that

$$\epsilon_1 \Delta x + \epsilon_2 \Delta y = \Delta f(x_0, y_0) - D_1 f(x_0, y_0)\Delta x - D_2 f(x_0, y_0)\Delta y \tag{1}$$

where $\epsilon_1 \to 0$ and $\epsilon_2 \to 0$ as $(\Delta x, \Delta y) \to (0, 0)$. We have

$$\begin{aligned} \Delta f(x_0, y_0) &= f(x_0 + \Delta x, y_0 + \Delta y) - f(x_0, y_0) \\ &= [3(x_0 + \Delta x)(y_0 + \Delta y)^2 - 4(x_0 + \Delta x)^2 + (y_0 + \Delta y)^2] \\ &\quad - [3x_0 y_0^2 - 4x_0^2 + y_0^2] \\ &= 6x_0 y_0 \Delta y + 3x_0(\Delta y)^2 + 3y_0^2 \Delta x + 6y_0 \Delta x \Delta y + 3\Delta x(\Delta y)^2 \\ &\quad - 8x_0 \Delta x - 4(\Delta x)^2 + 2y_0 \Delta y + (\Delta y)^2 \end{aligned} \tag{2}$$

Next, we find the following.

$$D_1 f(x_0, y_0)\Delta x = 3y_0^2 \Delta x - 8x_0 \Delta x \tag{3}$$

$$D_2 f(x_0, y_0)\Delta y = 6x_0 y_0 \Delta y + 2y_0 \Delta y \tag{4}$$

Substituting from (2), (3), and (4) into (1), we obtain

$$\epsilon_1 \Delta x + \epsilon_2 \Delta y = (6y_0 \Delta y - 4\Delta x)\Delta x + (3x_0 \Delta y + 3\Delta x \Delta y + \Delta y)\Delta y$$

Therefore, we take $\epsilon_1 = 6y_0 \Delta y - 4\Delta x$ and $\epsilon_2 = 3x_0 \Delta y + 3\Delta x \Delta y + \Delta y$. If $(\Delta x, \Delta y) \to (0, 0)$, then $\epsilon_1 \to 0$ and $\epsilon_2 \to 0$. Thus $f$ is differentiable at all points in $R^2$.

32. At a given instant, the length of one side of a rectangle is 6 ft and it is increasing at the rate of 1 ft/sec; the length of another side of the rectangle is 10 ft and it is decreasing at the rate of 2 ft/sec. Find the rate of change of the area of the rectangle at the given instant.

SOLUTION: Let $x$ ft and $y$ ft be the lengths of the sides of the rectangle and let $A$ sq ft be the area of the rectangle. Then $x$, $y$, and $A$ are functions of $t$, where $t$ is the number of seconds that have elapsed since some fixed moment in time. We are given that $dx/dt = 1$ and $dy/dt = -2$ at the moment when $x = 6$, $y = 10$. We want to find $dA/dt$ at this particular moment. We have

$$A = xy$$

Thus,

$$\frac{dA}{dt} = \frac{\partial A}{\partial x}\frac{dx}{dt} + \frac{\partial A}{\partial y}\frac{dy}{dt}$$

$$= y\frac{dx}{dt} + x\frac{dy}{dt}$$

Substituting the given values, we obtain

$$\frac{dA}{dt} = 10 \cdot 1 + 6(-2)$$

$$= -2$$

Therefore, the area is decreasing at the rate of 2 square feet per second at the moment under consideration.

**36.** Verify that $u(x, t) = A\cos(kat)\sin(kx)$, where $A$ and $k$ are arbitrary constants, satisfies the partial differential equation for a vibrating string:

$$\frac{\partial^2 u}{\partial t^2} = a^2\frac{\partial^2 u}{\partial x^2}$$

SOLUTION: Partial-differentiating $u$ with respect to $t$, we obtain

$$\frac{\partial u}{\partial t} = -kaA\sin(kat)\sin(kx)$$

$$\frac{\partial^2 u}{\partial t^2} = -k^2a^2A\cos(kat)\sin(kx) \tag{1}$$

Partial-differentiating $u$ with respect to $x$, we obtain

$$\frac{\partial u}{\partial x} = kA\cos(kat)\cos(kx)$$

$$\frac{\partial^2 u}{\partial x^2} = -k^2A\cos(kat)\sin(kx)$$

Thus,

$$a^2\frac{\partial^2 u}{\partial x^2} = -k^2a^2A\cos(kat)\sin(kx) \tag{2}$$

Comparing Eqs. (1) and (2), we obtain

$$\frac{\partial^2 u}{\partial t^2} = a^2\frac{\partial^2 u}{\partial x^2}$$

which is the desired result.

**40.** Let $f$ be the function defined by

$$f(x, y) = \begin{cases} \dfrac{e^{-1/x^2}y}{e^{-2/x^2} + y^2} & \text{if } x \neq 0 \\ 0 & \text{if } x = 0 \end{cases}$$

Prove that $f$ is discontinuous at the origin.

SOLUTION: We are given that $f(0, 0) = 0$. We use Theorem 19.2.8 to show that either $\lim\limits_{(x, y) \to (0, 0)} f(x, y)$ does not exist or that $\lim\limits_{(x, y) \to (0, 0)} f(x, y) \neq 0$. Let $S$ be the curve $y = e^{-1/x^2}$. Because

$$\lim_{x \to 0} e^{-1/x^2} = 0$$

then $(0, 0)$ is an accumulation point of $S$. Furthermore,

$$\begin{aligned}
\lim_{\substack{(x, y) \to (0, 0) \\ (P \text{ in } S)}} f(x, y) &= \lim_{x \to 0} f(x, e^{-1/x^2}) \\
&= \lim_{x \to 0} \frac{e^{-1/x^2} e^{-1/x^2}}{e^{-2/x^2} + e^{-2/x^2}} \\
&= \lim_{x \to 0} \frac{e^{-2/x^2}}{2e^{-2/x^2}} \\
&= \frac{1}{2}
\end{aligned}$$

Therefore, if $\lim\limits_{(x, y) \to (0, 0)} f(x, y)$ exists, it has value $\frac{1}{2}$. Hence, either $\lim\limits_{(x, y) \to (0, 0)} f(x, y)$ does not exist or else $\lim\limits_{(x, y) \to (0, 0)} f(x, y) \neq f(0, 0)$. Thus, $f$ is discontinuous at the origin.

# 20

# Directional derivatives, gradients, applications of partial derivatives, and line integrals

## 20.1 DIRECTIONAL DERIVATIVES AND GRADIENTS

**20.1.1 Definition** Let $f$ be a function of two variables $x$ and $y$. If $\mathbf{U}$ is the unit vector $\cos\theta\mathbf{i} + \sin\theta\mathbf{j}$, then the *directional derivative* of $f$ in the direction of $\mathbf{U}$, denoted by $D_{\mathbf{U}}f$, is given by

$$D_{\mathbf{U}}f(x, y) = \lim_{h \to 0} \frac{f(x + h\cos\theta, y + h\sin\theta) - f(x, y)}{h}$$

if this limit exists.

**20.1.2 Theorem** If $f$ is a differentiable function of $x$ and $y$, and $\mathbf{U} = \cos\theta\mathbf{i} + \sin\theta\mathbf{j}$, then

$$D_{\mathbf{U}}f(x, y) = f_x(x, y)\cos\theta + f_y(x, y)\sin\theta$$

**20.1.3 Definition** If $f$ is a function of two variables $x$ and $y$ and $f_x$ and $f_y$ exist, then the *gradient* of $f$, denoted by $\nabla f$ (read: "del $f$"), is defined by

$$\nabla f(x, y) = f_x(x, y)\mathbf{i} + f_y(x, y)\mathbf{j}$$

If $f$ is a differentiable function of $x$ and $y$ and $\mathbf{U}$ is a unit vector in $R^2$, then

$$D_{\mathbf{U}}f(x, y) = \mathbf{U} \cdot \nabla f(x, y)$$

Moreover, if $\mathbf{U}$ has the same direction as $\nabla f$, then $D_{\mathbf{U}}f$ has an absolute maximum value, and the absolute maximum value of $D_{\mathbf{U}}f$ is $|\nabla f|$, the magnitude of the gradient vector.

We extend the definition of a directional derivative to a function of three variables.

**20.1.4 Definition** Suppose that $f$ is a function of three variables $x, y$, and $z$. If $\mathbf{U}$ is the unit vector $\cos\alpha\mathbf{i} + \cos\beta\mathbf{j} + \cos\gamma\mathbf{k}$, then the directional derivative of $f$ in the direction of $\mathbf{U}$, denoted by $D_{\mathbf{U}}f$, is given by

$$D_{\mathbf{U}}f(x, y, z) = \lim_{h\to 0} \frac{f(x + h\cos\alpha, y + h\cos\beta, z + h\cos\gamma) - f(x, y, z)}{h}$$

if this limit exists.

**20.1.5 Theorem** If $f$ is a differentiable function of $x, y$, and $z$ and

$$\mathbf{U} = \cos\alpha\mathbf{i} + \cos\beta\mathbf{j} + \cos\gamma\mathbf{k}$$

then

$$D_{\mathbf{U}}f(x, y, z) = f_x(x, y, z)\cos\alpha + f_y(x, y, z)\cos\beta + f_z(x, y, z)\cos\gamma$$

**20.1.6 Definition** If $f$ is a function of three variables $x, y$, and $z$ and the first partial derivatives $f_x$, $f_y$, and $f_z$ exist, then the *gradient* of $f$, denoted by $\nabla f$, is defined by

$$\nabla f(x, y, z) = f_x(x, y, z)\mathbf{i} + f_y(x, y, z)\mathbf{j} + f_z(x, y, z)\mathbf{k}$$

If $\mathbf{U}$ is a unit vector in $R^3$, then

$$D_{\mathbf{U}}f(x, y, z) = \mathbf{U}\cdot\nabla f(x, y, z)$$

## *Exercises 20.1*

**4.** Find the directional derivative of the given function in the direction of the given unit vector $\mathbf{u}$ by using Definition 20.1.4, and then verify your result by applying Theorem 20.1.5.

$$f(x, y, z) = 6x^2 - 2xy + yz \qquad \mathbf{u} = \frac{3}{7}\mathbf{i} + \frac{2}{7}\mathbf{j} + \frac{6}{7}\mathbf{k}$$

SOLUTION: We use Definition 20.1.4 with

$$\cos\alpha = \frac{3}{7} \qquad \cos\beta = \frac{2}{7} \quad \text{and} \quad \cos\gamma = \frac{6}{7}$$

Thus,

$$D_{\mathbf{u}}f(x, y, z) = \lim_{h\to 0} \frac{f\left(x + \frac{3}{7}h, y + \frac{2}{7}h, z + \frac{6}{7}h\right) - f(x, y, z)}{h}$$

$$= \lim_{h\to 0} \frac{\left[6\left(x + \frac{3}{7}h\right)^2 - 2\left(x + \frac{3}{7}h\right)\left(y + \frac{2}{7}h\right) + \left(y + \frac{2}{7}h\right)\left(z + \frac{6}{7}h\right)\right] - [6x^2 - 2xy + yz]}{h}$$

$$= \lim_{h\to 0} \frac{6x^2 + \frac{36}{7}xh + \frac{54}{49}h^2 - 2xy - \frac{4}{7}xh - \frac{6}{7}yh - \frac{12}{7}h^2 + yz + \frac{6}{7}yh + \frac{2}{7}hz + \frac{12}{49}h^2 - 6x^2 + 2xy - yz}{h}$$

$$= \lim_{h\to 0} \frac{\frac{32}{7}xh - \frac{18}{49}h^2 + \frac{2}{7}hz}{h}$$

$$= \lim_{h \to 0} \frac{32}{7}x - \frac{18}{49}h + \frac{2}{7}z$$

$$= \frac{32}{7}x + \frac{2}{7}z$$

Next, we apply Theorem 20.1.5. We have

$$f_x(x, y, z) = 12x - 2y \qquad f_y(x, y, z) = -2x + z \qquad f_z(x, y, z) = y$$

Substituting into the formula of the theorem, we get

$$D_{\mathbf{u}}f(x, y, z) = (12x - 2y)\frac{3}{7} + (-2x + z)\frac{2}{7} + \frac{6}{7}y$$

$$= \frac{32}{7}x + \frac{2}{7}z$$

Therefore, the result of using Definition 20.1.4 agrees with the result of applying Theorem 20.1.5.

8. Find the value of the directional derivative at the particular point $P_0$ for the given function in the direction of $\mathbf{u}$.

$$f(x, y, z) = \ln(x^2 + y^2 + z^2) \qquad u = \frac{1}{\sqrt{3}}\mathbf{i} - \frac{1}{\sqrt{3}}\mathbf{j} - \frac{1}{\sqrt{3}}\mathbf{k} \qquad P_0 = (1, 3, 2)$$

SOLUTION: We have

$$f_x(x, y, z) = \frac{2x}{x^2 + y^2 + z^2} \qquad f_y(x, y, z) = \frac{2y}{x^2 + y^2 + z^2}$$

$$f_z(x, y, z) = \frac{2z}{x^2 + y^2 + z^2}$$

Thus,

$$f_x(1, 3, 2) = \frac{1}{7} \qquad f_y(1, 3, 2) = \frac{3}{7} \qquad f_z(1, 3, 2) = \frac{2}{7}$$

Applying Theorem 20.1.5, we have

$$D_{\mathbf{u}}(1, 3, 2) = f_x(1, 3, 2)\cos\alpha + f_y(1, 3, 2)\cos\beta + f_z(1, 3, 2)\cos\gamma$$

$$= \frac{1}{7} \cdot \frac{1}{\sqrt{3}} + \frac{3}{7}\left(-\frac{1}{\sqrt{3}}\right) + \frac{2}{7}\left(-\frac{1}{\sqrt{3}}\right)$$

$$= -\frac{4}{7\sqrt{3}}$$

12. A function $f$, a point $P$, and a unit vector $\mathbf{u}$ are given. Find: **(a)** the gradient of $f$ at $P$, and **(b)** the rate of change of the function in the direction of $\mathbf{u}$ at $P$.

$$f(x, y) = e^{2xy} \quad P = (2, 1) \quad \mathbf{u} = \frac{4}{5}\mathbf{i} - \frac{3}{5}\mathbf{j}$$

SOLUTION:

**(a)**
$$\nabla f(x, y) = f_x(x, y)\mathbf{i} + f_y(x, y)\mathbf{j}$$
$$= 2ye^{2xy}\mathbf{i} + 2xe^{2xy}\mathbf{j}$$

Thus,

$$\nabla f(2, 1) = 2e^4\mathbf{i} + 4e^4\mathbf{j}$$

(b) The rate of change of the function in the direction of $\mathbf{u}$ at $P$ is given by

$$\mathbf{u} \cdot \nabla f(2, 1) = \left(\frac{4}{5}\mathbf{i} - \frac{3}{5}\mathbf{j}\right) \cdot (2e^4\mathbf{i} + 4e^4\mathbf{j})$$

$$= \frac{8}{5}e^4 - \frac{12}{5}e^4$$

$$= -\frac{4}{5}e^4$$

**16.** Draw a contour map showing the level curves of the function of Exercise 12 at $e^8, e^4, 1, e^{-4}$, and $e^{-8}$. Also show the representation of $\nabla f(2, 1)$ having its initial point at $(2, 1)$.

SOLUTION: The function of Exercise 12 is defined by $f(x, y) = e^{2xy}$. If $f(x, y) = e^8$, then $e^{2xy} = e^8$, or equivalently, $xy = 4$. The graph of the equation $xy = 4$ is an equilateral hyperbola in the first and third quadrants with asymptotes the $x$ and $y$ axes which contains the point $(2, 2)$. Similarly, the level curve at $e^4$ is the equilateral hyperbola $xy = 2$ that contains the point $(\frac{1}{2}\sqrt{2}, \frac{1}{2}\sqrt{2})$. If $f(x, y) = 1$, then $e^{2xy} = 1$. Hence, $xy = 0$, and thus, either $x = 0$ or $y = 0$. The level curve at 1 consists of the $x$ and $y$ axes. If $f(x, y) = e^{-4}$, then $xy = -2$. The level curve at $e^{-4}$ is the equilateral hyperbola in the second and fourth quadrants that contains the point $(\frac{1}{2}\sqrt{2}, -\frac{1}{2}\sqrt{2})$. Similarly, the level curve at $e^{-8}$ is the equilateral hyperbola $xy = -4$ which contains the point $(2, -2)$. In Fig. 20.1.16 we show the contour map and the representation of a vector in the direction of $\nabla f(2, 1)$. We cannot show a representation of $\nabla f(2, 1)$ itself, because $|\nabla f(2, 1)| = 244$.

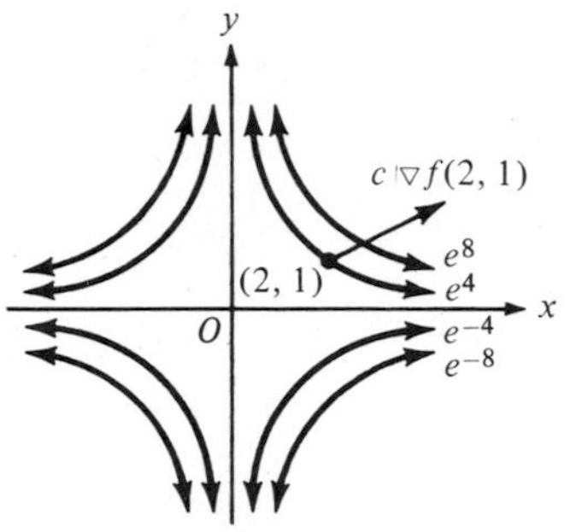

Figure 20.1.16

**20.** Find $D_{\mathbf{u}}f$ at the given point $P$ for which $\mathbf{u}$ is a unit vector in the direction of $\overrightarrow{PQ}$. Also at $P$ find $D_{\mathbf{u}}f$ if $\mathbf{u}$ is a unit vector for which $D_{\mathbf{u}}f$ is a maximum.

$f(x, y, z) = x^2 + y^2 - 4xz$; $P(3, 1, -2)$, $Q(-6, 3, 4)$

SOLUTION: We find the gradient vector. Thus,

$$\nabla f(x, y, z) = f_x(x, y, z)\mathbf{i} + f_y(x, y, z)\mathbf{j} + f_z(x, y, z)\mathbf{k}$$
$$= (2x - 4z)\mathbf{i} + 2y\mathbf{j} - 4x\mathbf{k}$$

Hence,

$$\nabla f(3, 1, -2) = 14\mathbf{i} + 2\mathbf{j} - 12\mathbf{k} \qquad \textbf{(1)}$$

Next, we find $\mathbf{u}$, a unit vector in the direction of $\overrightarrow{PQ}$. We have

$$\mathbf{V}(\overrightarrow{PQ}) = -9\mathbf{i} + 2\mathbf{j} + 6\mathbf{k}$$

Thus,

$$\mathbf{u} = \frac{\mathbf{V}(\overrightarrow{PQ})}{|\mathbf{V}(\overrightarrow{PQ})|} = \frac{1}{11}(-9\mathbf{i} + 2\mathbf{j} + 6\mathbf{k}) \qquad \textbf{(2)}$$

Therefore, from (1) and (2) we obtain

$$D_{\mathbf{u}}f(3, 1, -2) = \mathbf{u} \cdot \nabla f(3, 1, -2)$$

$$= \frac{1}{11}(-9\mathbf{i} + 2\mathbf{j} + 6\mathbf{k}) \cdot (14\mathbf{i} + 2\mathbf{j} - 12\mathbf{k})$$

$$= -\frac{194}{11}$$

which is the value of $D_{\mathbf{u}}f(3, 1, -2)$ in the direction of $\overrightarrow{PQ}$. $D_{\mathbf{u}}f(3, 1, -2)$ has a maximum value if $\mathbf{u}$ is in the direction of $\nabla f(3, 1, -2)$. Moreover, the maximum value of $D_{\mathbf{u}}(3, 1, -2)$ is given by $|\nabla f(3, 1, -2)|$. Thus, the maximum value of $D_{\mathbf{u}}f(3, 1, -2)$ is

$$|14\mathbf{i} + 2\mathbf{j} - 12\mathbf{k}| = 2\sqrt{86}$$

**24.** The temperature is $T$ degrees at any point $(x, y, z)$ in three-dimensional space and $T = 60/(x^2 + y^2 + z^2 + 3)$. Distance is measured in inches. Find: **(a)** the rate of change of the temperature at the point $(3, -2, 2)$ in the direction of the vector $-2\mathbf{i} + 3\mathbf{j} - 6\mathbf{k}$, and **(b)** the direction and magnitude of the greatest rate of change of $T$ at $(3, -2, 2)$.

SOLUTION: Let $T$ be the function defined by

$$T(x, y, z) = \frac{60}{x^2 + y^2 + z^2 + 3}$$

Then

$$T_x(x, y, z) = \frac{-120x}{(x^2 + y^2 + z^2 + 3)^2}$$

$$T_y(x, y, z) = \frac{-120y}{(x^2 + y^2 + z^2 + 3)^2}$$

$$T_z(x, y, z) = \frac{-120z}{(x^2 + y^2 + z^2 + 3)^2}$$

Thus,

$$T_x(3, -2, 2) = -\frac{9}{10}; \quad T_y(3, -2, 2) = \frac{3}{5}; \quad T_z(3, -2, 2) = -\frac{3}{5}$$

Hence,

$$\nabla T(3, -2, 2) = -\frac{3}{10}(3\mathbf{i} - 2\mathbf{j} + 2\mathbf{k}) \tag{1}$$

(a) Let $\mathbf{A} = -2\mathbf{i} + 3\mathbf{j} - 6\mathbf{k}$. A unit vector $\mathbf{u}$ in the direction of $\mathbf{A}$ is given by

$$\mathbf{u} = \frac{\mathbf{A}}{|\mathbf{A}|}$$

$$= \frac{1}{7}(-2\mathbf{i} + 3\mathbf{j} - 6\mathbf{k}) \tag{2}$$

Hence, from (2) and (1), we obtain

$$D_{\mathbf{u}}T(3, -2, 2) = \mathbf{u} \cdot \nabla T(3, -2, 2)$$

$$= \frac{1}{7}(-2\mathbf{i} + 3\mathbf{j} - 6\mathbf{k}) \cdot \left(-\frac{3}{10}\right)(3\mathbf{i} - 2\mathbf{j} + 2\mathbf{k})$$

$$= \frac{36}{35}$$

Therefore, the temperature is increasing at the rate of $\frac{36}{35}$ degrees per inch.

(b) The greatest rate of change of $T$ is in the direction of the gradient vector. From (1) we obtain the direction cosines of $\nabla T(3, -2, 2)$. They are

$$\cos\alpha = -\frac{3}{\sqrt{17}} \qquad \cos\beta = \frac{2}{\sqrt{17}} \qquad \cos\gamma = -\frac{2}{\sqrt{17}}$$

Furthermore, the magnitude of the greatest rate of change is given by $|\nabla T(3, -2, 2)| = \frac{3}{10}\sqrt{17}$. Thus, the greatest rate of change in the temperature is $\frac{3}{10}\sqrt{17}$ degrees per inch.

## 20.2 TANGENT PLANES AND NORMALS TO SURFACES

**20.2.1 Definition** A vector that is orthogonal to the unit tangent vector of every curve $C$ through a point $P_0$ on a surface $S$ is called a *normal vector* to $S$ at $P_0$.

**20.2.2 Theorem** If an equation of a surface $S$ is $F(x, y, z) = 0$, and $F_x, F_y$, and $F_z$ are continuous and not all zero at the point $P_0(x_0, y_0, z_0)$ on $S$, then a normal vector to $S$ at $P_0$ is $\nabla F(x_0, y_0, z_0)$.

**20.2.3 Definition** If an equation of a surface $S$ is $F(x, y, z) = 0$, then the *tangent plane* of $S$ at a point $P_0(x_0, y_0, z_0)$ is the plane through $P_0$ having as a normal vector $\nabla F(x_0, y_0, z_0)$.

An equation of the tangent plane of Definition 20.2.3 is

$$\nabla f(x_0, y_0, z_0) \cdot [(x - x_0)\mathbf{i} + (y - y_0)\mathbf{j} + (z - z_0)\mathbf{k}] = 0$$

or, equivalently,

$$f_x(x_0, y_0, z_0)(x - x_0) + f_y(x_0, y_0, z_0)(y - y_0) + f_z(x_0, y_0, z_0)(z - z_0) = 0$$

**20.2.4 Definition** The *normal line* to a surface $S$ at a point $P_0$ on $S$ is the line through $P_0$ having as a set of direction numbers the components of any normal vector to $S$ at $P_0$.

If an equation of the surface $S$ is $F(x, y, z) = 0$, then symmetric equations of the normal line to $S$ at $P_0(x_0, y_0, z_0)$ are

$$\frac{x - x_0}{F_x(x_0, y_0, z_0)} = \frac{y - y_0}{F_y(x_0, y_0, z_0)} = \frac{z - z_0}{F_z(x_0, y_0, z_0)}$$

**20.2.5 Definition** The *tangent line* to a curve $C$ at a point $P_0$ is the line through $P_0$ having as a set of direction numbers the components of the unit tangent vector to $C$ at $P_0$.

## *Exercises 20.2*

In Exercises 1-12, find an equation of the tangent plane and equations of the normal line to the given surface at the indicated point.

**4.** $x^2 + y^2 - z^2 = 6$; $(3, -1, 2)$

SOLUTION: Let $F$ be the function defined by

$$f(x, y, z) = x^2 + y^2 - z^2 - 6$$

Because the given surface is the graph of $F(x, y, z) = 0$, then an equation of the tangent plane to the surface at $(3, -1, 2)$ is given by

$$F_x(3, -1, 2)(x - 3) + F_y(3, -1, 2)(y + 1) + F_z(3, -1, 2)(z - 2) = 0 \tag{1}$$

We have

$$F_x(x, y, z) = 2x \qquad F_y(x, y, z) = 2y \qquad F_z(x, y, z) = -2z$$

Thus,

$$F_x(3, -1, 2) = 6 \qquad F_y(3, -1, 2) = -2 \qquad F_z(3, -1, 2) = -4 \tag{2}$$

Substituting from (2) into (1), we obtain

$$\begin{aligned} 6(x - 3) - 2(y + 1) - 4(z - 2) &= 0 \\ 3x - y - 2z - 6 &= 0 \end{aligned}$$

which is an equation of the tangent plane to the surface at the point $(3, -1, 2)$. Furthermore, symmetric equations of the normal line to the surface at the point $(3, -1, 2)$ are given by

$$\frac{x-3}{F_x(3,-1,2)} = \frac{y+1}{F_y(3,-1,2)} = \frac{z-2}{F_z(3,-1,2)} \tag{3}$$

Substituting from (2) into (3), we obtain

$$\frac{x-3}{6} = \frac{y+1}{-2} = \frac{z-2}{-4}$$

or, equivalently,

$$\frac{x-3}{3} = \frac{y+1}{-1} = \frac{z-2}{-2}$$

which are symmetric equations of the normal line to the surface at the point $(3, -1, 2)$.

8. $z = x^{1/2} + y^{1/2}$; $(1, 1, 2)$

SOLUTION: Let $F$ be the function defined by

$$F(x, y, z) = x^{1/2} + y^{1/2} - z$$

Thus, the given surface is the graph of $F(x, y, z) = 0$. We have

$$F_x(x, y, z) = \frac{1}{2}x^{-1/2} \qquad F_y(x, y, z) = \frac{1}{2}y^{-1/2} \qquad F_z(x, y, z) = -1$$

Thus,

$$F_x(1, 1, 2) = \frac{1}{2} \qquad F_y(1, 1, 2) = \frac{1}{2} \qquad F_z(1, 1, 2) = -1$$

Therefore, an equation of the tangent plane to the surface at $(1, 1, 2)$ is given by

$$\frac{1}{2}(x-1) + \frac{1}{2}(y-1) - (z-2) = 0$$

$$x + y - 2z + 2 = 0$$

Because $[1, 1, -2]$ is a set of direction numbers for the normal line to the surface at the point $(1, 1, 2)$, symmetric equations for the line are

$$\frac{x-1}{1} = \frac{y-1}{1} = \frac{z-2}{-2}$$

12. $x^{1/2} + z^{1/2} = 8$; $(25, 2, 9)$

SOLUTION: Let $F$ be the function defined by

$$F(x, y, z) = x^{1/2} + z^{1/2} - 8 = 0$$

The given surface is the graph of $F(x, y, z) = 0$. We have

$$F_x(25, 2, 9) = \frac{1}{10} \qquad F_y(25, 2, 9) = 0 \qquad F_z(25, 2, 9) = \frac{1}{6}$$

Therefore, an equation of the tangent plane to the surface at the point $(25, 2, 9)$ is

$$\frac{1}{10}(x-25) + 0(y-2) + \frac{1}{6}(z-9) = 0$$

$$3x + 5z - 120 = 0$$

Because a set of direction numbers for the line normal to the surface at the point $(25, 2, 9)$ is $[3, 0, 5]$, symmetric equations of the line are

$$\frac{x-25}{3} = \frac{z-9}{5} \qquad y = 2$$

In Exercises 13-18, if the two given surfaces intersect in a curve, find equations of the tangent line to the curve of intersection at the given point; if the two given surfaces are tangent at the given point, prove it.

**16.** $x = 2 + \cos \pi yz,\ y = 1 + \sin \pi xz;\quad (3, 1, 2)$

SOLUTION: Let $F$ and $G$ be the functions defined by

$$F(x, y, z) = 2 + \cos \pi yz - x$$
$$G(x, y, z) = 1 + \sin \pi xz - y$$

We have

$$\nabla F(x, y, z) = -\mathbf{i} - \pi z \sin \pi yz\,\mathbf{j} - \pi y \sin \pi yz\,\mathbf{k}$$
$$\nabla G(x, y, z) = \pi z \cos \pi xz\,\mathbf{i} - \mathbf{j} + \pi x \cos \pi xz\,\mathbf{k}$$

We take

$$\mathbf{N}_1 = \nabla F(3, 1, 2) = -\mathbf{i}$$
$$\mathbf{N}_2 = \nabla G(3, 1, 2) = 2\pi\mathbf{i} - \mathbf{j} + 3\pi\mathbf{k}$$

Because $\mathbf{N}_1$ is a normal vector to the surface $F(x, y, z) = 0$ at the point $(3, 1, 2)$ and $\mathbf{N}_2$ is a normal vector to the surface $G(x, y, z) = 0$ at the point $(3, 1, 2)$, then $\mathbf{N}_1 \times \mathbf{N}_2$ is a vector whose representations are parallel to the line of intersection of the two tangent planes. And this line is the tangent line to the curve of intersection of the two surfaces. We have

$$\mathbf{N}_1 \times \mathbf{N}_2 = (-\mathbf{i}) \times (2\pi\mathbf{i} - \mathbf{j} + 3\pi\mathbf{k})$$
$$= 3\pi\mathbf{j} + \mathbf{k}$$

Therefore, a set of direction numbers for the tangent line to the curve of intersection at the point $(3, 1, 2)$ is $[0, 3\pi, 1]$, and symmetric equations of the line are

$$x = 3 \qquad \frac{y-1}{3\pi} = \frac{z-2}{1}$$

**18.** $x^2 + y^2 + z^2 = 8,\ yz = 4;\quad (0, 2, 2)$

SOLUTION: Let $F$ and $G$ be the functions defined by

$$F(x, y, z) = x^2 + y^2 + z^2 - 8 \text{ and } G(x, y, z) = yz - 4$$

We have

$$\nabla F(x, y, z) = 2x\mathbf{i} + 2y\mathbf{j} + 2z\mathbf{k}$$
$$\nabla G(x, y, z) = z\mathbf{j} + y\mathbf{k}$$

We take

$$\mathbf{N}_1 = \nabla F(0, 2, 2) = 4\mathbf{j} + 4\mathbf{k}$$
$$\mathbf{N}_2 = \nabla G(0, 2, 2) = 2\mathbf{j} + 2\mathbf{k}$$

Because $\mathbf{N}_1$ is a normal vector to the surface $F(x, y, z) = 0$ at the point $(0, 2, 2)$ and $\mathbf{N}_2$ is a normal vector to the surface $G(x, y, z) = 0$ at the point $(0, 2, 2)$ and $\mathbf{N}_1$ is parallel to $\mathbf{N}_2$, we conclude that the two surfaces have a common tangent plane and are thus tangent surfaces at the point $(0, 2, 2)$.

## 20.3 EXTREMA OF FUNCTIONS OF TWO VARIABLES

**20.3.1 Definition** The function $f$ of two variables is said to have an *absolute maximum value* on a disk $B$ in the $xy$ plane if there is some point $(x_0, y_0)$ in $B$ such that $f(x_0, y_0) \geq f(x, y)$ for all points $(x, y)$ in $B$. In such a case, $f(x_0, y_0)$ is the absolute maximum value of $f$ on $B$.

**20.3.2 Definition** The function $f$ of two variables is said to have an *absolute minimum value* on a disk $B$ in the $xy$ plane if there is some point $(x_0, y_0)$ in $B$ such that $f(x_0, y_0) \leqslant f(x, y)$ for all points $(x, y)$ in $B$. In such a case, $f(x_0, y_0)$ is the absolute minimum value of $f$ on $B$.

**20.3.3 Theorem** (*The Extreme-Value Theorem for Functions of Two Variables*) Let $B$ be a closed disk in the $xy$ plane, and let $f$ be a function of two variables that is continuous on $B$. Then there is at least one point in $B$ where $f$ has an absolute maximum value and at least one point in $B$ where $f$ has an absolute minimum value.

**20.3.4 Definition** The function $f$ of two variables is said to have a *relative maximum value* at the point $(x_0, y_0)$ if there exists an open disk $B((x_0, y_0); r)$ such that $f(x, y) \leqslant f(x_0, y_0)$ for all $(x, y)$ in the open disk.

**20.3.5 Definition** The function $f$ of two variables is said to have a *relative minimum value* at the point $(x_0, y_0)$ if there exists an open disk $B((x_0, y_0); r)$ such that $f(x, y) \geqslant f(x_0, y_0)$ for all $(x, y)$ in the open disk.

**20.3.6 Theorem** If $f(x, y)$ exists at all points in some open disk $B((x_0, y_0); r)$ and if $f$ has a relative extremum at $(x_0, y_0)$, then if $f_x(x_0, y_0)$ and $f_y(x_0, y_0)$ exist,

$$f_x(x_0, y_0) = f_y(x_0, y_0) = 0$$

**20.3.7 Definition** A point $(x_0, y_0)$ for which both $f_x(x_0, y_0) = 0$ and $f_y(x_0, y_0) = 0$ is called a *critical point.*

**20.3.8 Theorem** (*Second-Derivative Test*) Let $f$ be a function of two variables such that $f$ and its first- and second-order partial derivatives are continuous on some open disk $B((a, b); r)$. Suppose further that $f_x(a, b) = f_y(a, b) = 0$. Then

**(i)** $f$ has a relative minimum value at $(a, b)$ if

$$f_{xx}(a, b) f_{yy}(a, b) - f_{xy}{}^2(a, b) > 0 \quad \text{and} \quad f_{xx}(a, b) > 0$$

**(ii)** $f$ has a relative maximum value at $(a, b)$ if

$$f_{xx}(a, b) f_{yy}(a, b) - f_{xy}{}^2(a, b) > 0 \quad \text{and} \quad f_{xx}(a, b) < 0$$

**(iii)** $f(a, b)$ is not a relative extremum if

$$f_{xx}(a, b) f_{yy}(a, b) - f_{xy}{}^2(a, b) < 0$$

**(iv)** We can make no conclusion if

$$f_{xx}(a, b) f_{yy}(a, b) - f_{xy}{}^2(a, b) = 0$$

If $f$ is a function of $x$, $y$, and $z$, then the critical numbers of $f$, subject to the constraint $g(x, y, z) = 0$, are the same as the critical numbers of the function $F$, defined as follows.

$$F(x, y, z, \lambda) = f(x, y, z) + \lambda g(x, y, z)$$

If we use $F$ to find the critical numbers of $f$, we are using the method of *Lagrange multipliers.* The method may be extended to include more than one constraint. Thus, the critical numbers of $f$ subject to the constraints $g(x, y, z) = 0$ and $h(x, y, z) = 0$ are the same as the critical numbers of $G$, defined as follows.

$$G(x, y, z, \lambda, \mu) = f(x, y, z) + \lambda g(x, y, z) + \mu h(x, y, z)$$

*Exercises 20.3*

**6.** Determine the relative extrema of $f$, if there are any.

$$f(x, y) = \frac{2x + 2y + 1}{x^2 + y^2 + 1}$$

SOLUTION: We have

$$f_x(x, y) = \frac{(x^2 + y^2 + 1)2 - (2x + 2y + 1)(2x)}{(x^2 + y^2 + 1)^2}$$

$$= -2\left[\frac{x^2 + 2xy - y^2 + x - 1}{(x^2 + y^2 + 1)^2}\right] \tag{1}$$

and

$$f_y(x, y) = \frac{(x^2 + y^2 + 1)(2) - (2x + 2y + 1)(2y)}{(x^2 + y^2 + 1)^2}$$

$$= 2\left[\frac{x^2 - 2xy - y^2 - y + 1}{(x^2 + y^2 + 1)^2}\right] \tag{2}$$

If $f_x(x, y) = 0$, then from (1) we obtain

$$x^2 + 2xy - y^2 + x - 1 = 0 \tag{3}$$

If $f_y(x, y) = 0$, then from (2) we obtain

$$x^2 - 2xy - y^2 - y + 1 = 0 \tag{4}$$

Adding the members of (3) and (4), we get

$$2x^2 - 2y^2 + x - y = 0$$
$$2(x + y)(x - y) + (x - y) = 0$$
$$(x - y)[2(x + y) + 1] = 0$$

Thus, by setting each factor equal to zero

$$y = x \quad \text{or} \quad y = -x - \frac{1}{2}$$

Setting $y = x$ in (3), we get

$$2x^2 + x - 1 = 0$$
$$(x + 1)(2x - 1) = 0$$

$$x = -1 \quad \text{or} \quad x = \frac{1}{2}$$

$$y = -1 \qquad y = \frac{1}{2}$$

Thus, $(-1, -1)$ and $(\frac{1}{2}, \frac{1}{2})$ are critical points.

Setting $y = -x - \frac{1}{2}$ in (3), we obtain

$$x^2 - 2x^2 - x - x^2 - x - \frac{1}{4} + x - 1 = 0$$

$$8x^2 + 4x + 5 = 0$$

$$x = \frac{-4 \pm \sqrt{-144}}{16}$$

Because there are no real solutions, we find no additional critical points. Next, we test each of the critical points we have found. Partial-differentiating with respect to $x$ on both sides of (1), we obtain

$$f_{xx}(x, y) = -2\left[\frac{(x^2+y^2+1)^2(2x+2y+1) - (x^2+2xy-y^2+x-1)(2)(x^2+y^2+1)(2x)}{(x^2+y^2+1)^4}\right]$$

$$= -2\left[\frac{(x^2+y^2+1)(2x+2y+1) - 4x(x^2+2xy-y^2+x-1)}{(x^2+y^2+1)^3}\right] \qquad (6)$$

Partial-differentiating with respect to $y$ on both sides of (2), we obtain

$$f_{yy}(x, y) = 2\left[\frac{(x^2+y^2+1)^2(-2x-2y-1) - (x^2-2xy-y^2-y+1)(2)(x^2+y^2+1)(2y)}{(x^2+y^2+1)^4}\right]$$

$$= -2\left[\frac{(x^2+y^2+1)(2x+2y+1) + 4y(x^2-2xy-y^2-y+1)}{(x^2+y^2+1)^3}\right] \qquad (7)$$

Partial-differentiating on both sides of (1) with respect to $y$, we obtain

$$f_{xy}(x, y) = -2\left[\frac{(x^2+y^2+1)^2(2x-2y) - (x^2+2xy-y^2+x-1)(2)(x^2+y^2+1)(2y)}{(x^2+y^2+1)^4}\right]$$

$$= -4\left[\frac{(x^2+y^2+1)(x-y) - 2y(x^2+2xy-y^2+x-1)}{(x^2+y^2+1)^3}\right] \qquad (8)$$

We test the critical point $(-1, -1)$. From (6), (7), and (8), we calculate

$$f_{xx}(-1, -1) = \frac{2}{3} \qquad f_{yy}(-1, -1) = \frac{2}{3} \qquad f_{xy}(-1, -1) = 0$$

Therefore,

$$f_{xx}(-1, -1) > 0$$

and

$$f_{xx}(-1, -1)f_{yy}(-1, -1) - f_{xy}^2(-1, -1) = \frac{2}{3}\cdot\frac{2}{3} - 0 = \frac{4}{9} > 0$$

By Theorem 20.3.8(i), we conclude that $f$ has a relative minimum value at $(-1, -1)$. Because $f(-1, -1) = -1$, the relative minimum value of $f$ is $-1$. Next, we test the critical point $(\frac{1}{2}, \frac{1}{2})$. From (6), (7), and (8) we calculate

$$f_{xx}\left(\frac{1}{2}, \frac{1}{2}\right) = -\frac{8}{3} \qquad f_{yy}\left(\frac{1}{2}, \frac{1}{2}\right) = -\frac{8}{3} \qquad f_{xy}\left(\frac{1}{2}, \frac{1}{2}\right) = 0$$

Therefore,

$$f_{xx}\left(\frac{1}{2}, \frac{1}{2}\right) < 0$$

and

$$f_{xx}\left(\frac{1}{2}, \frac{1}{2}\right) f_{yy}\left(\frac{1}{2}, \frac{1}{2}\right) - f_{xy}^2\left(\frac{1}{2}, \frac{1}{2}\right) = \left(-\frac{8}{3}\right)\left(-\frac{8}{3}\right) - 0$$

$$= \frac{64}{9} > 0$$

By Theorem 20.3.8(ii), we conclude that $f$ has a relative maximum value at $(\frac{1}{2}, \frac{1}{2})$. Because $f(\frac{1}{2}, \frac{1}{2}) = 2$, the relative maximum value of $f$ is 2.

In Exercises 7-12, use the method of Lagrange multipliers to find the critical points of the given function subject to the indicated constraint.

8. $f(x, y) = x^2 + xy + 2y^2 - 2x$ with constraint $x - 2y + 1 = 0$

SOLUTION: Let $g$ be the function defined by $g(x, y) = x - 2y + 1$. Let $F$ be the function defined by

$$F(x, y, \lambda) = f(x, y) + \lambda g(x, y)$$
$$= x^2 + xy + 2y^2 - 2x + \lambda(x - 2y + 1)$$

We find the critical points of $F$. Thus, we take

$$F_x(x, y, \lambda) = 2x + y - 2 + \lambda = 0 \qquad (1)$$
$$F_y(x, y, \lambda) = x + 4y - 2\lambda = 0 \qquad (2)$$
$$F_\lambda(x, y, \lambda) = x - 2y + 1 = 0 \qquad (3)$$

We solve the system of Eqs. (1), (2), and (3).

Solving Eq. (1) for $\lambda$ and substituting this value for $\lambda$ in Eq. (2), we obtain

$$5x + 6y - 4 = 0 \qquad (4)$$

Multiplying on both sides of Eq. (3) by 3 and adding the resulting members to the members of Eq. (4), we get

$$8x - 1 = 0$$
$$x = \frac{1}{8}$$

Substituting in (4), we obtain $y = \frac{9}{16}$. Thus, $(\frac{1}{8}, \frac{9}{16})$ is the only critical point.

**12.** $f(x, y, z) = x^2 + y^2 + z^2$ with constraint $y^2 - x^2 = 1$

SOLUTION: Let $g(x, y) = x^2 - y^2 + 1$. Let $F$ be the function defined by

$$F(x, y, z, \lambda) = f(x, y, z) + \lambda g(x, y)$$
$$= x^2 + y^2 + z^2 + \lambda x^2 - \lambda y^2 + \lambda$$

We take

$$F_x(x, y, z, \lambda) = 2x + 2\lambda x = 0 \qquad (1)$$
$$F_y(x, y, z, \lambda) = 2y - 2\lambda y = 0 \qquad (2)$$
$$F_z(x, y, z, \lambda) = 2z = 0 \qquad (3)$$
$$F_\lambda(x, y, z, \lambda) = x^2 - y^2 + 1 = 0 \qquad (4)$$

From (3) we have $z = 0$. From (2) we have either $\lambda = 1$ or $y = 0$. We reject $y = 0$ because then Eq. (4) would require that $x^2 + 1 = 0$ which is impossible. If $\lambda = 1$ in Eq. (1), we get $x = 0$. Substituting $x = 0$ in Eq. (4), we have $y = \pm 1$. Therefore, $(0, 1, 0)$ and $(0, -1, 0)$ are the critical points.

**18.** Prove that the box having the largest volume that can be placed inside a sphere is in the shape of a cube.

SOLUTION: Let $a$ units be the diameter of the sphere, where $a$ is a constant. We assume that the box is rectangular and that the box with largest volume is a box that is inscribed in the sphere. If $x, y$, and $z$ are the number of units in the lengths of the sides of the box and $V$ is the number of cubic units in the volume of the box, then $V = xyz$. Because the box is inscribed in the sphere, a diagonal of the box is a diameter of the sphere. Thus, $x^2 + y^2 + z^2 = a^2$. We want to find the maximum value of $V$ subject to the constraint $x^2 + y^2 + z^2 - a^2 = 0$. Thus, we let $F$ be the function defined by

$$F(x, y, z, \lambda) = xyz + \lambda(x^2 + y^2 + z^2 - a^2)$$

We take

$$F_x(x, y, z, \lambda) = yz + 2x\lambda = 0 \qquad (1)$$
$$F_y(x, y, z, \lambda) = xz + 2y\lambda = 0 \qquad (2)$$

$$F_z(x, y, z, \lambda) = xy + 2z\lambda = 0 \tag{3}$$

$$F_\lambda(x, y, z, \lambda) = x^2 + y^2 + z^2 - a^2 = 0 \tag{4}$$

Because $x > 0, y > 0$, and $z > 0$, we may solve Eqs. (1) and (2) for $\lambda$. Thus,

$$\lambda = -\frac{yz}{2x} \quad \text{and} \quad \lambda = -\frac{xz}{2y}$$

which gives

$$-\frac{yz}{2x} = -\frac{xz}{2y}$$

$$y^2 = x^2$$

$$y = x$$

Similarly, if we eliminate $\lambda$ from Eqs. (2) and (3), the result is

$$z^2 = y^2$$

$$z = y$$

Therefore, the only critical points occur when $x = y = z$, and thus, $V$ has an absolute maximum value when the box is a cube.

22. Find the points on the surface $y^2 - xz = 4$ that are closest to the origin and find the minimum distance.

SOLUTION: Let $F$ be the number of units in the distance between the origin and any point $(x, y, z)$ on the surface $y^2 - xz = 4$. We have $F = \sqrt{x^2 + y^2 + z^2}$ and $F > 0$ because the surface $y^2 - xz = 4$ does not contain the origin. Furthermore, because $y^2 = xz + 4$, then

$$F = \sqrt{x^2 + xz + 4 + z^2}$$

$$F^2 = x^2 + z^2 + xz + 4$$

Partial-differentiating on both sides, we obtain

$$2F\frac{\partial F}{\partial x} = 2x + z \tag{1}$$

$$2F\frac{\partial F}{\partial z} = 2z + x \tag{2}$$

Because $F \neq 0$, if $\partial F/\partial x = 0$ and $\partial F/\partial z = 0$, then

$$2x + z = 0 \quad \text{and} \quad 2z + x = 0$$

Therefore, $x = 0$ and $z = 0$ is a critical point of the function, and when $x = z = 0$, we have $F = 2$. We show that 2 is the absolute minimum value of $F$. Partial-differentiating with respect to $x$ on both sides of (1), we obtain

$$2F\frac{\partial^2 F}{\partial x^2} + \frac{\partial F}{\partial x} \cdot 2\frac{\partial F}{\partial x} = 2$$

Because $\partial F/\partial x = 0$ and $F = 2$ at the critical point, we have

$$\frac{\partial^2 F}{\partial x^2} = \frac{1}{2} \tag{3}$$

at the critical point. Partial-differentiating with respect to $z$ on both sides of (2), we obtain

$$2F\frac{\partial^2 F}{\partial z^2} + 2\frac{\partial F}{\partial z} \cdot \frac{\partial F}{\partial z} = 2$$

Thus,

$$\frac{\partial^2 F}{\partial z^2} = \frac{1}{2} \tag{4}$$

at the critical point. Partial-differentiating on both sides of (1) with respect to $z$, we obtain

$$2F \frac{\partial^2 F}{\partial z \partial x} + 2 \frac{\partial F}{\partial z} \cdot \frac{\partial F}{\partial x} = 1$$

Thus,

$$\frac{\partial^2 F}{\partial z \partial x} = \frac{1}{4} \tag{5}$$

at the critical point. From (3), (4), and (5) we have

$$\frac{\partial^2 F}{\partial x^2} \cdot \frac{\partial^2 F}{\partial z^2} - \left(\frac{\partial^2 F}{\partial z \partial x}\right)^2 = \frac{1}{2} \cdot \frac{1}{2} - \left(\frac{1}{4}\right)^2 = \frac{3}{16} > 0$$

Thus, by Theorem 20.3.8 the absolute minimum value of $F$ occurs at the critical point. Furthermore, if $x = 0$ and $z = 0$, we have $y^2 = 4$. Thus, the minimum distance of 2 units occurs at the points $(0, 2, 0)$ and $(0, -2, 0)$.

**26.** Suppose that $T$ degrees is the temperature at any point $(x, y, z)$ on the sphere $x^2 + y^2 + z^2 = 4$ and $T = 100xy^2z$. Find the points on the sphere where the temperature is the greatest and also the points where the temperature is the least. Also find the temperature at these points.

SOLUTION: Because $y^2 = 4 - x^2 - z^2$, $T$ is a function of $x$ and $z$ and

$$T(x, z) = 100xz(4 - x^2 - z^2) \tag{1}$$

Thus,

$$\begin{aligned} T_x(x, z) &= 100[xz(-2x) + z(4 - x^2 - z^2)] \\ &= -100z(3x^2 + z^2 - 4) \end{aligned}$$

and

$$\begin{aligned} T_z(x, z) &= 100[xz(-2z) + x(4 - x^2 - z^2)] \\ &= -100x(3z^2 + x^2 - 4) \end{aligned}$$

If $T_x(x, z) = 0$ and $T_z(x, z) = 0$, we have

$$z(3x^2 + z^2 - 4) = 0 \tag{2}$$

$$x(3z^2 + x^2 - 4) = 0 \tag{3}$$

If $z = 0$, then from Eq. (3) we get $x = 0$ or $x = \pm 2$.
If $x = 0$, then from Eq. (2) we get $z = 0$ or $z = \pm 2$.
If $x \neq 0$ and $z \neq 0$, then we have

$$3x^2 + z^2 - 4 = 0 \quad \text{and} \quad 3z^2 + x^2 - 4 = 0 \tag{4}$$

The solutions of Eqs. (4) are $x = \pm 1$ and $z = \pm 1$. We use (1) to find the value of $T$ at each critical point of $T$. Thus,

$$T(0, 2) = 0 \qquad T(0, -2) = 0 \qquad T(2, 0) = 0 \qquad T(-2, 0) = 0$$

and

$$T(1, 1) = T(-1, -1) = 200$$

$$T(1, -1) = T(-1, 1) = -200$$

Therefore, 200 degrees is the greatest temperature, and this temperature occurs at the points $(1, \pm\sqrt{2}, 1)$ and $(-1, \pm\sqrt{2}, -1)$. The least temperature is $-200$ degrees, which occurs at the points $(1, \pm\sqrt{2}, -1)$ and $(-1, \pm\sqrt{2}, 1)$.

## 20.4 SOME APPLICATIONS OF PARTIAL DERIVATIVES TO ECONOMICS

**20.4.1 Definition** Let $x$ units of a first commodity and $y$ units of a second commodity be demanded when the unit prices are $p$ dollars and $q$ dollars, respectively. If $f$ and $g$ are the respective demand functions of the commodities so that

$$x = f(p, q) \quad \text{and} \quad y = g(p, q)$$

then

(i) $\dfrac{\partial x}{\partial p}$ gives the (partial) *marginal demand of* $x$ *with respect to* $p$;

(ii) $\dfrac{\partial x}{\partial q}$ gives the (partial) *marginal demand of* $x$ *with respect to* $q$;

(iii) $\dfrac{\partial y}{\partial p}$ gives the (partial) *marginal demand of* $y$ *with respect to* $p$;

(iv) $\dfrac{\partial y}{\partial q}$ gives the (partial) *marginal demand of* $y$ *with respect to* $q$.

The two commodities in Definition 20.4.1 are *complementary* if and only if both $\partial x/\partial q$ and $\partial y/\partial p$ are negative, and the commodities are *substitutes* if and only if both $\partial x/\partial q$ and $\partial y/\partial p$ are positive.

If the cost of producing $x$ units of one commodity and $y$ units of another commodity is given by $C(x, y)$, then $C$ is called a *joint-cost function.* The partial derivatives of $C$ are called *marginal cost functions.* If a monopolist produces two related commodities whose demand equations are $x = f(p, q)$ and $y = g(p, q)$, and the joint-cost function is $C$, then if $S$ dollars is the profit, we have

$$S = px + qy - C(x, y)$$

## *Exercises 20.4*

In Exercises 1-9, demand equations for two related commodities are given. In each exercise, determine the four marginal demands. Determine if the commodities are complementary, substitutes, or neither. In Exercises 1-6, draw sketches of the two demand surfaces.

**4.** $x = 9 - 3p + q$ **(1)**

$y = 10 - 2p - 5q$ **(2)**

SOLUTION: The marginal demands are given by

$$\frac{\partial x}{\partial p} = -3 \qquad \frac{\partial x}{\partial q} = 1 \qquad \frac{\partial y}{\partial p} = -2 \qquad \frac{\partial y}{\partial q} = -5$$

Because $\partial x/\partial q > 0$ and $\partial y/\partial p < 0$, the commodities are neither complementary nor substitutes. To draw sketches of the demand surfaces, we first determine the permissible replacements of $p$ and $q$. Because $x \geqslant 0$ and $y \geqslant 0$, we have $9 - 3p + q \geqslant 0$ and $10 - 2p - 5q \geqslant 0$. Let $L_1$ be the line $9 - 3p + q = 0$ in the $pq$ plane and let $L_2$ be the line $10 - 2p - 5q = 0$ in the $pq$ plane. Then $L_1$ intersects the $p$-axis at the point $A$ where $p = 3$, and $L_2$ intersects the $q$-axis at the point $B$ where $q = 2$.

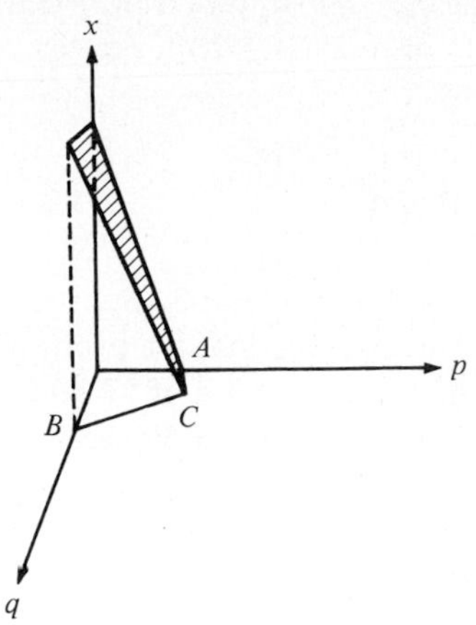


Figure 20.4.4(a)

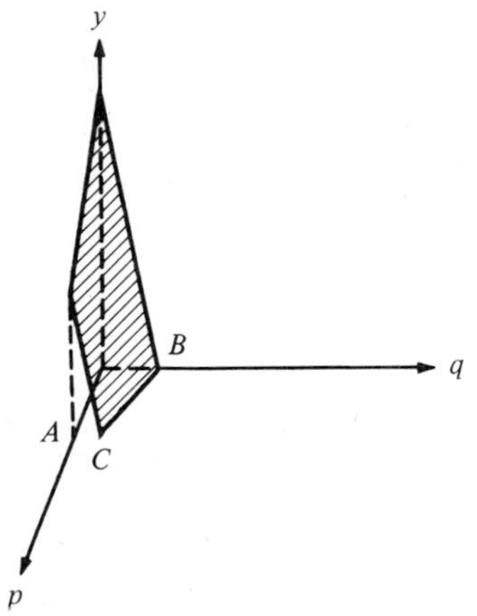


Figure 20.4.4(b)

Furthermore, $L_1$ intersects $L_2$ at point $C$ where $p = \frac{55}{17} = 3.24$ and $q = 0.71$. Because $p \geqslant 0$ and $q \geqslant 0$, then the domain of each demand function is the set of all points that lie either on the quadrilateral $OACB$ or in the interior of the quadrilateral. In Fig. 20.4.4(a) we show a sketch of the demand surface for the commodity having demand Eq. (1). In Fig. 20.4.4(b) we show a sketch of the demand surface for the commodity having demand Eq. (2).

**8.** $pqx = 4 \quad p^2qy = 16$

SOLUTION: Because

$$x = \frac{4}{pq} \quad \text{and} \quad y = \frac{16}{p^2q}$$

The marginal demands are given by

$$\frac{\partial x}{\partial p} = -\frac{4}{p^2q} \quad \frac{\partial x}{\partial q} = -\frac{4}{pq^2} \quad \frac{\partial y}{\partial p} = -\frac{32}{p^3q} \quad \frac{\partial y}{\partial q} = -\frac{16}{p^2q^2}$$

Because $\partial x/\partial q < 0$ and $\partial y/\partial p < 0$, then the commodities are complementary.

**14.** The demand equations for two commodities that are produced by a monopolist are

$$x = 6 - 2p + q \quad \text{and} \quad y = 7 + p - q$$

where $100x$ is the quantity of the first commodity demanded if the price is $p$ dollars per unit and $100y$ is the quantity of the second commodity demanded if the price is $q$ dollars per unit. If it costs \$2 to produce each unit of the first commodity and \$3 to produce each unit of the second commodity, find the quantities demanded and the prices of the two commodities in order to have the greatest profit. Take $x$ and $y$ as the independent variables.

SOLUTION: When $100x$ units of the first commodity and $100y$ units of the second commodity are produced and sold, the number of dollars in the total revenue is $100px + 100qy$, and the number of dollars in the total cost of production is $200x + 300y$. Hence, if $S$ dollars is the profit, we have

$$S = (100px + 100qy) - (200x + 300y) \tag{1}$$

We want to eliminate $p$ and $q$. Solving the demand equations for $p$ and $q$, we obtain

$$p = 13 - x - y \quad \text{and} \quad q = 20 - x - 2y \tag{2}$$

Substituting from Eqs. (2) into (1), we express $S$ as a function of $x$ and $y$. Thus,

$$\begin{aligned} S &= 100x(13 - x - y) + 100y(20 - x - 2y) - (200x + 300y) \\ &= -100x^2 - 200y^2 - 200xy + 1100x + 1700y \end{aligned}$$

and

$$\frac{\partial S}{\partial x} = -200x - 200y + 1100$$

$$\frac{\partial S}{\partial y} = -400y - 200x + 1700$$

Setting $\partial S/\partial x = 0$ and $\partial S/\partial y = 0$, we have

$$2x + 2y = 11 \quad \text{and} \quad 2x + 4y = 17$$

Thus, $x = 2.5$ and $y = 3$. Furthermore,

$$\frac{\partial^2 S}{\partial x^2} = -200 \qquad \frac{\partial^2 S}{\partial y^2} = -400 \qquad \frac{\partial^2 S}{\partial y \partial x} = -200$$

Thus, $\partial^2 S/\partial x^2 < 0$ and

$$\frac{\partial^2 S}{\partial x^2} \cdot \frac{\partial^2 S}{\partial y^2} - \left(\frac{\partial^2 S}{\partial y \partial x}\right)^2 = (-200)(-400) - (-200)^2 > 0$$

Therefore, $S$ has an absolute maximum value when $x = 2.5$ and $y = 3$. Substituting these values into (2), we get $p = 7.5$ and $q = 11.5$. We conclude that the profit is a maximum if 250 units of the first commodity are demanded at a price of \$7.50 each and 300 units of the second commodity are demanded at a price of \$11.50 each.

**16.** The production function $f$ for a certain commodity has function values

$$f(x, y) = x + \frac{5}{2}y - \frac{1}{8}x^2 - \frac{1}{4}y^2 - \frac{9}{8}$$

The amounts of the two inputs are given by $100x$ and $100y$, whose prices per unit are respectively, \$4 and \$8, and the amount of output is given by $100z$, whose price per unit is \$16. Determine the greatest profit by two methods: **(a)** without using Lagrange multipliers, and **(b)** using Lagrange multipliers.

SOLUTION:

**(a)** If $P$ dollars is the profit, then

$$\begin{aligned} P &= 16(100z) - 4(100x) - 8(100y) \\ &= 400[4z - x - 2y] \qquad (1) \end{aligned}$$

Because $z = f(x, y)$, we have

$$\begin{aligned} P &= 400\left[4\left(x + \frac{5}{2}y - \frac{1}{8}x^2 - \frac{1}{4}y^2 - \frac{9}{8}\right) - x - 2y\right] \\ &= 400\left[3x + 8y - \frac{1}{2}x^2 - y^2 - \frac{9}{2}\right] \qquad (2) \end{aligned}$$

Thus,

$$\frac{\partial P}{\partial x} = 400(3 - x) \quad \text{and} \quad \frac{\partial P}{\partial y} = 400(8 - 2y)$$

If $\partial P/\partial x = 0$ and $\partial P/\partial y = 0$, we have $x = 3$ and $y = 4$. Furthermore,

$$\frac{\partial^2 P}{\partial x^2} = -400 \qquad \frac{\partial^2 P}{\partial y^2} = -800 \qquad \frac{\partial^2 P}{\partial x \partial y} = 0$$

Because

$$\frac{\partial^2 P}{\partial x^2} \cdot \frac{\partial^2 P}{\partial y^2} - \left(\frac{\partial^2 P}{\partial x \partial y}\right)^2 = (-400)(-800) - 0 > 0$$

and $\partial^2 P/\partial x^2 < 0$, we conclude that the maximum value of $P$ occurs when $x = 3$ and $y = 4$. With $x = 3$ and $y = 4$ in Eq. (2), we have $P = 6400$. Therefore, the maximum profit is \$6400.

**(b)** We want to find the maximum value of $P$ which is defined by Eq. (1) subject to the constraint $z = f(x, y)$. We let $g$ be the function defined by

$$g(x, y, z) = f(x, y) - z = 0$$

and take $F$ to be the function defined by

$$F(x, y, z, \lambda) = P(x, y, z) + \lambda g(x, y, z)$$

$$= 400(4z - x - 2y) + \lambda\left(x + \frac{5}{2}y - \frac{1}{8}x^2 - \frac{1}{4}y^2 - \frac{9}{8} - z\right)$$

We have

$$F_x(x, y, z, \lambda) = -400 + \lambda\left(1 - \frac{1}{4}x\right) = 0 \tag{3}$$

$$F_y(x, y, z, \lambda) = -800 + \lambda\left(\frac{5}{2} - \frac{1}{2}y\right) = 0 \tag{4}$$

$$F_z(x, y, z, \lambda) = 1600 - \lambda = 0 \tag{5}$$

$$F_\lambda(x, y, z, \lambda) = x + \frac{5}{2}y - \frac{1}{8}x^2 - \frac{1}{4}y^2 - \frac{9}{8} - z = 0 \tag{6}$$

From (5) we get $\lambda = 1600$. Substituting into (3) and (4), we obtain $x = 3$ and $y = 4$. And from (6) we get $z = \frac{27}{4}$. Substituting for $x, y$, and $z$ into (1), we obtain $P = 6400$.

## 20.5 OBTAINING A FUNCTION FROM ITS GRADIENT

**20.5.1 Theorem** Suppose that $M$ and $N$ are functions of two variables $x$ and $y$ defined on an open disk $B((x_0, y_0); r)$ in $R^2$, and $M_y$ and $N_x$ are continuous on $B$. Then the vector

$$M(x, y)\mathbf{i} + N(x, y)\mathbf{j}$$

is a gradient on $B$ if and only if

$$M_y(x, y) = N_x(x, y)$$

at all points in $B$.

**20.5.2 Theorem** Let $M, N$, and $R$ be functions of three variables $x, y$, and $z$ defined on an open ball $B((x_0, y_0, z_0); r)$ in $R^3$, and $M_y, M_z, N_x, N_z, R_x$, and $R_y$ are continuous on $B$. Then the vector $M(x, y, z)\mathbf{i} + N(x, y, z)\mathbf{j} + R(x, y, z)\mathbf{k}$ is a gradient on $B$ if and only if $M_y(x, y, z) = N_x(x, y, z), M_z(x, y, z) = R_x(x, y, z)$, and $N_z(x, y, z) = R_y(x, y, z)$.

## *Exercises 20.5*

In the following exercises, determine if the vector is a gradient. If it is, find a function having the given gradient.

4. $(4y^2 + 6xy - 2)\mathbf{i} + (3x^2 + 8xy + 1)\mathbf{j}$

SOLUTION: We apply Theorem 20.5.1 with

$$M(x, y) = 4y^2 + 6xy - 2 \quad \text{and} \quad N(x, y) = 3x^2 + 8xy + 1$$

We have

$$M_y(x, y) = 8y + 6x \quad \text{and} \quad N_x(x, y) = 6x + 8y$$

Because $M_y(x, y) = N_x(x, y)$, the given vector is a gradient of the function $f$, where

$$f_x(x, y) = 4y^2 + 6xy - 2 \tag{1}$$

$$f_y(x, y) = 3x^2 + 8xy + 1 \tag{2}$$

Integrating both members of (1) with respect to $x$, we get

$$f(x, y) = 4xy^2 + 3x^2y - 2x + C(y) \tag{3}$$

where $C(y)$ is independent of $x$. Partial-differentiating both members of (3) with respect to $y$, we obtain

$$f_y(x, y) = 8xy + 3x^2 + C'(y) \tag{4}$$

Equating the right members of (2) and (4), we get

$$8xy + 3x^2 + C'(y) = 3x^2 + 8xy + 1$$
$$C'(y) = 1$$
$$C(y) = y + C$$

Substituting for $C(y)$ into Equation (3), we obtain

$$f(x, y) = 4xy^2 + 3x^2y - 2x + y + C$$

which is the desired function.

8. $\dfrac{2x-1}{y}\mathbf{i} + \dfrac{x-x^2}{y^2}\mathbf{j}$

SOLUTION: Let

$$M(x, y) = \frac{2x-1}{y} \quad \text{and} \quad N(x, y) = \frac{x-x^2}{y^2}$$

We have

$$M_y(x, y) = \frac{-2x+1}{y^2} \quad \text{and} \quad N_x(x, y) = \frac{1-2x}{y^2}$$

Because $M_y(x, y) = N_x(x, y)$, the given vector is the gradient of a function $f$ such that

$$f_x(x, y) = \frac{2x-1}{y} \tag{1}$$

$$f_y(x, y) = \frac{x-x^2}{y^2} \tag{2}$$

Integrating both members of (1) with respect to $x$, we get

$$f(x, y) = \frac{x^2-x}{y} + C(y) \tag{3}$$

Partial-differentiating with respect to $y$, we have

$$f_y(x, y) = \frac{x-x^2}{y^2} + C'(y) \tag{4}$$

From (2) and (4), we have

$$\frac{x-x^2}{y^2} + C'(y) = \frac{x-x^2}{y^2}$$

Thus,

$$C'(y) = 0$$
$$C(y) = C$$

From (3), we have

$$f(x, y) = \frac{x^2-x}{y} + C$$

12. $(\sin 2x - \tan y)\mathbf{i} - x\sec^2 y\,\mathbf{j}$

SOLUTION: Let $M(x, y) = \sin 2x - \tan y$ and $N(x, y) = -x\sec^2 y$. We have

$$M_y(x, y) = -\sec^2 y \quad \text{and} \quad N_x(x, y) = -\sec^2 y$$

Because $M_y(x, y) = N_x(x, y)$, the given vector is the gradient of a function $f$ such that

$$f_x(x, y) = \sin 2x - \tan y \tag{1}$$

$$f_y(x, y) = -x \sec^2 y \tag{2}$$

Integrating on both sides of (2) with respect to $y$, we get

$$f(x, y) = -x \tan y + C(x) \tag{3}$$

Partial-differentiating with respect to $x$ on both sides of (3), we obtain

$$f_x(x, y) = -\tan y + C'(x) \tag{4}$$

Substituting from (4) into (1), we obtain

$$-\tan y + C'(x) = \sin 2x - \tan y$$

$$C'(x) = \sin 2x$$

$$C(x) = -\frac{1}{2} \cos 2x + C$$

Substituting for $C(x)$ into (3), we obtain

$$f(x, y) = -x \tan y - \frac{1}{2} \cos 2x + C$$

**16.** $(2xy + 7z^3)\mathbf{i} + (x^2 + 2y^2 - 3z)\mathbf{j} + (21xz^2 - 4y)\mathbf{k}$

SOLUTION: We apply Theorem 20.5.2. Let

$$M(x, y, z) = 2xy + 7z^3$$

$$N(x, y, z) = x^2 + 2y^2 - 3z$$

$$R(x, y, z) = 21xz^2 - 4y$$

We have

$$M_y(x, y, z) = 2x \quad \text{and} \quad N_x(x, y, z) = 2x$$

$$M_z(x, y, z) = 21z^2 \quad \text{and} \quad R_x(x, y, z) = 21z^2$$

$$N_z(x, y, z) = -3 \quad \text{and} \quad R_y(x, y, z) = -4$$

Because $N_z(x, y, z) \neq R_y(x, y, z)$, the given vector is not a gradient.

**20.** $e^z \cos x\mathbf{i} + z \sin y\mathbf{j} + (e^z \sin x - \cos y)\mathbf{k}$

SOLUTION: Let

$$M(x, y, z) = e^z \cos x$$

$$N(x, y, z) = z \sin y$$

$$R(x, y, z) = e^z \sin x - \cos y$$

We have

$$M_y(x, y, z) = 0 \quad \text{and} \quad N_x(x, y, z) = 0$$

$$M_z(x, y, z) = e^z \cos x \quad \text{and} \quad R_x(x, y, z) = e^z \cos x$$

$$N_z(x, y, z) = \sin y \quad \text{and} \quad R_y(x, y, z) = \sin y$$

Because $M_y(x, y, z) = N_x(x, y, z)$, $M_z(x, y, z) = R_x(x, y, z)$, and $N_z(x, y, z) = R_y(x, y, z)$, then the given vector is the gradient of the function $f$ such that

$$f_x(x, y, z) = e^z \cos x \tag{1}$$

$$f_y(x, y, z) = z \sin y \tag{2}$$

$$f_z(x, y, z) = e^z \sin x - \cos y \tag{3}$$

Integrating on both sides of (1) with respect to $x$, we get

$$f(x, y, z) = e^z \sin x + g(y, z) \tag{4}$$

where $g(y, z)$ is independent of $x$. Partial-differentiating with respect to $y$ on both sides of (4), we obtain

$$f_y(x, y, z) = g_y(y, z) \tag{5}$$

From (2) and (5), we have

$$g_y(y, z) = z \sin y \tag{6}$$

Integrating on both sides of (6) with respect to $y$, we obtain

$$g(y, z) = -z \cos y + h(z) \tag{7}$$

where $h(z)$ is independent of $y$. Substituting from (7) into (4), we obtain

$$f(x, y, z) = e^z \sin x - z \cos y + h(z) \tag{8}$$

Partial-differentiating with respect to $z$ on both sides of (8), we get

$$f_z(x, y, z) = e^z \sin x - \cos y + h'(z) \tag{9}$$

Substituting for $f_z(x, y, z)$ from Eq. (9) into Eq. (3), we obtain

$$e^z \sin x - \cos y + h'(z) = e^z \sin x - \cos y$$

$$h'(z) = 0$$

Thus,

$$h(z) = C$$

Substituting for $h(z)$ into Eq. (8), we get

$$f(x, y, z) = e^z \sin x - z \cos y + C$$

## 20.6 LINE INTEGRALS

**20.6.2 Definition** Let $M$ and $N$ be functions of two variables $x$ and $y$ such that they are continuous on an open disk $B$ in $R^2$. Let $C$ be a curve lying in $B$ and having parametric equations

$$x = f(t) \qquad y = g(t) \qquad a \leqslant t \leqslant b$$

such that $f'$ and $g'$ are continuous on $[a, b]$. Then the *line integral* of $M(x, y)\, dx + N(x, y)\, dy$ over $C$ is given by

$$\int_C M(x, y)dx + N(x, y)dy = \int_a^b [M(f(t), g(t))f'(t) + N(f(t), g(t))g'(t)]\, dt$$

or, equivalently, by using vector notation,

$$\int_C M(x, y)dx + N(x, y)dy = \int_a^b \langle M(f(t), g(t)), N(f(t), g(t))\rangle \cdot \langle f'(t), g'(t)\rangle\, dt$$

**20.6.3 Definition** Let the curve $C$ consist of smooth arcs $C_1, C_2, \ldots, C_n$. Then the *line integral* of $M(x, y)dx + N(x, y)dy$ over $C$ is defined by the following equation:

$$\int_C M(x, y)dx + N(x, y)dy = \sum_{i=1}^{n} \left( \int_{C_1} M(x, y)dx + N(x, y)dy \right)$$

Line integrals are used to calculate the total work done by a variable force that causes a particle to move along a curve.

**20.6.1 Definition** Let $C$ be a curve lying in an open disk $B$ in $R^2$ for which a vector equation of $C$ is $\mathbf{R}(t) = f(t)\mathbf{i} + g(t)\mathbf{j}$, where $f'$ and $g'$ are continuous on $[a, b]$. Furthermore, let a force field on $B$ be defined by $\mathbf{F}(x, y) = M(x, y)\mathbf{i} + N(x, y)\mathbf{j}$, where $M$ and $N$ are continuous on $B$. Then if $W$ is the measure of the work done by $\mathbf{F}$ in moving an object along $C$ from $(f(a), g(a))$ to $(f(b), g(b))$,

$$W = \int_a^b [M(f(t), g(t))f'(t) + N(f(t), g(t))g'(t)]\, dt$$

or, equivalently, by using vector notation,

$$W = \int_a^b \langle M(f(t), g(t)), N(f(t), g(t))\rangle \cdot \langle f'(t), g'(t)\rangle\, dt$$

or, equivalently,

$$W = \int_a^b \mathbf{F}(f(t), g(t)) \cdot \mathbf{R}'(t)\, dt$$

The following definition extends the concept of a line integral to functions of three variables.

**20.6.4 Definition** Let $M, N$, and $R$ be functions of three variables, $x, y$, and $z$ such that they are continuous on an open disc $B$ in $R^3$. Let $C$ be a curve, lying in $B$, and having parametric equations

$$x = f(t) \qquad y = g(t) \qquad z = h(t) \qquad a \leqslant t \leqslant b$$

such that $f', g'$, and $h'$ are continuous on $[a, b]$. Then the *line integral* of $M(x, y, z)dx + N(x, y, z)dy + R(x, y, z)dz$ over $C$ is given by

$$\int_C M(x, y, z)dx + N(x, y, z)dy + R(x, y, z)dz$$

$$= \int_a^b [M(f(t), g(t), h(t))f'(t) + N(f(t), g(t), h(t))g'(t) + R(f(t), g(t), h(t))h'(t)]\, dt$$

or, equivalently, by using vector notation

$$\int_C M(x, y, z)dx + N(x, y, z)dy + N(x, y, z)dz$$

$$= \int_a^b \langle M(f(t), g(t), h(t)), N(f(t), g(t), h(t)), R(f(t), g(t), h(t))\rangle \cdot \langle f'(t), g'(t), h'(t)\rangle dt$$

*Exercises 20.6*

In Exercises 1-20, evaluate the line integral over the given curve.

**4.** $\displaystyle\int_C yx^2dx + (x+y)dy$

$C$: the line $y = -x$ from the origin to the point $(1, -1)$

SOLUTION: We let $x$ be the parameter, and replace $y$ by $-x$ and $dy$ by $-dx$ in the line integral, with $0 \leqslant x \leqslant 1$. Thus, we have

$$\int_C yx^2dx + (x+y)dy = \int_0^1 -x^3dx + (x-x)(-dx)$$
$$= -\int_0^1 x^3dx$$
$$= -\frac{1}{4}x^4\Big]_0^1$$
$$= -\frac{1}{4}$$

**6.** The line integral of Exercise 4; $C$: the $y$-axis from the origin to $(0, -1)$; and the line $y = -1$ from $(0, -1)$ to $(1, -1)$

SOLUTION: We apply Definition 20.6.3, with $C_1$ the $y$-axis from the origin to $(0, -1)$ and $C_2$ the line $y = -1$ from $(0, -1)$ to $(1, -1)$. Thus

$$\int_C yx^2\,dx + (x+y)dy = \int_{C_1} yx^2dx + (x+y)dy + \int_{C_2} yx^2dx + (x+y)dy \qquad (1)$$

Parametric equations of $C_1$ are

$x = 0$ and $y = -t \qquad 0 \leqslant t \leqslant 1$

Thus, $dx = 0$ and $dy = -dt$. We have

$$\int_{C_1} yx^2dx + (x+y)dy = \int_0^1 (-t)(-dt)$$
$$= \frac{1}{2}t^2\Big]_0^1$$
$$= \frac{1}{2} \qquad (2)$$

For curve $C_2$ we let $x$ be the parameter. We have $x = x$ and $y = -1$, and $0 \leqslant x \leqslant 1$. Thus, $dy = 0$, and hence,

$$\int_{C_2} yx^2\,dx + (x+y)dy = \int_0^1 -x^2dx$$
$$= -\frac{1}{3}x^3\Big]_0^1$$

$$= -\frac{1}{3} \tag{3}$$

Substituting from (2) and (3) into (1), we have

$$\int_C yx^2 dx + (x+y)dy = \frac{1}{2} - \frac{1}{3} = \frac{1}{6}$$

**10.** $\displaystyle\int_C xydx - y^2 dy$

$C$: $\mathbf{R}(t) = t^2\mathbf{i} + t^3\mathbf{j}$ from the point $(1, 1)$ to the point $(4, -8)$

SOLUTION: Parametric equations of the curve $C$ are

$$x = t^2 \qquad y = t^3 \qquad \text{from } t = 1 \text{ to } t = -2$$

or equivalently,

$$x = t^2 \qquad y = -t^3 \qquad \text{for } -1 \leqslant t \leqslant 2$$

We have $dx = 2tdt$ and $dy = -3t^2 dt$. Thus,

$$\int_C xydx - y^2dy = \int_{-1}^{2} t^2(-t^3)(2tdt) - (-t^3)^2(-3t^2)dt$$

$$= \int_{-1}^{2} (-2t^6 + 3t^8)dt$$

$$= -\frac{2}{7}t^7 + \frac{1}{3}t^9\Big]_{-1}^{2}$$

$$= -\frac{258}{7} + 171$$

$$= \frac{939}{7}$$

**14.** $\displaystyle\int_C (xy - z)dx + e^x dy + ydz$

$C$: the line segment from $(1, 0, 0)$ to $(3, 4, 8)$

SOLUTION: A set of direction numbers for the line segment is $[1, 2, 4]$. Thus, parametric equations of $C$ are

$$x = 1 + t \qquad y = 2t \qquad z = 4t \qquad \text{for } 0 \leqslant t \leqslant 2$$

Because $dx = dt$, $dy = 2dt$, and $dz = 4dt$, we have

$$\int_C (xy - z)dx + e^x dy + ydz = \int_0^2 [(1+t)(2t) - 4t]\, dt + e^{1+t}(2dt) + (2t)(4dt)$$

$$= \int_0^2 (6t + 2t^2 + 2e^{t+1})dt$$

$$= 3t^2 + \frac{2}{3}t^3 + 2e^{t+1}\Big]_0^2$$

$$= \frac{52}{3} + 2(e^3 - e)$$

**18.** $\int_C 2xydx + (6y^2 - xz)dy + 10zdz$

$C$: the twisted cubic $\mathbf{R}(t) = t\mathbf{i} + t^2\mathbf{j} + t^3\mathbf{k}, 0 \leq t \leq 1$

SOLUTION: Parametric equations of $C$ are

$$x = t \quad y = t^2 \quad z = t^3 \quad \text{for } 0 \leq t \leq 1$$

Thus,

$$\int_C 2xydx + (6y^2 - xz)dy + 10zdz = \int_0^1 2t^3dt + (6t^4 - t^4)(2tdt) + 10t^3(3t^2)dt$$

$$= \int_0^1 (2t^3 + 40t^5)dt$$

$$= \frac{1}{2}t^4 + \frac{20}{3}t^6\bigg]_0^1$$

$$= \frac{43}{6}$$

In Exercises 21-34, find the total work done in moving an object along the given arc $C$ if the motion is caused by the given force field. Assume the arc is measured in inches and the force is measured in pounds.

**22.** $\mathbf{F}(x, y) = 2xy\mathbf{i} + (x^2 + y^2)\mathbf{j}$

$C$: the arc of the parabola $y^2 = x$ from the origin to the point $(1, 1)$

SOLUTION: A vector equation of $C$ is

$$\mathbf{R}(t) = t^2\mathbf{i} + t\mathbf{j} \quad \text{for } 0 \leq t \leq 1$$

We apply Definition 20.6.1. Thus, we have

$$f(t) = t^2 \text{ and } g(t) = t$$

Therefore,

$$\mathbf{F}(f(t), g(t)) = \mathbf{F}(t^2, t) = 2t^3\mathbf{i} + (t^4 + t^2)\mathbf{j}$$

and

$$\mathbf{R}'(t) = 2t\mathbf{i} + \mathbf{j}$$

Thus,

$$\begin{aligned} \mathbf{F}(f(t), g(t)) \cdot \mathbf{R}'(t) &= [2t^3\mathbf{i} + (t^4 + t^2)\mathbf{j}] \cdot [2t\mathbf{i} + \mathbf{j}] \\ &= 4t^4 + (t^4 + t^2) \\ &= 5t^4 + t^2 \end{aligned}$$

Therefore

$$W = \int_0^1 (5t^4 + t^2)dt$$

$$= \left[t^5 + \frac{1}{3}t^3\right]_0^1$$

$$= \frac{4}{3}$$

Thus, the total work done is $\frac{4}{3}$ inch-pounds.

26. $\mathbf{F}(x, y) = -x^2y\mathbf{i} + 2y\mathbf{j}$

$C$: the line segment from $(a, 0)$ to $(0, a)$

SOLUTION: A Cartesian equation of $C$ is $x + y = a$. Thus, a vector equation of $C$ is

$$\mathbf{R}(t) = (-t + a)\mathbf{i} + t\mathbf{j} \quad 0 \leqslant t \leqslant a$$

We apply Definition 20.6.1 with $f(t) = -t + a$ and $g(t) = t$. We have

$$\begin{aligned} \mathbf{F}(f(t), g(t)) = \mathbf{F}(-t + a, t) &= -(-t + a)^2 t\mathbf{i} + 2t\mathbf{j} \\ &= (-t^3 + 2at^2 - a^2t)\mathbf{i} + 2t\mathbf{j} \end{aligned}$$

$$\mathbf{R}'(t) = -\mathbf{i} + \mathbf{j}$$

and

$$\begin{aligned} \mathbf{F}(f(t), g(t)) \cdot \mathbf{R}'(t) &= [(-t^3 + 2at^2 - a^2t)\mathbf{i} + 2t\mathbf{j}] \cdot [-\mathbf{i} + \mathbf{j}] \\ &= t^3 - 2at^2 + a^2t + 2t \end{aligned}$$

Thus,

$$W = \int_0^a (t^3 - 2at^2 + a^2t + 2t)dt$$

$$= \frac{1}{4}t^4 - \frac{2}{3}at^3 + \frac{1}{2}a^2t^2 + t^2\Big]_0^a$$

$$= \frac{1}{12}a^4 + a^2$$

Therefore, the total work done is $\frac{1}{12}a^4 + a^2$ inch-pounds.

32. $\mathbf{F}(x, y, z) = (xyz + x)\mathbf{i} + (x^2z + y)\mathbf{j} + (x^2y + z)\mathbf{k}$

$C$: $\mathbf{R}(t) = t\mathbf{i} + t^2\mathbf{j} + t^3\mathbf{k}$, $0 \leqslant t \leqslant 2$

SOLUTION: Thus, we have

$$\begin{aligned} \mathbf{F}(t, t^2, t^3) &= (t^6 + t)\mathbf{i} + (t^5 + t^2)\mathbf{j} + (t^4 + t^3)\mathbf{k} \\ \mathbf{R}'(t) &= \mathbf{i} + 2t\mathbf{j} + 3t^2\mathbf{k} \end{aligned}$$

$$\begin{aligned} \mathbf{F}(t, t^2, t^3) \cdot \mathbf{R}'(t) &= (t^6 + t) + 2t(t^5 + t^2) + 3t^2(t^4 + t^3) \\ &= 6t^6 + 3t^5 + 2t^3 + t \end{aligned}$$

Thus,

$$W = \int_0^2 (6t^6 + 3t^5 + 2t^3 + t)dt$$

$$= \frac{6}{7}t^7 + \frac{1}{2}t^6 + \frac{1}{2}t^4 + \frac{1}{2}t^2\Big]_0^2$$

$$= \frac{1062}{7}$$

Therefore, the total work done is $\frac{1067}{7}$ inch-pounds.

## 20.7 LINE INTEGRALS INDEPENDENT OF THE PATH

**20.7.1 Theorem** Suppose that $M$ and $N$ are functions of two variables $x$ and $y$ defined on an open disk $B((x_0, y_0); r)$ in $R^2$, $M_y$ and $N_x$ are continuous on $B$, and

$$\nabla\phi(x, y) = M(x, y)\mathbf{i} + N(x, y)\mathbf{j}$$

Suppose that $C$ is any sectionally smooth curve in $B$ from the point $(x_1, y_1)$ to the point $(x_2, y_2)$. Then the line integral

$$\int_C M(x, y)dx + N(x, y)dy$$

is independent of the path $C$ and

$$\int_C M(x, y)dx + N(x, y)dy = \phi(x_2, y_2) - \phi(x_1, y_1)$$

If $\mathbf{F}$ is the vector-valued function defined by

$$\mathbf{F}(x, y) = M(x, y)\mathbf{i} + N(x, y)\mathbf{j}$$

and there is some function $\phi$ such that

$$\mathbf{F}(x, y) = \nabla\phi(x, y)$$

then $\mathbf{F}$ is called a *gradient field;* $\phi$ is called a *potential function;* and $\phi(x, y)$ is the *potential* of $\mathbf{F}$ at $(x, y)$. Furthermore, $\mathbf{F}$ is said to be a *conservative* force field and the expression $M(x, y)dx + N(x, y)dy$ is called an *exact differential.*

**20.7.2 Theorem** Suppose that $M, N$, and $R$ are functions of three variables $x, y$, and $z$ defined on an open ball $B((x_0, y_0, z_0); r)$ in $R^3$; $M_y, M_z, N_x, N_z, R_x$, and $R_y$ are continuous on $B$; and

$$\nabla\phi(x, y, z) = M(x, y, z)\mathbf{i} + N(x, y, z)\mathbf{j} + R(x, y, z)\mathbf{k}$$

Suppose that $C$ is any sectionally smooth curve in $B$ from the point $(x_1, y_1, z_1)$ to the point $(x_2, y_2, z_2)$. Then the line integral

$$\int_C M(x, y, z)dx + N(x, y, z)dy + R(x, y, z)dz$$

is independent of the path $C$ and

$$\int_C M(x, y, z)dx + N(x, y, z)dy + R(x, y, z)dz = \phi(x_2, y_2, z_2) - \phi(x_1, y_1, z_1)$$

## *Exercises 20.7*

In Exercises 1-10, prove that the given force field is conservative and find a potential function.

4. $\mathbf{F}(x, y) = (\sin y \sinh x + \cos y \cosh x)\mathbf{i} + (\cos y \cosh x - \sin y \sinh x)\mathbf{j}$

SOLUTION: $\mathbf{F}$ is conservative if and only if $\mathbf{F}$ is the gradient of some function $\phi$. We let

$$M(x, y) = \sin y \sinh x + \cos y \cosh x$$
$$N(x, y) = \cos y \cosh x - \sin y \sinh x$$

Then

$$M_y(x, y) = \cos y \sinh x - \sin y \cosh x$$
$$N_x(x, y) = \cos y \sinh x - \sin y \cosh x$$

Because $M_y(x, y) = N_x(x, y)$, then **F** is the gradient of some function $\phi$, and thus, **F** is a conservative field. Furthermore, we have

$$\phi_x(x, y) = \sin y \sinh x + \cos y \cosh x \quad (1)$$

$$\phi_y(x, y) = \cos y \cosh x - \sin y \sinh x \quad (2)$$

Integrating with respect to $x$ on both sides of (1), we have

$$\phi(x, y) = \sin y \cosh x + \cos y \sinh x + C(y) \quad (3)$$

Partial-differentiating with respect to $y$ on both sides of (3), we have

$$\phi_y(x, y) = \cos y \cosh x - \sin y \sinh x + C'(y) \quad (4)$$

Equating the right members of (2) and (4), we get

$$\cos y \cosh x - \sin y \sinh x + C'(y) = \cos y \cosh x - \sin y \sinh x$$
$$C'(y) = 0$$
$$C(y) = C$$

Substituting for $C(y)$ into (3), we obtain

$$\phi(x, y) = \sin y \cosh x + \cos y \sinh x + C$$

where $\phi$ is the required potential function.

8. $\mathbf{F}(x, y, z) = (2y^3 - 8xz^2)\mathbf{i} + (6xy^2 + 1)\mathbf{j} - (8x^2z + 3z^2)\mathbf{k}$

SOLUTION: We have

$$M(x, y, z) = 2y^3 - 8xz^2 \qquad N(x, y, z) = 6xy^2 + 1 \qquad R(x, y, z) = -(8x^2z + 3z^2)$$

Thus,

$$M_y(x, y, z) = 6y^2 \quad \text{and} \quad N_x(x, y, z) = 6y^2$$

$$M_z(x, y, z) = -16xz \quad \text{and} \quad R_x(x, y, z) = -16xz$$

$$N_z(x, y, z) = 0 \quad \text{and} \quad R_y(x, y, z) = 0$$

Because $M_y(x, y, z) = N_x(x, y, z)$, $M_z(x, y, z) = R_x(x, y, z)$, and $N_z(x, y, z) = R_y(x, y, z)$, then **F** is the gradient of some function $\phi$, and thus, **F** is a conservative force field. Moreover, we have

$$\phi_x(x, y, z) = 2y^3 - 8xz^2 \quad (1)$$

$$\phi_y(x, y, z) = 6xy^2 + 1 \quad (2)$$

$$\phi_z(x, y, z) = -(8x^2z + 3z^2) \quad (3)$$

Integrating with respect to $x$ on both sides of (1), we get

$$\phi(x, y, z) = 2xy^3 - 4x^2z^2 + g(y, z) \quad (4)$$

Partial-differentiating with respect to $y$ on both sides of (4), we have

$$\phi_y(x, y, z) = 6xy^2 + g_y(y, z) \quad (5)$$

Equating the right members of (2) and (5), we get

$$g_y(y, z) = 1$$

Thus,

$$g(y, z) = y + h(z)$$

Substituting the value for $g(y, z)$ into (4), we have

$$\phi(x, y, z) = 2xy^3 - 4x^2z^2 + y + h(z) \tag{6}$$

Partial-differentiating with respect to $z$, we obtain

$$\phi_z(x, y, z) = -8x^2z + h'(z) \tag{7}$$

Equating the right members of (3) and (7), we have

$$-8x^2z + h'(z) = -8x^2z - 3z^2$$
$$h'(z) = -3z^2$$

Thus,

$$h(z) = -z^3 + C$$

Substituting the value for $h(z)$ into (6), we obtain

$$\phi(x, y, z) = 2xy^3 - 4x^2z^2 + y - z^3 + C$$

where $\phi$ is a potential function.

In Exercises 11-20, use the results of the indicated exercise to prove that the value of the given line integral is independent of the path. Then evaluate the line integral by applying either Theorem 20.7.1 or Theorem 20.7.2 and using the potential function found in the indicated exercise. In each exercise, $C$ is any sectionally smooth curve from the point $A$ to the point $B$.

**14.** $\int_C (\sin y \sinh x + \cos y \cosh x)dx + (\cos y \cosh x - \sin y \sinh x)dy$

$A$ is $(1, 0)$ and $B$ is $(2, \pi)$; Exercise 4.

SOLUTION: In Exercise 4 we found a potential function $\phi$ defined by

$$\phi(x, y) = \sin y \cosh x + \cos y \sinh x$$

such that

$$\nabla\phi(x, y) = (\sin y \sinh x + \cos y \cosh x)\mathbf{i} + (\cos y \cosh x - \sin y \sinh x)\mathbf{j}$$

Thus, by Theorem 20.7.1, the given line integral is independent of the path. Furthermore, the value of the given line integral is given by

$$\phi(2, \pi) - \phi(1, 0) = -\sinh 2 - \sinh 1$$

**18.** $\int_C (2y^3 - 8xz^2)dx + (6xy^2 + 1)dy - (8x^2z + 3z^2)dz$

$A$ is $(2, 0, 0)$ and $B$ is $(3, 2, 1)$; Exercise 8.

SOLUTION: In Exercise 8 we found a potential function $\phi$, defined by

$$\phi(x, y, z) = 2xy^3 - 4x^2z^2 + y - z^3$$

such that

$$\nabla\phi(x, y, z) = (2y^3 - 8xz^2)\mathbf{i} + (6xy^2 + 1)\mathbf{j} + (-8x^2z - 3z^2)\mathbf{j}$$

Therefore, by Theorem 20.7.2, the given line integral is independent of the path. Moreover, the value of the line integral is given by

$$\begin{aligned}\phi(3, 2, 1) - \phi(2, 0, 0) &= [2 \cdot 3 \cdot 2^3 - 4 \cdot 3^2 \cdot 1^2 + 2 - 1^3] - 0 \\ &= 13\end{aligned}$$

In Exercises 21-30, show that the value of the line integral is independent of the path and compute the value in any convenient manner. In each exercise, $C$ is any sectionally smooth curve from the point $A$ to the point $B$.

**24.** $\displaystyle\int_C \sin y dx + (\sin y + x\cos y)dy$

$A$ is $(-2, 0)$ and $B$ is $(2, \frac{1}{6}\pi)$.

SOLUTION: Let

$$M(x, y) = \sin y \quad \text{and} \quad N(x, y) = \sin y + x \cos y$$

we have

$$M_y(x, y) = \cos y \quad \text{and} \quad N_x(x, y) = \cos y$$

Because $M_y(x, y) = N_x(x, y)$, the integrand of the given line integral is an exact differential, and thus, the integral is independent of the path. Let the path $C$ consist of the two parts $C_1$ and $C_2$ where $C_1$ is the $x$-axis from $(-2, 0)$ to $(2, 0)$ and $C_2$ is the line $x = 2$ from $(2, 0)$ to $(2, \frac{1}{6}\pi)$. For $C_1$ we let $x$ be the parameter. Thus, for $C_1$ we have $y = 0$ and $-2 \leqslant x \leqslant 2$. Therefore, $\sin y = 0$ and $dy = 0$, and hence

$$\int_{C_1} \sin y\, dx + (\sin y + x\cos y)dy = \int_{-2}^{2} 0dx + x \cdot 0 = 0$$

For $C_2$ we let $y$ be the parameter. Thus, for $C_2$ we have $x = 2$ and $0 \leqslant y \leqslant \frac{1}{6}\pi$. Therefore, $dx = 0$, and hence

$$\begin{aligned}\int_{C_2} \sin y\, dx + (\sin y + x\cos y)dy &= \int_0^{\pi/6} (\sin y + 2\cos y)dy \\ &= -\cos y + 2\sin y\Big]_0^{\pi/6} \\ &= \left(-\frac{1}{2}\sqrt{3} + 1\right) - (-1) \\ &= 2 - \frac{1}{2}\sqrt{3}\end{aligned}$$

Therefore,

$$\begin{aligned}\int_C \sin y\, dx + (\sin y + x\cos y)dy &= \int_{C_1} \sin y\, dx + (\sin y + x\cos y)dy \\ &\quad + \int_{C_2} \sin y\, dx + (\sin y + x\cos y)dy \\ &= 0 + \left(2 - \frac{1}{2}\sqrt{3}\right) \\ &= 2 - \frac{1}{2}\sqrt{3}\end{aligned}$$

**28.** $\displaystyle\int_C (yz + x)dx + (xz + y)dy + (xy + z)dz$

$A$ is $(0, 0, 0)$ and $B$ is $(1, 1, 1)$.

SOLUTION: Let

$$M(x, y, z) = yz + x \qquad N(x, y, z) = xz + y \qquad R(x, y, z) = xy + z$$

We have

$$M_y(x, y, z) = z \quad \text{and} \quad N_x(x, y, z) = z$$
$$M_z(x, y, z) = y \quad \text{and} \quad R_x(x, y, z) = y$$
$$N_z(x, y, z) = x \quad \text{and} \quad R_y(x, y, z) = x$$

Therefore, the given integrand is an exact differential, and hence, the line integral is independent of the path. We let $C$ be the line segment from $(0, 0, 0)$ to $(1, 1, 1)$. Parametric equations for $C$ are

$$x = t \qquad y = t \qquad z = t \qquad 0 \leq t \leq 1$$

Therefore,

$$\int_C (yz + x)dx + (xz + y)dy + (xy + z)dz = \int_0^1 (t^2 + t)dt + (t^2 + t)dt + (t^2 + t)dt$$

$$= 3\int_0^1 (t^2 + t)dt$$

$$= t^3 + \frac{3}{2}t^2\Big]_0^1$$

$$= \frac{5}{2}$$

## *Review Exercises*

4. Find the rate of change of the function value in the direction of **u** at $P$.

$$f(x, y, z) = yz - y^2 - xz \qquad \mathbf{u} = \frac{6}{7}\mathbf{i} + \frac{3}{7}\mathbf{j} + \frac{2}{7}\mathbf{k} \qquad P = (1, 2, 3)$$

SOLUTION:

$$\nabla f(x, y, z) = -z\mathbf{i} + (z - 2y)\mathbf{j} + (y - x)\mathbf{k}$$
$$\nabla f(1, 2, 3) = -3\mathbf{i} - \mathbf{j} + \mathbf{k}$$

We have

$$D_{\mathbf{u}}f(1, 2, 3) = \nabla f(1, 2, 3) \cdot \mathbf{u}$$

$$= (-3\mathbf{i} - \mathbf{j} + \mathbf{k}) \cdot \left(\frac{6}{7}\mathbf{i} + \frac{3}{7}\mathbf{j} + \frac{2}{7}\mathbf{k}\right)$$

$$= -\frac{19}{7}$$

8. Find an equation of the tangent plane and equations of the normal line to the given surface at the indicated point.

$$z = x^2 + 2xy \qquad (1, 3, 7)$$

SOLUTION: Let $F$ be the function defined by

$$F(x, y, z) = x^2 + 2xy - z$$

Thus, the surface is the graph of $F(x, y, z) = 0$. We have

$$\nabla F(x, y, z) = (2x + 2y)\mathbf{i} + 2x\mathbf{j} - \mathbf{k}$$
$$\nabla F(1, 3, 7) = 8\mathbf{i} + 2\mathbf{j} - \mathbf{k}$$

Thus, an equation of the tangent plane to the surface at $(1, 3, 7)$ is

$$8(x-1) + 2(y-3) - (z-7) = 0$$
$$8x + 2y - z - 7 = 0$$

and equations of the normal line to the surface at $(1, 3, 7)$ are

$$\frac{x-1}{8} = \frac{y-3}{2} = \frac{z-7}{-1}$$

12. Evaluate the line integral over the given curve.

$$\int_C xe^y\,dx - xe^z\,dy + e^z\,dz$$

$C$: $\mathbf{R}(t) = t\mathbf{i} + t^2\mathbf{j} + t^3\mathbf{k}, 0 \leqslant t \leqslant 1$

SOLUTION: Parametric equations of $C$ are

$$x = t \qquad y = t^2 \qquad z = t^3 \quad \text{for } 0 \leqslant t \leqslant 1$$

Therefore,

$$\begin{aligned}\int_C xe^y\,dx - xe^z\,dy + e^z\,dz &= \int_0^1 te^{t^2}\,dt - \int_0^1 te^{t^3}(2t\,dt) + \int_0^1 e^{t^3}(3t^2\,dt)\\ &= \left[\frac{1}{2}e^{t^2}\right]_0^1 - \left[\frac{2}{3}e^{t^3}\right]_0^1 + \left[e^{t^3}\right]_0^1\\ &= \frac{1}{2}(e-1) + \frac{1}{3}(e-1)\\ &= \frac{5}{6}(e-1)\end{aligned}$$

16. Prove that the value of the given line integral is independent of the path, and compute the value in any convenient manner. $C$ is any sectionally smooth curve from the point $A$ to the point $B$.

$$\int_C z\sin y\,dx + xz\cos y\,dy + x\sin y\,dz$$

$A$ is $(0, 0, 0)$ and $B$ is $(2, 3, \frac{1}{2}\pi)$.

SOLUTION: We let

$$M(x, y, z) = z\sin y \qquad N(x, y, z) = xz\cos y \qquad R(x, y, z) = x\sin y$$

Thus,

$$M_y(x, y, z) = z\cos y \quad \text{and} \quad N_x(x, y, z) = z\cos y$$
$$M_z(x, y, z) = \sin y \quad \text{and} \quad R_x(x, y, z) = \sin y$$
$$N_z(x, y, z) = x\cos y \quad \text{and} \quad R_y(x, y, z) = x\cos y$$

Therefore, the value of the line integral is independent of the path. Let $C$ be the path that consists of $C_1$, $C_2$, and $C_3$ where $C_1$ is the segment from $(0, 0, 0)$ to $(2, 0, 0)$,

$C_2$ is the segment from $(2, 0, 0)$ to $(2, 3, 0)$, and $C_3$ is the segment from $(2, 3, 0)$ to $(2, 3, \frac{1}{2}\pi)$.

For $C_1$, we let $x$ be the parameter, with $y = 0, z = 0$, and $0 \leqslant x \leqslant 2$. Thus, $dy = dz = 0$, and hence,

$$\int_{C_1} z \sin y\, dx + xz \cos y\, dy + x \sin y\, dz = \int_0^2 0\, dz = 0 \tag{1}$$

For $C_2$, we let $y$ be the parameter, and we have $x = 2, z = 0$, and $0 \leqslant y \leqslant 3$. Thus, $dx = dz = 0$, and hence,

$$\int_{C_2} z \sin y\, dx + xz \cos y\, dy + x \sin y\, dz = \int_0^3 0\, dy = 0 \tag{2}$$

For $C_3$, we let $z$ be the parameter, and we have $x = 2, y = 3$, and $0 \leqslant z \leqslant \frac{1}{2}\pi$. Thus, $dx = dy = 0$, and hence,

$$\begin{aligned}\int_{C_3} z \sin y\, dx + xz \cos y\, dy + x \sin y\, dz &= \int_0^{\pi/2} 2 \sin 3\, dz \\ &= (2 \sin 3)z\Big]_0^{\pi/2} \\ &= \pi \sin 3 \end{aligned} \tag{3}$$

Adding the results obtained in Eqs. (1), (2), and (3), we have

$$\begin{aligned}\int_C z \sin y\, dx + xz \cos y\, dy + x \sin y\, dz &= 0 + 0 + \pi \sin 3 \\ &= \pi \sin 3\end{aligned}$$

**18.** If $f(x, y, z) = \sinh(x + z) \cosh y$, find the rate of change of $f(x, y, z)$ with respect to distance in $R^3$ at the point $P(1, 1, 0)$ in the direction $\overrightarrow{PQ}$ if $Q$ is the point $(-1, 0, 2)$.

SOLUTION: We want to find $D_{\mathbf{u}}f(1, 1, 0)$ where $\mathbf{u}$ is a unit vector in the direction of $\overrightarrow{PQ}$. We have

$$\mathbf{V}(PQ) = -2\mathbf{i} - \mathbf{j} + 2\mathbf{k}$$

Thus,

$$\mathbf{u} = \frac{\mathbf{V}(PQ)}{|\mathbf{V}(PQ)|} = -\frac{2}{3}\mathbf{i} - \frac{1}{3}\mathbf{j} + \frac{2}{3}\mathbf{k}$$

Furthermore,

$$\nabla f(x, y, z) = \cosh(x + z) \cosh y\, \mathbf{i} + \sinh(x + z) \sinh y\, \mathbf{j} + \cosh(x + z) \cosh y\, \mathbf{k}$$

Thus,

$$\nabla f(1, 1, 0) = \cosh^2 1\mathbf{i} + \sinh^2 1\mathbf{j} + \cosh^2 1\mathbf{k}$$

Therefore,

$$\begin{aligned} D_{\mathbf{u}}f(1, 1, 0) &= \mathbf{u} \cdot \nabla f(1, 1, 0) \\ &= \left\langle -\frac{2}{3}, -\frac{1}{3}, \frac{2}{3} \right\rangle \cdot \langle \cosh^2 1, \sinh^2 1, \cosh^2 1 \rangle \\ &= -\frac{2}{3} \cosh^2 1 - \frac{1}{3} \sinh^2 1 + \frac{2}{3} \cosh^2 1 \end{aligned}$$

$$= -\frac{1}{3}\sinh^2 1$$

22. The temperature is $T$ degrees at any point $(x, y)$ of the curve $4x^2 + 12y^2 = 1$ and $T = 4x^2 + 24y^2 - 2x$. Find the points on the curve where the temperature is the greatest and where it is the least. Also find the temperature at these points.

SOLUTION: We use the method of Lagrange multipliers. We want to find the extrema of $T$ subject to the constraint $4x^2 + 12y^2 - 1 = 0$. Thus, we let $F$ be the function defined by

$$F(x, y, \lambda) = 4x^2 + 24y^2 - 2x + \lambda(4x^2 + 12y^2 - 1)$$

We take

$$F_x(x, y, \lambda) = 8x - 2 + 8x\lambda = 0 \qquad (1)$$

$$F_y(x, y, \lambda) = 48y + 24y\lambda = 0 \qquad (2)$$

$$F_\lambda(x, y, \lambda) = 4x^2 + 12y^2 - 1 = 0 \qquad (3)$$

From (2) we have either $y = 0$ or $\lambda = -2$. If $y = 0$, from Eq. (3) we obtain $x = \pm\frac{1}{2}$. If $\lambda = -2$, from Eq. (1) we get $x = -\frac{1}{4}$, and from (3) we get $y = \pm\frac{1}{4}$. Thus, the critical points are $(\pm\frac{1}{2}, 0)$ and $(-\frac{1}{4}, \pm\frac{1}{4})$. Let

$$T(x, y) = 4x^2 + 24y^2 - 2x$$

Then

$$T\left(\frac{1}{2}, 0\right) = 0 \qquad T\left(-\frac{1}{2}, 0\right) = 2 \qquad T\left(-\frac{1}{4}, \frac{1}{4}\right) = \frac{9}{4} \qquad T\left(-\frac{1}{4}, -\frac{1}{4}\right) = \frac{9}{4}$$

Therefore, the maximum temperature is $\frac{9}{4}°$ which occurs at the points $(-\frac{1}{4}, \pm\frac{1}{4})$ and the minimum temperature is $0°$ at the point $(\frac{1}{2}, 0)$.

26. A piece of wire $L$ feet long is cut into three pieces. One piece is bent into the shape of a circle; a second piece is bent into the shape of a square; and the third piece is bent into the shape of an equilateral triangle. How should the wire be cut so that: **(a)** the combined area of the three figures is as small as possible, and **(b)** the combined area of the three figures is as large as possible?

SOLUTION: Let

$r$ = the number of feet in the radius of the circle
$x$ = the number of feet in the side of the square
$y$ = the number of feet in the side of the triangle
$A$ = the number of sq. feet in the combined area of the three figures

We have

$$A = \pi r^2 + x^2 + \frac{1}{4}\sqrt{3}\,y^2 \qquad (1)$$

Because the number of feet in the circumference of the circle is $2\pi r$, the number of feet in the perimeter of the square is $4x$, and the number of feet in the perimeter of the triangle is $3y$, we also have

$$L = 2\pi r + 4x + 3y \qquad (2)$$

We want to find the extrema of $A$ subject to the constraint $2\pi r + 4x + 3y - L = 0$. We use the method of Lagrange multipliers to find the critical points. Let $F$ be the function defined by

$$F(r, x, y, \lambda) = \pi r^2 + x^2 + \frac{1}{4}\sqrt{3}\, y^2 + \lambda(2\pi r + 4x + 3y - L)$$

We take

$$F_r(r, x, y, \lambda) = 2\pi r + 2\pi\lambda = 0 \tag{3}$$

$$F_x(r, x, y, \lambda) = 2x + 4\lambda = 0 \tag{4}$$

$$F_y(r, x, y, \lambda) = \frac{1}{2}\sqrt{3}\, y + 3\lambda = 0 \tag{5}$$

$$F_\lambda(r, x, y, \lambda) = 2\pi r + 4x + 3y - L = 0 \tag{6}$$

Solving (3) for $\lambda$, we have $\lambda = -r$. Solving (4) for $x$, we get $x = -2\lambda = 2r$. Solving (5) for $y$, we obtain $y = -2\sqrt{3}\lambda = 2\sqrt{3}\, r$. Substituting the values obtained for $x$ and $y$ into (6), we have

$$2\pi r + 4(2r) + 3(2\sqrt{3}\, r) - L = 0$$

Solving for $r$ gives

$$r = \frac{L}{2\pi + 8 + 6\sqrt{3}} \approx 0.04L$$

From this value for $r$ we obtain

$$x = 2r \approx 0.08L$$

$$y = 2\sqrt{3}\, r \approx 0.14L$$

Substituting the values obtained for $r, x,$ and $y$ into (1), we have

$$\begin{aligned} A &= \pi r^2 + (2r)^2 + \frac{1}{4}\sqrt{3}\,(2\sqrt{3}\, r)^2 \\ &= r^2(\pi + 4 + 3\sqrt{3}) \\ &= \left[\frac{L}{2\pi + 8 + 6\sqrt{3}}\right]^2 (\pi + 4 + 3\sqrt{3}) \\ &= \frac{L^2}{4(\pi + 4 + 3\sqrt{3}} \\ &\approx 0.02L^2 \end{aligned} \tag{7}$$

We compare the value of $A$ obtained at the critical point with the value of $A$ at each endpoint. If $x = 0$ and $y = 0$, then from Eq. (2) we have $r = L/2\pi$. Substituting these values into Eq. (1), we obtain

$$A = \pi\left(\frac{L}{2\pi}\right)^2 \approx 0.08L^2 \tag{8}$$

If $r = 0$ and $y = 0$, then from Eq. (2) we have $x = \frac{1}{4}L$. Thus, from (1) we get

$$A = \left(\frac{1}{4}L\right)^2 \approx 0.06L^2 \tag{9}$$

And if $r = 0$ and $x = 0$, then $y = \frac{1}{3}L$ and

$$A = \frac{1}{4}\sqrt{3}\left(\frac{1}{3}L\right)^2 \approx 0.05L^2 \tag{10}$$

Comparing the values of $A$ given in Eqs. (7), (8), (9), and (10), we conclude that the minimum combined area is $0.02L^2$ square feet and that it occurs when the radius is $0.04L$ feet, the length of a side of the square is $0.08L$ feet, and the length of a side of the triangle is $0.14L$ feet. The maximum combined area is $0.08L^2$ sq. feet, and it occurs when all the wire is used to make the circle.

**28.** In parts (a), (b), and (c), demand equations of two related commodities are given. In each part, determine if the commodities are complementary, substitutes, or neither.

(a) $x = -4p + 2q + 6 \quad y = 5p - q + 10$
(b) $x = 6 - 3p - 2q \quad y = 4 + 2p - q$
(c) $x = -7q - p + 7 \quad y = 18 - 3q - 9p$

SOLUTION:

(a) $\dfrac{\partial x}{\partial q} = 2$ and $\dfrac{\partial y}{\partial p} = 5$

Because $\partial x/\partial q > 0$ and $\partial y/\partial p > 0$, the commodities are substitutes.

(b) $\dfrac{\partial x}{\partial q} = -2$ and $\dfrac{\partial y}{\partial p} = 2$

Because $\partial x/\partial q$ and $\partial y/\partial p$ are neither both positive nor both negative, the commodities are neither complementary nor substitutes.

(c) $\dfrac{\partial x}{\partial q} = -7$ and $\dfrac{\partial y}{\partial p} = -9$

Because $\partial x/\partial q < 0$ and $\partial y/\partial p < 0$, the commodities are complementary.

# 21 Multiple integration

## 21.1 THE DOUBLE INTEGRAL

**21.1.1 Definition** Let $f$ be a function defined on a closed rectangular region $R$. The number $L$ is said to be the *limit* of sums of the form $\sum_{i=1}^{n} f(\xi_i, \gamma_i)\,\Delta_i A$ if $L$ satisfies the property that for any $\epsilon > 0$, there exists a $\delta > 0$ such that

$$\left| \sum_{i=1}^{n} f(\xi_i, \gamma_i)\,\Delta_i A - L \right| < \epsilon$$

for every partition $\Delta$ for which $\|\Delta\| < \delta$ and for all possible selections of the point $(\xi_i, \gamma_i)$ in the $i$th rectangle, $i = 1, 2, \ldots, n$. If such a number $L$ exists, we write

$$\lim_{\|\Delta\| \to 0} \sum_{i=1}^{n} f(\xi_i, \gamma_i)\,\Delta_i A = L$$

**21.1.2 Definition** A function $f$ of two variables is said to be *integrable* on a rectangular region $R$ if $f$ is defined on $R$ and the number $L$ of Definition 21.1.1 exists. This number $L$ is called the *double integral* of $f$ on $R$, and we write

$$\lim_{\|\Delta\| \to 0} \sum_{i=1}^{n} f(\xi_i, \gamma_i)\,\Delta_i A = \iint_R f(x, y)\, dA$$

**21.1.3 Theorem** If a function $f$ of two variables is continuous on a closed rectangular region $R$, then $f$ is integrable on $R$.

**21.1.4 Theorem** Let $f$ be a function of two variables that is continuous on a closed region $R$ in the $xy$ plane and $f(x, y) \geq 0$ for all $(x, y)$ in $R$. If $V(S)$ is the measure of the volume of the solid $S$ having the region $R$ as its base and having an altitude of measure $f(x, y)$ at the point $(x, y)$ in $R$, then

$$V(S) = \lim_{\|\Delta\| \to 0} \sum_{i=1}^{n} f(\xi_i, \gamma_i)\, \Delta_i A = \iint_R f(x, y)\, dA$$

**21.1.5 Theorem** If $c$ is a constant and the function $f$ is integrable on a closed region $R$, then $cf$ is integrable on $R$ and

$$\iint_R cf(x, y)\, dA = c \iint_R f(x, y)\, dA$$

**21.1.6 Theorem** If the functions $f$ and $g$ are integrable on a closed region $R$, then the function $f + g$ is integrable on $R$ and

$$\iint_R [f(x, y) + g(x, y)]\, dA = \iint_R f(x, y)\, dA + \iint_R g(x, y)\, dA$$

**21.1.9 Theorem** Suppose that the function $f$ is continuous on the closed region $R$ and that region $R$ is composed of the two subregions $R_1$ and $R_2$ which have no points in common except for points on parts of their boundaries. Then

$$\iint_R f(x, y)\, dA = \iint_{R_1} f(x, y)\, dA + \iint_{R_2} f(x, y)\, dA$$

In Table 21(a) we show a flowchart that can be used to calculate a sum that approximates the double integral of a function $f$ over a rectangle $R$ in the $xy$ plane with relative error less than 0.001. The boundaries of $R$ are the lines $x = a, x = b$, $y = c$, and $y = d$, with $a < b$ and $c < d$. We divide the interval $[a, b]$ into $n$ subintervals and the interval $[c, d]$ into $n$ subintervals, which results in a partition of $R$ into $n^2$ subrectangles $R_i, i = 1, 2, \ldots, n^2$. We take $(\bar{x}_i, \bar{y}_i)$ to be the midpoint of $R_i$ and approximate the double integral by the sum

$$\sum_{i=1}^{n^2} f(\bar{x}_i, \bar{y}_i)\, \Delta_i A$$

Next, we double $n$ and repeat the above. If the difference between the new approximation and the old approximation has absolute value less than 0.001 times the new approximation, we terminate the process. Otherwise, we double $n$ and repeat the above.

In Table 21(b) we show an actual computer program written in BASIC that uses the flowchart of Table 21(a) to approximate the double integral

$$\iint_R x^2 y^3\, dA$$

where $R$ is the rectangle bounded by the lines $x = 0, x = 1, y = 0$, and $y = 2$.

**Table 21(a)**

$$S = \int_a^b \int_c^d f(x, y)\,dy\,dx \quad \text{(Relative error is less than .001.)}$$

$N^2$ = number of rectangles

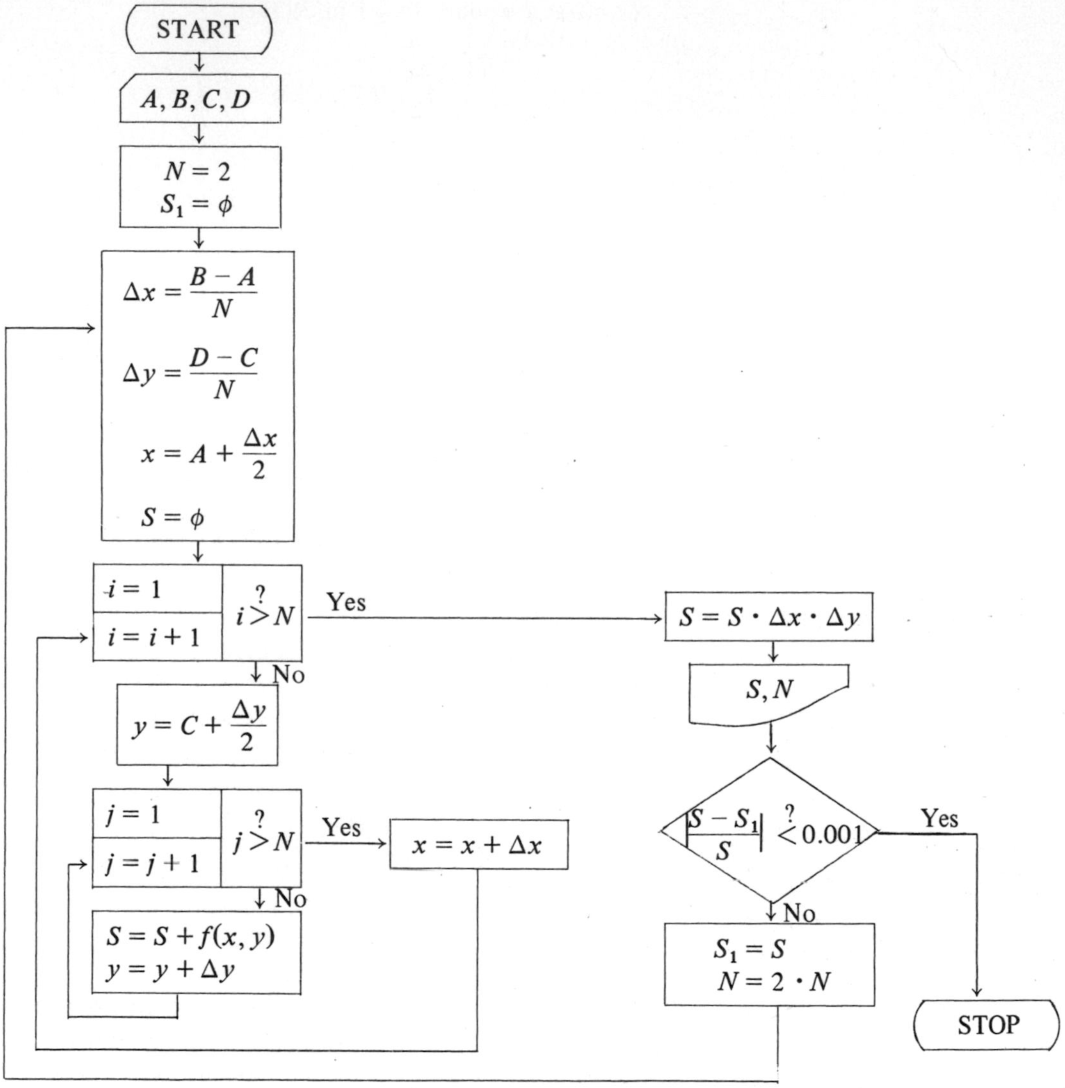


**Table 21(b)**

```
LIST

10  DATA 0,1,0,2
20  READ A,B,C,D
30  LET S1=0
40  FOR K=1 TO 10
50  LET N=2↑K
60  LET D1=(B-A)/N
70  LET D2=(D-C)/N
80  LET S=0
90  LET X=A+D1/2
100  FOR I=1 TO N
```

```
110 LET Y=C+D2/2
120 FOR J=1 TO N
130 LET S=S+(X↑2)*(Y↑3)
140 LET Y=Y+D2
150 NEXT J
160 LET X=X+D1
170 NEXT I
180 LET S=S*D1*D2
190 PRINT "S = ";S,"N = ";N
200 IF ABS((S-S1)/S)<.001 THEN 999
210 LET S1=S
220 NEXT K
999 END

RUN

S = 1.09375          N = 2
S = 1.27148          N = 4
S = 1.31775          N = 8
S = 1.32943          N = 16
S = 1.33236          N = 32
S = 1.33309          N = 64
```

## *Exercises 21.1*

**4.** Find an approximate value of the given double integral where $R$ is the rectangular region having the vertices $P$ and $Q$, $\Delta$ is a regular partition of $R$, and $(\xi_i, \gamma_i)$ is the midpoint of each subregion.

$$\iint_R (xy + 3y^2)\,dA;\ P(0, -2);\ Q(6, 4)$$

$$\Delta:\ x_1 = 0,\ x_2 = 2,\ x_3 = 4,\ y_1 = -2, y_2 = 0,\ y_3 = 2$$

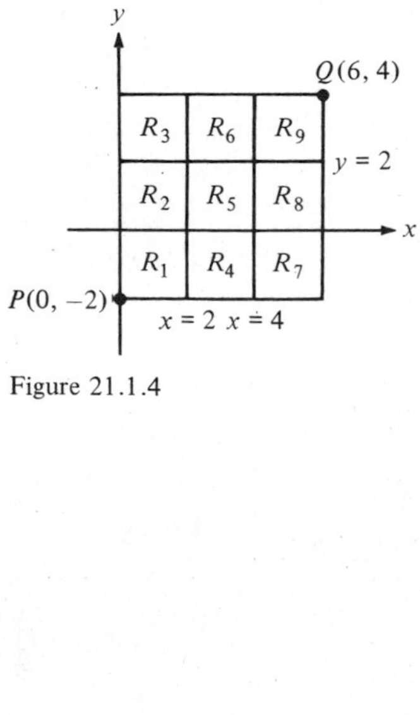


Figure 21.1.4

SOLUTION: In Fig. 21.1.4 we show the region $R$ which is partitioned into 9 subregions $R_i, i = 1, 2, \ldots, 9$. Let $\Delta_i A$ be the number of square units in the area of subregion $R_i$ and let $(\xi_i, \gamma_i)$ be the midpoint of subregion $R_i$. If $f$ is the function defined by $f(x, y) = xy + 3y^2$, then we have, approximately,

$$\iint_R (xy + 3y^2)\,dA = \sum_{i=1}^{9} f(\xi_i, \gamma_i)\Delta_i A$$

$$= 4\sum_{i=1}^{9} f(\xi_i, \gamma_i) \qquad \textbf{(1)}$$

because $\Delta_i A \doteq 4$ for each $i = 1, 2, \ldots, 9$. In Table 4 we show the calculations for finding the sum in (1). Substituting the value for the sum into (1), we obtain

$$\iint_R (xy + 3y^2)\,dA = 4(126)$$

$$= 504$$

Table 4

| $i$ | $(\xi_i, \gamma_i)$ | $f(\xi_i, \gamma_i)$ |
|---|---|---|
| 1 | $(1, -1)$ | 2 |
| 2 | $(1, 1)$ | 4 |
| 3 | $(1, 3)$ | 30 |
| 4 | $(3, -1)$ | 0 |
| 5 | $(3, 1)$ | 6 |
| 6 | $(3, 3)$ | 36 |
| 7 | $(5, -1)$ | $-2$ |
| 8 | $(5, 1)$ | 8 |
| 9 | $(5, 3)$ | 42 |

$$\sum_{i=1}^{9} f(\xi_i, \gamma_i) = 126$$

8. Find an approximate value of the given double integral where $R$ is the rectangular region having the vertices $P$ and $Q$, $\Delta$ is a regular partition of $R$, and $(\xi_i, \gamma_i)$ is an arbitrary point in each subregion.

$$\iint_R (2 - x - y)\, dA;\; P(0, 0);\; Q(6, 4)$$

$\Delta$: $x_1 = 0,\ x_2 = 2,\ x_3 = 4,\ y_1 = 0,\ y_2 = 2$

$$(\xi_1, \gamma_1) = \left(\frac{1}{2}, \frac{3}{2}\right) \qquad (\xi_2, \gamma_2) = (3, 1) \qquad (\xi_3, \gamma_3) = \left(\frac{11}{2}, \frac{1}{2}\right)$$

$$(\xi_4, \gamma_4) = (2, 2) \qquad (\xi_5, \gamma_5) = (2, 2) \qquad (\xi_6, \gamma_6) = (5, 3)$$

SOLUTION: Let $f$ be the function defined by $f(x, y) = 2 - x - y$, and let $\Delta_i A$ square units be the area of the $i$th subrectangle. Because each subrectangle is 2 units by 2 units, we have $\Delta_i A = 4$ for $i = 1, 2, \ldots, 6$. Thus, approximately,

$$\iint_R (2 - x - y)\, dA = \sum_{i=1}^{6} f(\xi_i, \gamma_i)\Delta_i A$$

$$= 4\sum_{i=1}^{6} f(\xi_i, \gamma_i)$$

$$= 4\left[f\left(\frac{1}{2}, \frac{3}{2}\right) + f(3, 1) + f\left(\frac{11}{2}, \frac{1}{2}\right) + f(2, 2) + f(2, 2)\right.$$

$$\left. + f(5, 3)\right]$$

$$= 4[0 - 2 - 4 - 2 - 2 - 6]$$

$$= -64$$

12. Approximate the volume of the solid bounded by the surface $100z = 300 - 25x^2 - 4y^2$; the planes $x = -1$, $x = 3$, $y = -3$, and $y = 5$; and the $xy$ plane. To find an approximate value of the double integral, take a partition of the region in the $xy$

plane by drawing the lines $x = 1$, $x = 2$, $y = 1$, and $y = 2$, and take $(\xi_i, \gamma_i)$ at the center of the $i$th subregion.

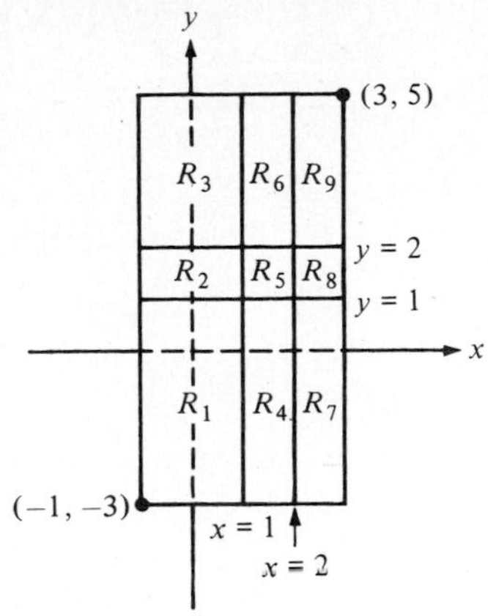


Figure 21.1.12

SOLUTION: In Fig. 21.1.12 we show a sketch of the region $R$ in the $xy$ plane which is the base of the solid. $R$ is divided into 9 subrectangles. Let $\Delta_i A$ be the number of square units in the area of $R_i$ and let $(\xi_i, \gamma_i)$ be the midpoint of $R_i$. Solving the equation of the surface for $z$ and setting $z = f(x, y)$, we define the function $f$ by

$$f(x, y) = 3 - \frac{1}{4}x^2 - \frac{1}{25}y^2$$

The approximate number of cubic units in the volume of the solid is

$$\iint_R \left(3 - \frac{1}{4}x^2 - \frac{1}{25}y^2\right) dA = \sum_{i=1}^{9} f(\xi_i, \gamma_i)\Delta_i x \tag{1}$$

The calculations for (1) are shown in Table 12. We conclude that the volume is approximately 72.2 cubic units.

**Table 12**

| $i$ | $\Delta_i A$ | $(\xi_i, \gamma_i)$ | $f(\xi_i, \gamma_i)$ | $f(\xi_i, \gamma_i)\,\Delta_i A$ |
|---|---|---|---|---|
| 1 | 8 | (0, −1) | 2.96 | 23.68 |
| 2 | 2 | (0, 1.5) | 2.91 | 5.82 |
| 3 | 6 | (0, 3.5) | 2.51 | 15.06 |
| 4 | 4 | (1.5, −1) | 2.40 | 9.60 |
| 5 | 1 | (1.5, 1.5) | 2.35 | 2.35 |
| 6 | 3 | (1.5, 3.5) | 1.95 | 5.85 |
| 7 | 4 | (2.5, −1) | 1.40 | 5.60 |
| 8 | 1 | (2.5, 1.5) | 1.35 | 1.35 |
| 9 | 3 | (2.5, 3.5) | 0.95 | 2.85 |

$$\sum_{i=1}^{9} f(\xi_i, \gamma_i)\,\Delta_i A = 72.16$$

**14.** Prove Theorem 21.1.6

SOLUTION: We apply Definitions 21.1.2 and 21.1.1. Because $f$ and $g$ are integrable on $R$, the numbers $L_1$ and $L_2$ exist, where

$$L_1 = \iint_R f(x, y)\,dA = \lim_{\|\Delta\| \to 0} \sum_{i=1}^{n} f(\xi_i, \gamma_i)\,\Delta_i A$$

and

$$L_2 = \iint_R g(x, y)\,dA = \lim_{\|\Delta\| \to 0} \sum_{i=1}^{n} g(\xi_i, \gamma_i)\,\Delta_i A$$

Thus, for any $\epsilon > 0$ there exist numbers $\delta_1 > 0$ and $\delta_2 > 0$ such that

$$\left|\sum_{i=1}^{n} f(\xi_i, \gamma_i)\Delta_i A - L_1\right| < \frac{1}{2}\epsilon$$

and

$$\left| \sum_{i=1}^{n} g(\xi_i, \gamma_i)\,\Delta_i A - L_2 \right| < \frac{1}{2}\epsilon$$

for all partitions of $R$ with $\|\Delta\| < \delta_1$ and $\|\Delta\| < \delta_2$. Thus, if $\delta = \min(\delta_1, \delta_2)$, then for all partitions of $R$ with $\|\Delta\| < \delta$, we have

$$\begin{aligned} \left| \sum_{i=1}^{n} [f(\xi_i, \gamma_i) + g(\xi_i, \gamma_i)] - [L_1 + L_2] \right| &\leqslant \left| \sum_{i=1}^{n} f(\xi_i, \gamma_i) - L_1 \right| \\ &\quad + \left| \sum_{i=1}^{n} g(\xi_i, \gamma_i) - L_2 \right| \\ &\leqslant \frac{1}{2}\epsilon + \frac{1}{2}\epsilon = \epsilon \end{aligned}$$

Therefore, $f + g$ is integrable and

$$\begin{aligned} \iint_R [f(x, y) + g(x, y)]\, dA &= L_1 + L_2 \\ &= \iint_R f(x, y)\, dA + \iint_R g(x, y)\, dA \end{aligned}$$

## 21.2 EVALUATION OF DOUBLE INTEGRALS AND ITERATED INTEGRALS

Let $f$ be a continuous function of two variables $x$ and $y$. Let $R$ be a region in the $xy$ plane that is bounded on the left by the line $x = x_1$, bounded on the right by the line $x = x_2$, bounded below by the smooth curve $y = g_1(x)$, and bounded above by the smooth curve $y = g_2(x)$. Then the double integral of $f$ on $R$ is equivalent to an iterated integral, and

$$\iint_R f(x, y)\, dA = \int_{x_1}^{x_2} \int_{g_1(x)}^{g_2(x)} f(x, y)\, dy\, dx$$

Let $R$ be a region in the $xy$ plane that is bounded below by the line $y = y_1$, bounded above by the line $y = y_2$, bounded on the left by the smooth curve $x = g_1(y)$, and bounded on the right by the smooth curve $x = g_2(y)$. Then

$$\iint_R f(x, y)\, dA = \int_{y_1}^{y_2} \int_{g_1(y)}^{g_2(y)} f(x, y)\, dx\, dy$$

Let $S$ be a solid that is bounded below by some region $R$ in the $xy$ plane, bounded above by the surface $z = f(x, y)$, and bounded on the sides by a cylindrical surface whose directrix is the boundary of the region $R$ in the $xy$ plane and whose rulings are perpendicular to the $xy$ plane. That is, every nonempty intersection of a line perpendicular to the $xy$ plane with the solid $S$ is a line segment that has one endpoint in the region $R$ and the other endpoint in the surface $z = f(x, y)$. Then the number of cubic units in the volume of $S$ is given by

$$V = \iint_R f(x, y)\, dA$$

Furthermore, the number of square units in the area of $R$ is given by

$$A = \iint_R dA$$

*Exercises 21.2*

In Exercises 1-8, evaluate the given iterated integral.

**4.** $\displaystyle\int_{-1}^{1}\int_{1}^{e^x} \frac{1}{xy}\, dy\, dx$

SOLUTION:

$$\int_{-1}^{1}\int_{1}^{e^x} \frac{1}{xy}\, dy\, dx = \int_{-1}^{1}\left[\frac{1}{x}\ln y\right]_{y=1}^{y=e^x} dx$$

$$= \int_{-1}^{1} \frac{1}{x}(\ln e^x - \ln 1)\, dx$$

$$= \int_{-1}^{1} dx$$

$$= x\Big]_{-1}^{1}$$

$$= 2$$

**8.** $\displaystyle\int_{0}^{\pi}\int_{0}^{y^2} \sin\frac{x}{y}\, dx\, dy$

SOLUTION:

$$\int_{0}^{\pi}\int_{0}^{y^2} \sin\frac{x}{y}\, dx\, dy = \int_{0}^{\pi}\left[-y\cos\frac{x}{y}\right]_{x=0}^{x=y^2}$$

$$= \int_{0}^{\pi} [-y\cos y + y]\, dy \qquad (1)$$

We use integration by parts for the integral in (1). Thus,

$$\int_{0}^{\pi}\int_{0}^{y^2} \sin\frac{x}{y}\, dx\, dy = \left[-y\sin y - \cos y + \frac{1}{2}y^2\right]_{0}^{\pi}$$

$$= 2 + \frac{1}{2}\pi^2$$

**12.** Find the exact value of the double integral.

$$\iint_R \cos(x+y)\, dA$$

$R$ is the region bounded by the lines $y = x$ and $x = \pi$, and the $x$-axis.

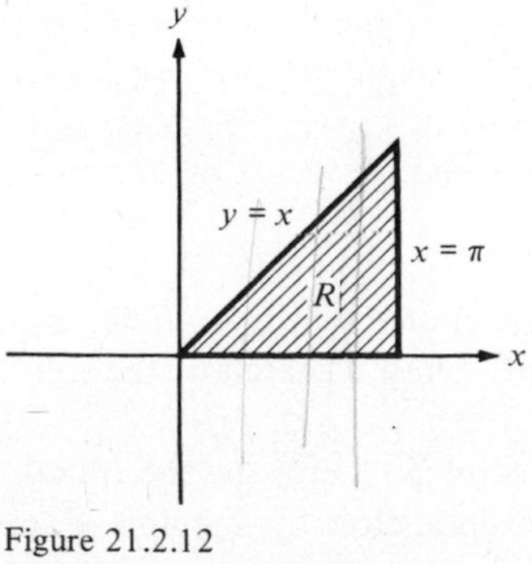

Figure 21.2.12

SOLUTION: In Fig. 21.2.12, we show a sketch of the region $R$, which is bounded below by the line $y = 0$, bounded above by the line $y = x$, bounded on the left by the line $x = 0$, and bounded on the right by the line $x = \pi$. Therefore, we have

$$\iint_R \cos(x+y)\, dA = \int_{0}^{\pi}\int_{0}^{x} \cos(x+y)\, dy\, dx$$

$$= \int_0^{\pi} [\sin(x+y)]_{y=0}^{y=x}\, dx$$

$$= \int_0^{\pi} (\sin 2x - \sin x)dx$$

$$= -\frac{1}{2}\cos 2x + \cos x\Big]_0^{\pi}$$

$$= -2$$

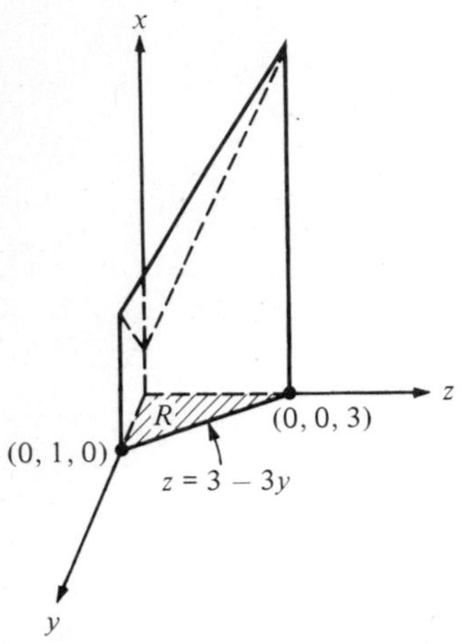

Figure 21.2.16

**16.** Find the volume of the solid bounded by the planes $x = y + 2z + 1$, $x = 0$, $y = 0$, $z = 0$, and $3y + z - 3 = 0$. Draw a sketch of the solid.

SOLUTION: Because the plane $3y + z - 3 = 0$ is perpendicular to the $yz$ plane, we let the $yz$ plane be the horizontal plane. Thus, the triangular region $R$ in the $yz$ plane bounded by the $y$-axis, the $z$-axis, and the line $3y + z - 3 = 0$ is the base of the solid $S$. See Fig. 21.2.16. If $f$ is the function defined by $f(y, z) = y + 2z + 1$, then the solid $S$ is bounded below by the region $R$ in the $yz$ plane and bounded above by the plane $x = f(y, z)$. Therefore, the number of cubic units in the volume of $S$ is given by

$$V = \lim_{\|\Delta\| \to 0} \sum_{i=1}^{n} f(\bar{y}_i, \bar{z}_i)\Delta_i A$$

$$= \iint_R f(y, z)\, dA \qquad (1)$$

Because the region $R$ in the $yz$ plane is bounded by the lines $y = 0$, $y = 1$, and the lines $z = 0$ and $z = 3 - 3y$, we may replace the double integral in (1) by an equivalent iterated integral. Thus,

$$V = \int_0^1 \int_0^{3-3y} (y + 2z + 1)dz\, dy$$

$$= \int_0^1 [yz + z^2 + z]_{z=0}^{z=3-3y}\, dy$$

$$= \int_0^1 [y(3-3y) + (3-3y)^2 + (3-3y)]\, dy$$

$$= \int_0^1 (6y^2 - 18y + 12)dy$$

$$= 2y^3 - 9y^2 + 12y\Big]_0^1$$

$$= 5$$

Thus, the volume is 5 cubic units.

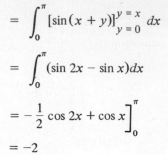

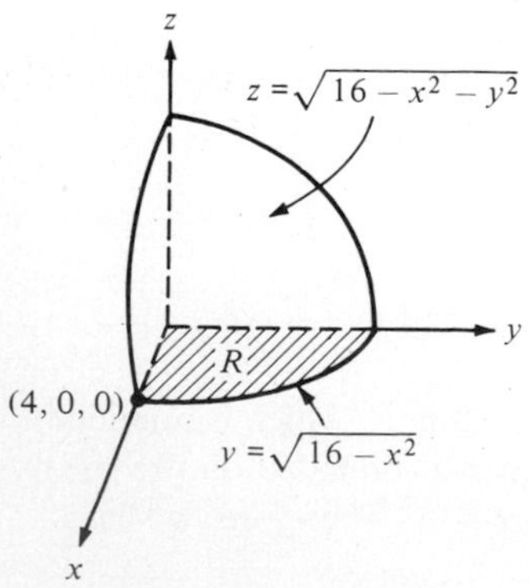

Figure 21.2.20

**20.** Find by double integration the volume of the portion of the solid bounded by the sphere $x^2 + y^2 + z^2 = 16$ which lies in the first octant. Draw a sketch of the solid.

SOLUTION: In Fig. 21.2.20, we show a sketch of the solid $S$. Let $f$ be the function defined by $f(x, y) = \sqrt{16 - x^2 - y^2}$. Then $S$ is bounded below by a region $R$ in the $xy$ plane and bounded above by the surface $z = f(x, y)$. Therefore,

$$V = \lim_{\|\Delta\| \to 0} \sum_{i=1}^{n} f(\bar{x}_i, \bar{y}_i)\, \Delta_i A$$

$$= \iint_R f(x, y)\, dA \qquad (1)$$

Because the region $R$ is bounded by the lines $x = 0$, $x = 4$, and the curves $y = 0$ and $y = \sqrt{16 - x^2}$, the double integral in (1) is equivalent to an iterated integral. Thus,

$$V = \int_0^4 \int_0^{\sqrt{16 - x^2}} \sqrt{16 - x^2 - y^2}\, dy\, dx \qquad (2)$$

To evaluate the "inner" integral in (2), we regard $x$ as a constant. We let $y = \sqrt{16 - x^2} \sin \theta$. Then $dy = \sqrt{16 - x^2} \cos \theta\, d\theta$, and $\sqrt{16 - x^2 - y^2} = \sqrt{16 - x^2} \cos \theta$. Furthermore, when $y = 0$, then $\theta = 0$, and when $y = \sqrt{16 - x^2}$, then $\theta = \frac{1}{2}\pi$. With these replacements in the right side of (2), we obtain

$$V = \int_0^4 \int_0^{\pi/2} (16 - x^2) \cos^2 \theta\, d\theta\, dx$$

$$= \frac{1}{2} \int_0^4 (16 - x^2) \int_0^{\pi/2} (1 + \cos 2\theta)\, d\theta\, dx$$

$$= \frac{1}{2} \int_0^4 (16 - x^2) \left[\theta + \frac{1}{2} \sin 2\theta\right]_0^{\pi/2} dx$$

$$= \frac{1}{2} \int_0^4 (16 - x^2)\, \frac{1}{2}\pi\, dx$$

$$= \frac{1}{4}\pi \left[16x - \frac{1}{3}x^3\right]_0^4$$

$$= \frac{32}{3}\pi$$

Thus, the volume of the solid is $\frac{32}{3}\pi$ cubic units.

**24.** Use double integrals to find the area of the region bounded by the given curves in the $xy$ plane. Draw a sketch of the region.

$$x^2 + y^2 = 16 \quad \text{and} \quad y^2 = 6x$$

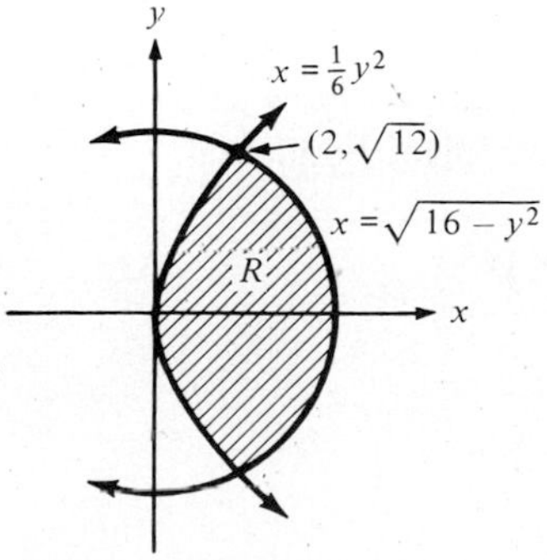

Figure 21.2.24

SOLUTION: In Fig. 21.2.24 we show a sketch of the region $R$ which is bounded by the lines $y = \pm 2\sqrt{3}$, and by the curves $x = \frac{1}{6}y^2$ and $x = \sqrt{16 - y^2}$. Thus,

$$A = \iint_R dx\, dy$$

$$= \int_{-2\sqrt{3}}^{2\sqrt{3}} \int_{\frac{1}{6}y^2}^{\sqrt{16 - y^2}} dx\, dy$$

$$= \int_{-2\sqrt{3}}^{2\sqrt{3}} [x]_{x = \frac{1}{6}y^2}^{x = \sqrt{16 - y^2}}$$

$$= \int_{-2\sqrt{3}}^{2\sqrt{3}} \left[\sqrt{16 - y^2} - \frac{1}{6}y^2\right] dy$$

$$= \int_{-2\sqrt{3}}^{2\sqrt{3}} \sqrt{16 - y^2}\, dy - \left[\frac{1}{18}y^3\right]_{-2\sqrt{3}}^{2\sqrt{3}}$$

$$= \int_{-2\sqrt{3}}^{2\sqrt{3}} \sqrt{16 - y^2}\, dy - \frac{8}{3}\sqrt{3} \qquad \textbf{(1)}$$

For the integral remaining in (1), we take $y = 4 \sin \theta$. Thus,

$$A = \int_{-\pi/3}^{\pi/3} 16 \cos^2 \theta\, d\theta - \frac{8}{3}\sqrt{3}$$

$$= 8 \int_{-\pi/3}^{\pi/3} (1 + \cos 2\theta) d\theta - \frac{8}{3}\sqrt{3}$$

$$= 8\left[\theta + \frac{1}{2} \sin 2\theta\right]_{-\pi/3}^{\pi/3} - \frac{8}{3}\sqrt{3}$$

$$= \frac{16}{3}\pi + \frac{4}{3}\sqrt{3}$$

Therefore, the area is $\frac{4}{3}(4\pi + \sqrt{3})$ square units.

**26.** Given the iterated integral

$$\int_0^a \int_0^x \sqrt{a^2 - x^2}\, dy\, dx$$

(a) Draw a sketch of the solid the measure of whose volume is represented by the given iterated integral.
(b) Evaluate the iterated integral.
(c) Write the iterated integral that gives the measure of the volume of the same solid with the order of the integration reversed.

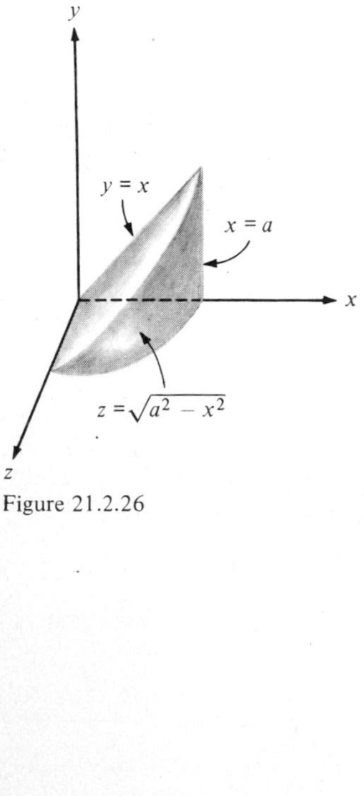

Figure 21.2.26

SOLUTION:

(a) The base of the solid is the triangle bounded by the lines $x = 0$ and $x = a$, and by the lines $y = 0$ and $y = x$. And the solid is bounded by the cylindrical surface $z = \sqrt{a^2 - x^2}$. In Fig. 21.2.26, we show a sketch of the solid. For a better perspective we take the $xz$ plane horizontal.

(b) $$\int_0^a \int_0^x \sqrt{a^2 - x^2}\, dy\, dx = \int_0^a [y\sqrt{a^2 - x^2}]_{y=0}^{y=x}\, dx$$

$$= \int_0^a x\sqrt{a^2 - x^2}\, dx$$

$$= -\frac{1}{3}(a^2 - x^2)^{3/2}\Big]_0^a$$

$$= \frac{1}{3}a^3$$

**(c)** We may regard the triangular region in the $xy$ plane as being bounded by the lines $y = 0, y = a$, and the lines $x = y$ and $x = a$. Therefore,

$$\int_0^a \int_0^x \sqrt{a^2 - x^2}\, dy\, dx = \int_0^a \int_y^a \sqrt{a^2 - x^2}\, dx\, dy$$

**30.** The iterated integral cannot be evaluated exactly in terms of elementary functions by the given order of integration. Reverse the order of integration and perform the computation.

$$\int_0^1 \int_y^1 e^{x^2}\, dx\, dy$$

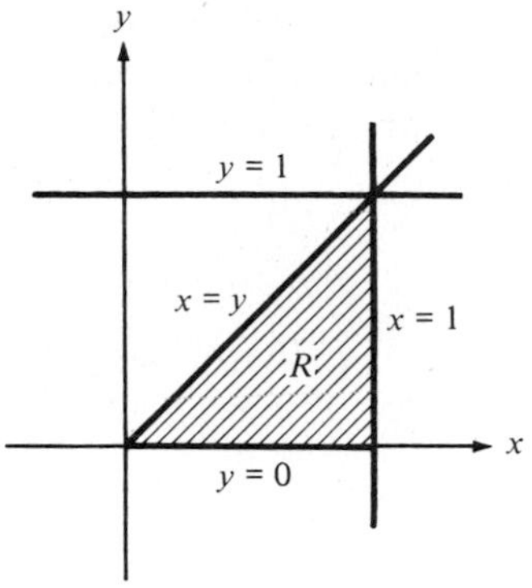

Figure 21.2.30

SOLUTION: The region $R$ over which the double integral is being taken is bounded by the lines $y = 0$, $y = 1$, and by the "curves" $x = y$ and $x = 1$, as illustrated in Fig. 21.2.30. We may also regard $R$ as being bounded by the lines $x = 0$, $x = 1$, and by the "curves" $y = 0$ and $y = x$. Therefore,

$$\int_0^1 \int_y^1 e^{x^2}\, dx\, dy = \int_0^1 \int_0^x e^{x^2}\, dy\, dx$$

$$= \int_0^1 [ye^{x^2}]_{y=0}^{y=x}\, dx$$

$$= \int_0^1 xe^{x^2}\, dx$$

$$= \frac{1}{2} e^{x^2} \Big]_0^1$$

$$= \frac{1}{2}(e - 1)$$

## 21.3 CENTER OF MASS AND MOMENTS OF INERTIA

If a lamina has the shape of a closed region $R$ in the $xy$ plane and $\rho(x, y)$ is the measure of the area density of the lamina at any point $(x, y)$ of $R$, where $\rho$ is continuous on $R$, then the measure of the total mass of the lamina is given by

$$M = \lim_{\|\Delta\| \to 0} \sum_{i=1}^{n} \rho(\bar{x}_i, \bar{y}_i)\, \Delta_i A$$

$$= \iint_R \rho(x, y)\, dA$$

Furthermore, if $(\bar{x}, \bar{y})$ is the center of mass of the lamina, we have

$$M \cdot \bar{x} = \lim_{\|\Delta\| \to 0} \sum_{i=1}^{n} \rho(\bar{x}_i, \bar{y}_i) \bar{x}_i\, \Delta_i A$$

$$= \iint_R \rho(x, y) x\, dA$$

and

$$M \cdot \bar{y} = \lim_{\|\Delta\| \to 0} \sum_{i=1}^{n} \rho(\bar{x}_i, \bar{y}_i)\bar{y}_i \, \Delta_i A$$

$$= \iint_R \rho(x, y) y \, dA$$

**21.3.1 Definition** The *moment of inertia* of a particle, whose mass is $m$ slugs, about an axis is defined to be $mr^2$ slug-ft$^2$, where $r$ ft is the perpendicular distance from the particle to the axis.

**21.3.2 Definition** Suppose that we are given a continuous distribution of mass occupying a region $R$ in the $xy$ plane, and suppose that the measure of area density of this distribution at the point $(x, y)$ is $\rho(x, y)$ slugs/ft$^2$, where $\rho$ is continuous on $R$. Then the moment of inertia $I_x$ slug-ft$^2$ about the $x$-axis of this distribution of mass is determined by

$$I_x = \lim_{\|\Delta\| \to 0} \sum_{i=1}^{n} \gamma_i^2 \rho(\xi_i, \gamma_i) \, \Delta_i A = \iint_R y^2 \rho(x, y) dA$$

Similarly, the measure $I_y$ of the moment of inertia about the $y$-axis is given by

$$I_y = \lim_{\|\Delta\| \to 0} \sum_{i=1}^{n} \xi_i^2 \rho(\xi_i, \gamma_i) \, \Delta_i A = \iint_R x^2 \rho(x, y) dA$$

and the measure $I_0$ of the moment of inertia about the origin, or the $z$-axis, is given by

$$I_0 = \lim_{\|\Delta\| \to 0} \sum_{i=1}^{n} (\xi_i^2 + \gamma_i^2)\rho(\xi_i, \gamma_i) \, \Delta_i A = \iint_R (x^2 + y^2)\rho(x, y) dA$$

The number $I_0$ is the measure of what is called the *polar moment of inertia.*

**21.3.3 Definition** If $I$ is the measure of the moment of inertia about an axis $L$ of a distribution of mass in a plane and $M$ is the measure of the total mass of the distribution, then the *radius of gyration* of the distribution about $L$ has measure $r$, where

$$r^2 = \frac{I}{M}$$

## *Exercises 21.3*

In Exercises 1-10, find the mass and the center of mass of the given lamina if the area density is as indicated. Mass is measured in slugs and distance is measured in feet.

2. A lamina in the shape of the region in the first quadrant bounded by the parabola $y = x^2$, the line $y = 1$, and the $y$-axis. The area density at any point is $(x + y)$ slugs/ft$^2$.

SOLUTION: A sketch of the region $R$ is shown in Fig. 21.3.2. Let $\rho$ be the area density function. Then $\rho(x, y) = x + y$, and the number of slugs in the mass of the lamina is given by

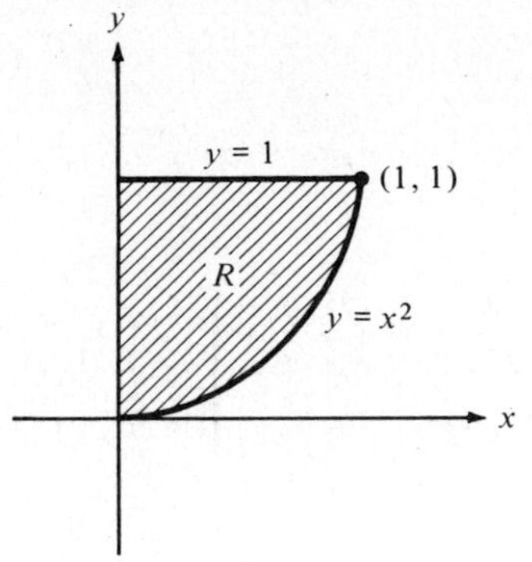


Figure 21.3.2

$$M = \lim_{\|\Delta\| \to 0} \sum_{i=1}^{n} \rho(\bar{x}_i, \bar{y}_i)\, \Delta_i A$$

$$= \iint_R \rho(x, y)\, dA$$

$$= \int_0^1 \int_{x^2}^1 (x + y)\, dy\, dx$$

$$= \int_0^1 \left[ xy + \frac{1}{2} y^2 \right]_{y=x^2}^{y=1} dx$$

$$= \int_0^1 \left( \frac{1}{2} + x - x^3 - \frac{1}{2} x^4 \right) dx$$

$$= \frac{1}{2}x + \frac{1}{2}x^2 - \frac{1}{4}x^4 - \frac{1}{10}x^5 \Big]_0^1$$

$$= \frac{13}{20} \tag{1}$$

Thus, the mass is $\frac{13}{20}$ slugs. Furthermore, if $(\bar{x}, \bar{y})$ is the center of mass, we have

$$M \cdot \bar{x} = \lim_{\|\Delta\| \to 0} \sum_{i=1}^{n} \rho(\bar{x}_i, \bar{y}_i)\bar{x}_i\, \Delta_i A$$

$$= \iint_R \rho(x, y) x\, dA$$

$$= \int_0^1 \int_{x^2}^1 (x + y) x\, dy\, dx$$

$$= \int_0^1 x \left[ xy + \frac{1}{2} y^2 \right]_{y=x^2}^{y=1} dx$$

$$= \int_0^1 \left( \frac{1}{2}x + x^2 - x^4 - \frac{1}{2} x^5 \right) dx$$

$$= \frac{1}{4}x^2 + \frac{1}{3}x^3 - \frac{1}{5}x^5 - \frac{1}{12}x^6 \Big]_0^1$$

$$= \frac{3}{10} \tag{2}$$

and

$$M \cdot \bar{y} = \lim_{\|\Delta\| \to 0} \sum_{i=1}^{n} \rho(\bar{x}_i, \bar{y}_i)\bar{y}_i\, \Delta_i A$$

$$= \iint_R \rho(x, y) y\, dA$$

$$= \int_0^1 \int_{x^2}^1 (x+y)y \, dy \, dx$$

$$= \int_0^1 \left[\frac{1}{2}xy^2 + \frac{1}{3}y^3\right]_{y=x^2}^{y=1} dx$$

$$= \int_0^1 \left(\frac{1}{3} + \frac{1}{2}x - \frac{1}{2}x^5 - \frac{1}{3}x^6\right) dx$$

$$= \frac{1}{3}x + \frac{1}{4}x^2 - \frac{1}{12}x^6 - \frac{1}{21}x^7\Big]_0^1$$

$$= \frac{19}{42} \tag{3}$$

Substituting from (1) into (2) and into (3), we have

$$\frac{13}{20}\bar{x} = \frac{3}{10} \quad \text{and} \quad \frac{13}{20}\bar{y} = \frac{19}{42}$$

$$\bar{x} = \frac{6}{13} \quad \text{and} \quad \bar{y} = \frac{190}{273}$$

Therefore, the center of mass is at $(\frac{6}{13}, \frac{190}{273})$.

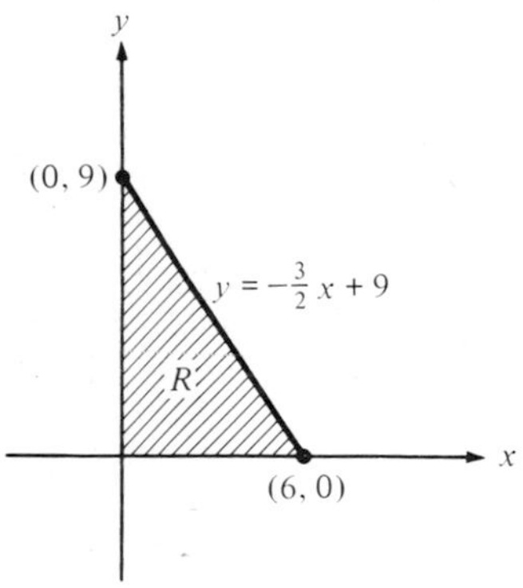

Figure 21.3.6

6. A lamina in the shape of the region bounded by the triangle whose sides are segments of the coordinate axes and the line $3x + 2y = 18$. The area density varies as the product of the distances from the coordinate axes.

SOLUTION: A sketch of the region $R$ is shown in Fig. 21.3.6. If $\rho(x, y)$ is the number of slugs/ft² in the area density, then $\rho(x, y) = kxy$. The number of slugs in the total mass is given by

$$M = \lim_{\|\Delta\| \to 0} \sum_{i=1}^{n} \rho(\bar{x}_i, \bar{y}_i) \, \Delta_i A$$

$$= \iint_R \rho(x, y) dA$$

$$= \int_0^6 \int_0^{-3x/2+9} kxy \, dy \, dx$$

$$= \frac{1}{2}k \int_0^6 x[y^2]_0^{-3x/2+9} dx$$

$$= \frac{1}{2}k \int_0^6 \left(\frac{9}{4}x^3 - 27x^2 + 81x\right) dx$$

$$= \frac{1}{2}k\left[\frac{9}{16}x^4 - 9x^3 + \frac{81}{2}x^2\right]_0^6$$

$$= \frac{243}{2}k \tag{1}$$

Thus, the mass is $\frac{243}{2}k$ slugs. If $(\bar{x}, \bar{y})$ is the center of mass, we have

$$\begin{aligned}
M \cdot \bar{x} &= \lim_{\|\Delta\| \to 0} \sum_{i=1}^{n} \rho(\bar{x}_i, \bar{y}_i)\bar{x}_i \, \Delta_i A \\
&= \iint_R \rho(x, y)x \, dA \\
&= \int_0^6 \int_0^{-3x/2+9} kx^2 y \, dy \, dx \\
&= \frac{1}{2}k \int_0^6 x^2 \big[y^2\big]_0^{-3x/2+9} dx \\
&= \frac{1}{2}k \int_0^6 \left(\frac{9}{4}x^4 - 27x^3 + 81x^2\right) dx \\
&= \frac{1}{2}k \left[\frac{9}{20}x^5 - \frac{27}{4}x^4 + 27x^3\right]_0^6 \\
&= \frac{1458}{5}k
\end{aligned} \tag{2}$$

and

$$\begin{aligned}
M \cdot \bar{y} &= \lim_{\|\Delta\| \to 0} \sum_{i=1}^{n} \rho(\bar{x}_i, \bar{y}_i)\bar{y}_i \, \Delta_i A \\
&= \iint_R kxy^2 \, dA
\end{aligned} \tag{3}$$

Because it is easier to integrate first with respect to $x$, we express the double integral in (3) as an iterated integral of the form $dx\,dy$. Thus, we regard the region $R$ as being bounded by the lines $y = 0, y = 9$, and the curves $x = 0, x = -\frac{2}{3}y + 6$. Hence, from (3) we obtain

$$\begin{aligned}
M \cdot \bar{y} &= \int_0^9 \int_0^{-2y/3+6} kxy^2 \, dx \, dy \\
&= \frac{1}{2}k \int_0^9 y^2 \big[x^2\big]_0^{-2y/3+6} dy \\
&= \frac{1}{2}k \int_0^9 \left(\frac{4}{9}y^4 - 8y^3 + 36y^2\right) dy \\
&= \frac{1}{2}k \left[\frac{4}{45}y^5 - 2y^4 + 12y^3\right]_0^9 \\
&= \frac{2187}{5}k
\end{aligned} \tag{4}$$

Substituting from (1) into (2) and (4), we obtain

$$\frac{243}{2}k\bar{x} = \frac{1458}{5}k \quad \text{and} \quad \frac{243}{2}k\bar{y} = \frac{2187}{5}k$$

$$\bar{x} = \frac{12}{5} \quad \text{and} \quad \bar{y} = \frac{18}{5}$$

Therefore, the center of mass is $(\frac{12}{5}, \frac{18}{5})$.

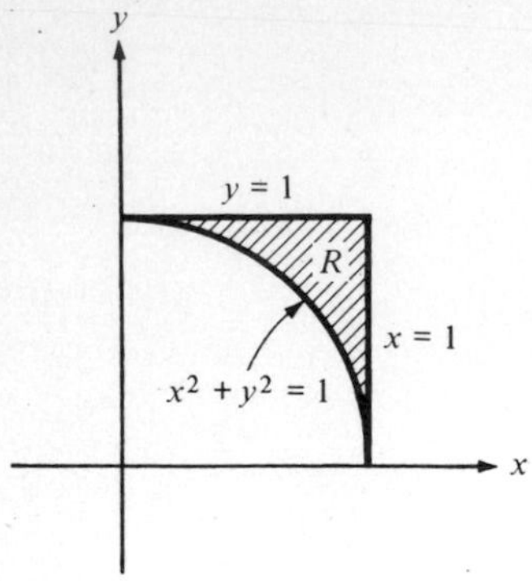

Figure 21.3.10

**10.** A lamina in the shape of the region bounded by the circle $x^2 + y^2 = 1$ and the lines $x = 1$ and $y = 1$. The area density at any point is $xy$ slugs/ft².

SOLUTION: Figure 21.3.10 shows a sketch of the region $R$. The number of slugs in the mass is given by

$$M = \lim_{\|\Delta\| \to 0} \sum_{i=1}^{n} \bar{x}_i \bar{y}_i \Delta_i x$$

$$= \iint_R xy \, dA$$

$$= \int_0^1 \int_{\sqrt{1-x^2}}^1 xy \, dy \, dx$$

$$= \frac{1}{2} \int_0^1 x[y^2]_{\sqrt{1-x^2}}^1 \, dx$$

$$= \frac{1}{2} \int_0^1 x[1 - (1 - x^2)] \, dx$$

$$= \frac{1}{2} \int_0^1 x^3 \, dx$$

$$= \frac{1}{8} \tag{1}$$

Thus, the mass is $\frac{1}{8}$ slugs. If $(\bar{x}, \bar{y})$ is the center of mass, then

$$M \cdot \bar{x} = \lim_{\|\Delta\| \to 0} \sum_{i=1}^{n} (\bar{x}_i \bar{y}_i) \bar{x}_i \Delta_i x$$

$$= \iint_R x^2 y \, dA$$

$$= \int_0^1 \int_{\sqrt{1-x^2}}^1 x^2 y \, dy \, dx$$

$$= \frac{1}{2} \int_0^1 x^2 [y^2]_{\sqrt{1-x^2}}^1 \, dx$$

$$= \frac{1}{2} \int_0^1 x^4 dx$$

$$= \frac{1}{10} \tag{2}$$

and

$$M \cdot \bar{y} = \lim_{\|\Delta\| \to 0} \sum_{i=1}^{n} (\bar{x}_i \bar{y}_i) \bar{y}_i \Delta_i x$$

$$= \iint_R xy^2 \, dA$$

$$= \int_0^1 \int_{\sqrt{1-y^2}}^1 xy^2 \, dx \, dy$$

$$= \frac{1}{2} \int_0^1 y^2 [x^2]_{\sqrt{1-y^2}}^1 \, dy$$

$$= \frac{1}{2} \int_0^1 y^4 dy$$

$$= \frac{1}{10} \tag{3}$$

Substituting from (1) into (2) and (3), we get

$$\frac{1}{8}\bar{x} = \frac{1}{10} \quad \text{and} \quad \frac{1}{8}\bar{y} = \frac{1}{10}$$

$$\bar{x} = \frac{4}{5} \quad \text{and} \quad \bar{y} = \frac{4}{5}$$

Therefore, the center of mass is $(\frac{4}{5}, \frac{4}{5})$.

**14.** Find the moment of inertia of the given homogeneous lamina about the indicated axis if the area density is $\rho$ slugs/ft$^2$ and the distance is measured in feet. A lamina in the shape of the region bounded by the parabola $x^2 = 4 - 4y$ and the $x$-axis, about the $x$-axis.

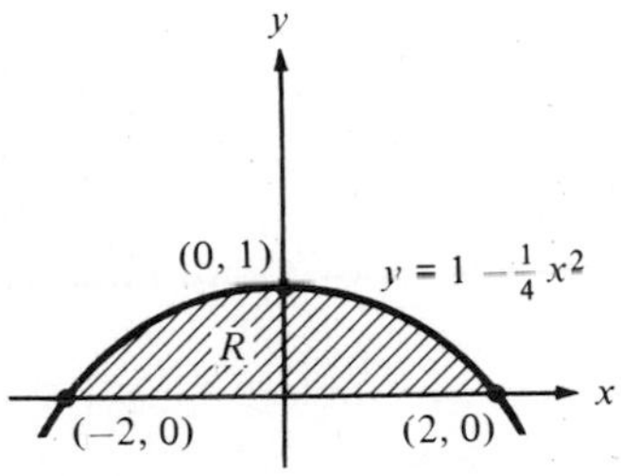

Figure 21.3.14

SOLUTION: Fig. 21.3.14 shows a sketch of the region $R$. The number of slug-ft$^2$ in the moment of inertia about the $x$-axis is given by

$$I_x = \lim_{\|\Delta\| \to 0} \sum_{i=1}^{n} \rho \, \bar{y}_i^2 \, \Delta_i A$$

$$= \iint_R \rho \, y^2 \, dA$$

$$= \rho \int_{-2}^{2} \int_0^{1 - \frac{1}{4}x^2} y^2 \, dy \, dx$$

$$= \frac{1}{3}\rho \int_{-2}^{2} [y^3]_0^{1 - \frac{1}{4}x^2} dx$$

$$= \frac{1}{3}\rho \int_{-2}^{2} \left(1 - \frac{3}{4}x^2 + \frac{3}{16}x^4 - \frac{1}{64}x^6\right) dx$$

$$= \frac{1}{3}\rho \left[x - \frac{1}{4}x^3 + \frac{3}{80}x^5 - \frac{1}{448}x^7\right]_{-2}^{2}$$

$$= \frac{64}{105}\rho$$

Thus, the moment of inertia about the $x$-axis is $\frac{64}{105}\rho$ slug-ft$^2$.

**18.** Find for the given lamina each of the following:
(a) the moment of inertia about the $x$-axis
(b) the moment of inertia about the $y$-axis
(c) the radius of gyration about the $x$-axis
(d) the polar moment of inertia.

The lamina of Exercise 2.

SOLUTION: Refer to Fig. 21.3.2 which shows a sketch of the region $R$. In Exercise 2 we are given that $\rho(x, y) = x + y$. We apply Definition 21.3.2. Thus,

(a)
$$I_x = \lim_{\|\Delta\| \to 0} \sum_{i=1}^{n} \rho(\bar{x}_i, \bar{y}_i)\bar{y}_i^2 \, \Delta_i A$$

$$= \iint_R (x + y)y^2 \, dA$$

$$= \int_0^1 \int_0^{\sqrt{y}} (x + y)y^2 \, dx \, dy$$

$$= \int_0^1 y^2 \left[\frac{1}{2}x^2 + xy\right]_{x=0}^{x=\sqrt{y}} dy$$

$$= \int_0^1 y^2 \left[\frac{1}{2}y + y^{3/2}\right] dy$$

$$= \int_0^1 \left(\frac{1}{2}y^3 + y^{7/2}\right) dy$$

$$= \frac{25}{72}$$

Thus, the moment of inertia about the $x$-axis is $\frac{25}{72}$ slug-ft$^2$.

(b) The measure of the moment of inertia about the $y$-axis is given by

$$I_y = \lim_{\|\Delta\| \to 0} \sum_{i=1}^{n} \rho(\bar{x}_i, \bar{y}_i)\bar{x}_i^2 \, \Delta_i A$$

$$= \iint_R (x + y)x^2 \, dA$$

$$= \int_0^1 \int_{x^2}^{1} (x + y)x^2 \, dy \, dx$$

$$= \int_0^1 x^2 \left[xy + \frac{1}{2}y^2\right]_{y=x^2}^{y=1} dx$$

$$= \int_0^1 x^2 \left[x + \frac{1}{2} - x^3 - \frac{1}{2}x^4\right] dx$$

$$= \int_0^1 \left(\frac{1}{2}x^2 + x^3 - x^5 - \frac{1}{2}x^6\right) dx$$

$$= \left[\frac{1}{6}x^3 + \frac{1}{4}x^4 - \frac{1}{6}x^6 - \frac{1}{14}x^7\right]_0^1$$

$$= \frac{5}{28}$$

Thus, the moment of inertia about the $y$-axis is $\frac{5}{28}$ slug-ft$^2$.

**(c)** If $r$ feet is the radius of gyration about the $x$-axis, then

$$r^2 = \frac{I_x}{M} \tag{1}$$

In part (a) we found $I_x = \frac{25}{72}$, and in Exercise 2 we found $M = \frac{13}{20}$. Substituting these values in (1), we obtain

$$r^2 = \frac{125}{234}$$

$$r = \frac{5}{78}\sqrt{130}$$

Thus, the radius of gyration about the $x$-axis is $\frac{5}{78}\sqrt{130}$ feet.

**(d)** The measure of the polar moment of inertia is given by

$$I_0 = \lim_{\|\Delta\| \to 0} \sum_{i=1}^{n} \rho(\bar{x}_i, \bar{y}_i)(\bar{x}_i^2 + \bar{y}_i^2)\, \Delta_i A$$

$$= \iint_R (x + y)(x^2 + y^2)\, dA$$

$$= \int_0^1 \int_{x^2}^1 (x + y)(x^2 + y^2)\, dy\, dx$$

$$= \int_0^1 \int_{x^2}^1 (x^3 + x^2 y + xy^2 + y^3)\, dy\, dx$$

$$= \int_0^1 \left[x^3 y + \frac{1}{2}x^2 y^2 + \frac{1}{3}xy^3 + \frac{1}{4}y^4\right]_{y=x^2}^{y=1} dx$$

$$= \int_0^1 \left[\left(x^3 + \frac{1}{2}x^2 + \frac{1}{3}x + \frac{1}{4}\right) - \left(x^5 + \frac{1}{2}x^6 + \frac{1}{3}x^7 + \frac{1}{4}x^8\right)\right] dx$$

$$= \left[\frac{1}{4}x^4 + \frac{1}{6}x^3 + \frac{1}{6}x^2 + \frac{1}{4}x - \frac{1}{6}x^6 - \frac{1}{14}x^7 - \frac{1}{24}x^8 - \frac{1}{36}x^9\right]_0^1$$

$$= \frac{265}{504}$$

Thus, the polar moment of inertia is $\frac{265}{504}$ slug-ft$^2$.

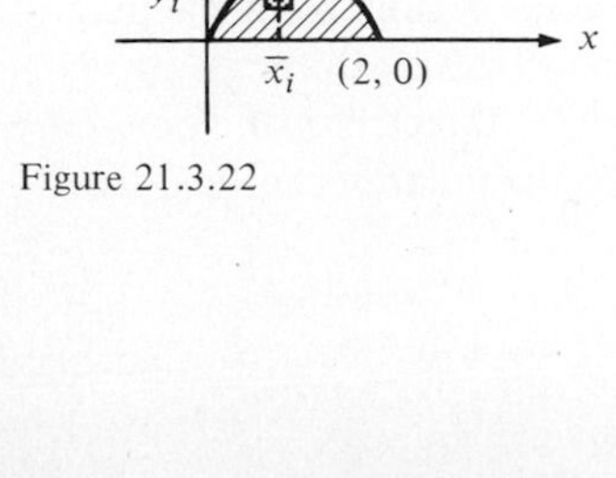

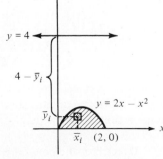

Figure 21.3.22

22. A lamina is in the shape of the region enclosed by the parabola $y = 2x - x^2$ and the $x$-axis. Find the moment of inertia of the lamina about the line $y = 4$ if the area density varies as its distance from the line $y = 4$. Mass is measured in slugs and distance is measured in feet.

SOLUTION: Fig. 21.3.22 shows a sketch of the region $R$. Let $(\bar{x}_i, \bar{y}_i)$ be the center of mass of the element of mass. Then $r_i = 4 - \bar{y}_i$ is the number of feet in the element of distance to the axis $y = 4$, and the number of slugs/ft$^2$ in the element of density is given by $\rho_i = k(4 - \bar{y}_i)$. Thus, the measure of the moment of inertia about the line $y = 4$ is given by

$$I = \lim_{\|\Delta\| \to 0} \sum_{i=1}^{n} \rho_i r_i^2 \Delta_i A$$

$$= \iint_R k(4 - y)(4 - y)^2 dA$$

$$= k \int_0^2 \int_0^{2x - x^2} (4 - y)^3 dy\, dx$$

$$= -\frac{1}{4} k \int_0^2 [(4 - y)^4]_0^{2x - x^2} dx$$

$$= -\frac{1}{4} k \int_0^2 [(4 - 2x + x^2)^4 - 4^4]\, dx$$

$$= -\frac{1}{4} k \int_0^2 ([3 + (x - 1)^2]^4 - 4^4) dx \qquad (1)$$

We let $z = x - 1$ in Eq. (1); thus,

$$I = -\frac{1}{4} k \int_{-1}^{1} [(3 + z^2)^4 - 4^4]\, dz$$

$$= -\frac{1}{4} k \int_{-1}^{1} [z^8 + 12z^6 + 54z^4 + 108z^2 + 81 - 256]\, dz$$

$$= -\frac{1}{4} k \left[\frac{1}{9} z^9 + \frac{12}{7} z^7 + \frac{54}{5} z^5 + 36z^3 - 175z\right]_{-1}^{1}$$

$$= \frac{19904}{315} k$$

## 21.4 THE DOUBLE INTEGRAL IN POLAR COORDINATES

Let $f$ be a function of two variables $r$ and $\theta$. Let $R$ be a region in the polar plane bounded by the rays $\theta = \theta_1, \theta = \theta_2$, and bounded by the curves $r = g_1(\theta)$ and $r = g_2(\theta)$, with $0 \leqslant g_1(\theta) \leqslant g_2(\theta)$ for all $\theta_1 \leqslant \theta \leqslant \theta_2$. Then the double integral of $f$ on $R$ is equivalent to an iterated integral, and

$$\iint_R f(r, \theta) dA = \iint_R f(r, \theta) r\, dr\, d\theta$$

$$= \int_{\theta_1}^{\theta_2} \int_{g_1(\theta)}^{g_2(\theta)} f(r, \theta) r\, dr\, d\theta$$

Double integrals in polar coordinates are used to calculate area, mass, volume, moments, and centers of mass, just as were double integrals in Cartesian coordinates. For example, if $S$ is a solid that is bounded below by some region $R$ in the polar plane, bounded above by the surface $z = f(r, \theta)$, where $f(r, \theta) \geqslant 0$ for all $(r, \theta)$ in $R$, and bounded on the sides by a cylindrical surface whose directrix is the boundary of the region $R$ and whose rulings are perpendicular to the polar plane, then the measure of the volume of $S$ is given by

$$V = \lim_{\|\Delta\| \to 0} \sum_{i=1}^{n} f(\bar{r}_i, \bar{\theta}_i)\, \Delta_i A$$

$$= \iint_R f(r, \theta) r\, dr\, d\theta$$

## Exercises 21.4

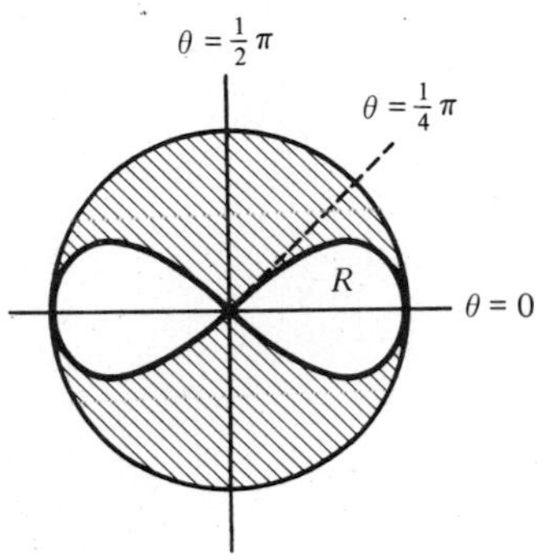

Figure 21.4.4

**4.** Use a double integral to find the area of the region inside the circle $r = 1$ and outside the lemniscate $r^2 = \cos 2\theta$.

SOLUTION: Fig. 21.4.4 shows the region. Let $R$ be the first quadrant part of the region inside the lemniscate. We take the area of the circle minus four times the area of $R$. Because $\cos 2\theta = 0$ when $\theta = \frac{1}{4}\pi$, the region $R$ is bounded by the rays $\theta = 0$ and $\theta = \frac{1}{4}\pi$. Furthermore, $R$ is bounded by the curves $r = 0$ and $r = \sqrt{\cos 2\theta}$ for $0 \leqslant \theta \leqslant \frac{1}{4}\pi$. Thus, the measure of the area of $R$ is given by

$$A = \lim_{\|\Delta\| \to 0} \sum_{i=1}^{n} \Delta_i A$$

$$= \iint_R r\, dr\, d\theta$$

$$= \int_0^{\pi/4} \int_0^{\sqrt{\cos 2\theta}} r\, dr\, d\theta$$

$$= \frac{1}{2} \int_0^{\pi/4} [r^2]_0^{\sqrt{\cos 2\theta}}\, d\theta$$

$$= \frac{1}{2} \int_0^{\pi/4} \cos 2\theta\, d\theta$$

$$= \frac{1}{4} \sin 2\theta \Big]_0^{\pi/4}$$

$$= \frac{1}{4}$$

Hence,

$$4A = 1$$

Because the area of the circle is $\pi$ square units, then the area of the region outside the lemniscate and inside the circle is $\pi - 1$ square units.

In Exercises 7-12, find the volume of the given solid.

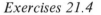

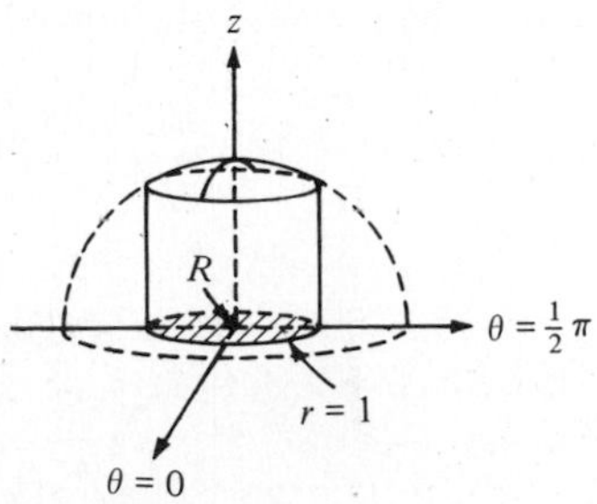

Figure 21.4.8

**8.** The solid cut out of the sphere $z^2 + r^2 = 4$ by the cylinder $r = 1$.

SOLUTION: A sketch of the half of the solid that lies above the polar plane is shown in Fig. 21.4.8. The base of the solid is the region $R$ in the polar plane. Solving the equation of the sphere for $z$, we obtain $z = \pm\sqrt{4 - r^2}$. We take $f(r, \theta) = \sqrt{4 - r^2}$.

Thus, if $V$ is the measure of the volume of the entire solid, we have

$$\frac{1}{2}V = \lim_{\|\Delta\|\to 0} \sum_{i=1}^{n} f(\bar{r}_i, \bar{\theta}_i)\,\Delta_i A$$

$$= \iint_R f(r, \theta) r\, dr\, d\theta$$

$$= \int_0^{2\pi} \int_0^1 \sqrt{4 - r^2}\, r\, dr\, d\theta$$

$$= -\frac{1}{3} \int_0^{2\pi} (4 - r^2)^{3/2}\Big]_0^1 d\theta$$

$$= -\frac{1}{3} \int_0^{2\pi} (3^{3/2} - 8) d\theta$$

$$= \frac{2}{3}\pi(8 - 3\sqrt{3})$$

Thus,

$$V = \frac{4}{3}\pi(8 - 3\sqrt{3})$$

Therefore, the volume of the solid is $\frac{4}{3}\pi(8 - 3\sqrt{3})$ cubic units.

**12.** The solid above the paraboloid $z = r^2$ and below the plane $z = 2r \sin\theta$.

SOLUTION: We find the intersection of the paraboloid and the plane. Eliminating $z$ from the two given equations, we have

$$r^2 - 2r \sin\theta = 0$$

$$r(r - 2\sin\theta) = 0$$

$$r = 0 \quad \text{and} \quad r = 2\sin\theta$$

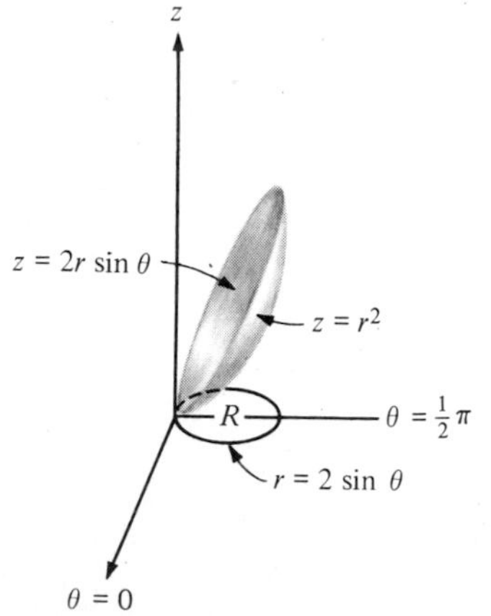

Figure 21.4.12

Thus, the given surfaces intersect at the pole and in a curve whose projection onto the polar plane is the circle $r = 2\sin\theta$. Fig. 21.4.12 shows a sketch of the solid and its projection $R$ in the polar plane. Because the solid is bounded above by the plane $z = 2r\sin\theta$ and bounded below by the paraboloid $z = r^2$, the measure of an element of volume is given by

$$(2\bar{r}_i \sin\bar{\theta}_i - \bar{r}_i^2)\,\Delta_i A$$

where $\Delta_i A$ is the measure of an element of area in the region $R$ and $(\bar{r}_i, \bar{\theta}_i)$ is a point in the curved rectangular element of area. Therefore, the measure of the volume of the solid is given by

$$V = \lim_{\|\Delta\|\to 0} \sum_{i=1}^{n} (2\bar{r}_i \sin\bar{\theta}_i - \bar{r}_i^2)\,\Delta_i A$$

$$= \iint_R (2r\sin\theta - r^2) dA$$

$$= \int_0^{\pi} \int_0^{2 \sin \theta} (2r \sin \theta - r^2) r \, dr \, d\theta$$

$$= \int_0^{\pi} \left[ \frac{2}{3} r^3 \sin \theta - \frac{1}{4} r^4 \right]_{r=0}^{r=2 \sin \theta} d\theta$$

$$= \int_0^{\pi} \left( \frac{16}{3} \sin^4 \theta - 4 \sin^4 \theta \right) d\theta$$

$$= \frac{4}{3} \int_0^{\pi} \sin^4 \theta \, d\theta$$

$$= \frac{1}{3} \int_0^{\pi} (1 - 2 \cos 2\theta + \cos^2 2\theta) \, d\theta$$

$$= \frac{1}{3} \int_0^{\pi} \left( \frac{3}{2} - 2 \cos 2\theta + \frac{1}{2} \cos 4\theta \right) d\theta$$

$$= \frac{1}{3} \left[ \frac{3}{2} \theta - \sin 2\theta + \frac{1}{8} \sin 4\theta \right]_0^{\pi}$$

$$= \frac{1}{2} \pi$$

Thus, the volume is $\frac{1}{2}\pi$ cubic units.

**16.** Find the mass and center of mass of a lamina in the shape of the region bounded by the limacon $r = 2 + \cos \theta, 0 \leqslant \theta \leqslant \pi$, and the polar axis with distance measured in feet. The area density at any point is $k \sin \theta$ slugs/ft$^2$.

SOLUTION: The measure of the mass is given by

$$M = \lim_{\|\Delta\| \to 0} \sum_{i=1}^{n} (k \sin \bar{\theta}_i) \, \Delta_i A$$

$$= \iint_R k \sin \theta \, dA$$

$$= k \int_0^{\pi} \int_0^{2 + \cos \theta} \sin \theta \, r \, dr \, d\theta$$

$$= \frac{1}{2} k \int_0^{\pi} \sin \theta \, [r^2]_0^{2 + \cos \theta} \, d\theta$$

$$= \frac{1}{2} k \int_0^{\pi} (2 + \cos \theta)^2 \sin \theta \, d\theta$$

$$= -\frac{1}{6} k (2 + \cos \theta)^3 ]_0^{\pi}$$

$$= \frac{13}{3}k \tag{1}$$

Thus, the mass is $\frac{13}{3}k$ slugs.

If $(\bar{x}, \bar{y})$ is the center of mass, then

$$M \cdot \bar{x} = \lim_{\|\Delta\| \to 0} \sum_{i=1}^{n} (k \sin \bar{\theta}_i)\bar{x}_i \, \Delta_i \theta$$

$$= \lim_{\|\Delta\| \to 0} \sum_{i=1}^{n} (k \sin \bar{\theta}_i)(\bar{r}_i \cos \bar{\theta}_i) \, \Delta_i A$$

$$= \iint_R (k \sin \theta)(r \cos \theta) \, dA$$

$$= k \int_0^{\pi} \int_0^{2+\cos\theta} \sin \theta \cos \theta \, r^2 \, dr \, d\theta$$

$$= \frac{1}{3}k \int_0^{\pi} \sin \theta \cos \theta (2 + \cos \theta)^3 \, d\theta$$

$$= \frac{1}{3}k \int_0^{\pi} [8 \cos \theta + 12 \cos^2 \theta + 6 \cos^3 \theta + \cos^4 \theta] \sin \theta \, d\theta$$

$$= -\frac{1}{3}k \left[4 \cos^2 \theta + 4 \cos^3 \theta + \frac{3}{2} \cos^4 \theta + \frac{1}{5} \cos^5 \theta\right]_0^{\pi}$$

$$= \frac{14}{5}k \tag{2}$$

and

$$M \cdot \bar{y} = \lim_{\|\Delta\| \to 0} \sum_{i=1}^{n} (k \sin \bar{\theta}_i)\bar{y}_i \, \Delta_i A$$

$$= \lim_{\|\Delta\| \to 0} \sum_{i=1}^{n} (k \sin \bar{\theta}_i)\bar{r}_i \sin \bar{\theta}_i \, \Delta_i A$$

$$= \iint_R kr \sin^2 \theta \, dA$$

$$= k \int_0^{\pi} \int_0^{2+\cos\theta} r \sin^2 \theta \, r \, dr \, d\theta$$

$$= \frac{1}{3}k \int_0^{\pi} \sin^2 \theta (2 + \cos \theta)^3 \, d\theta$$

$$= \frac{1}{3}k \int_0^{\pi} \sin^2 \theta (8 + 12 \cos \theta + 6 \cos^2 \theta + \cos^3 \theta) \, d\theta$$

$$= \frac{1}{3}k\left[8\int_0^\pi \sin^2\theta\, d\theta + 12\int_0^\pi \sin^2\theta\cos\theta\, d\theta + 6\int_0^\pi \sin^2\theta\cos^2\theta\, d\theta\right.$$

$$\left. + \int_0^\pi \sin^2\theta\cos^3\theta\, d\theta\right]$$

$$= \frac{1}{3}k\left[4\int_0^\pi (1-\cos 2\theta)\, d\theta + 12\int_0^\pi \sin^2\theta\cos\theta\, d\theta\right.$$

$$\left. + \frac{3}{4}\int_0^\pi (1-\cos 4\theta)\, d + \int_0^\pi (\sin^2\theta - \sin^4\theta)\cos\theta\, d\theta\right]$$

$$= \frac{1}{3}k\left[4\left(\theta - \frac{1}{2}\sin 2\theta\right) + 4\sin^3\theta + \frac{3}{4}\left(\theta - \frac{1}{4}\sin 4\theta\right)\right.$$

$$\left. + \left(\frac{1}{3}\sin^3\theta - \frac{1}{5}\sin^5\theta\right)\right]_0^\pi$$

$$= \frac{19}{12}\pi k \qquad (3)$$

Substituting from (1) into (2) and (3), we have

$$\frac{13}{3}k\bar{x} = \frac{14}{5}k \quad \text{and} \quad \frac{13}{3}k\bar{y} = \frac{19}{12}\pi k$$

$$\bar{x} = \frac{42}{65} \qquad \bar{y} = \frac{19}{52}\pi$$

Therefore, the center of mass is $(\frac{42}{65}, \frac{19}{52}\pi)$.

In Exercises 20-24, find the moment of inertia of the given lamina about the indicated axis or point if the area density is as indicated. Mass is measured in slugs, and distance is measured in feet.

**20.** A lamina in the shape of the region enclosed by the circle $r = \sin\theta$, about the $\frac{1}{2}\pi$ axis. The area density at any point is $k$ slugs/ft$^2$.

SOLUTION: Let $(\bar{r}_i, \bar{\theta}_i)$ be the center of mass of the element of mass. Then the number of feet in the distance to the $\frac{1}{2}\pi$ axis is given by $\bar{x}_i = \bar{r}_i\cos\bar{\theta}_i$. Thus, the measure of the moment of inertia about the $\frac{1}{2}\pi$ axis is given by

$$I = \lim_{\|\Delta\|\to 0}\sum_{i=1}^{n} k\bar{x}_i^2\,\Delta_i A$$

$$= \iint_R k(r\cos\theta)^2\, dA$$

$$= k\int_0^\pi\int_0^{\sin\theta} r^3\cos^2\theta\, dr\, d\theta$$

$$= \frac{1}{4}k\int_0^\pi \sin^4\theta\cos^2\theta\, d\theta$$

$$= \frac{1}{32}k \int_0^{\pi} (1 - \cos 2\theta)^2 (1 + \cos 2\theta) d\theta$$

$$= \frac{1}{32}k \int_0^{\pi} (1 - \cos 2\theta - \cos^2 2\theta + \cos^3 2\theta) d\theta$$

$$= \frac{1}{32}k \int_0^{\pi} \left[1 - \cos 2\theta - \frac{1}{2} - \frac{1}{2}\cos 4\theta + (1 - \sin^2 2\theta)\cos 2\theta\right] d\theta$$

$$= \frac{1}{32}k \left[\frac{1}{2}\theta - \frac{1}{2}\sin 2\theta - \frac{1}{8}\sin 4\theta + \frac{1}{2}\sin 2\theta - \frac{1}{6}\sin^3 2\theta\right]_0^{\pi}$$

$$= \frac{1}{64}\pi k$$

**28.** Evaluate by polar coordinates the double integral

$$\iint_R \frac{x}{\sqrt{x^2 + y^2}}\, dA$$

where $R$ is the region in the first quadrant bounded by the circle $x^2 + y^2 = 1$ and the coordinate axis.

SOLUTION:

$$\iint_R \frac{x}{\sqrt{x^2 + y^2}}\, dA = \int_0^{\pi/2} \int_0^1 \frac{r \cos \theta}{r} r\, dr\, d\theta$$

$$= \frac{1}{2} \int_0^{\pi/2} \cos \theta\, d\theta$$

$$= \frac{1}{2} \sin \theta \Big]_0^{\pi/2}$$

$$= \frac{1}{2}$$

## 21.5 AREA OF A SURFACE

**21.5.1 Theorem** Suppose that $f$ and its first partial derivatives are continuous on the closed region $R$ in the $xy$ plane. Then if $\sigma$ is the measure of the area of the surface $z = f(x, y)$, which lies over $R$,

$$\sigma = \iint_R \sqrt{f_x^2(x, y) + f_y^2(x, y) + 1}\, dx\, dy$$

**21.5.2 Theorem** Suppose that the function $f$ is positive on $[a, b]$ and $f'$ is continuous on $[a, b]$. Then if $\sigma$ is the measure of the area of the surface of revolution obtained by revolving the curve $y = f(x)$, with $a \leqslant x \leqslant b$, about the $x$-axis,

$$\sigma = 2\pi \int_a^b f(x) \sqrt{[f'(x)]^2 + 1}\, dx$$

*Exercises 21.5*

**4.** Find the area of the surface in the first octant which is cut from the cone $x^2 + y^2 = z^2$ by the plane $x + y = 4$.

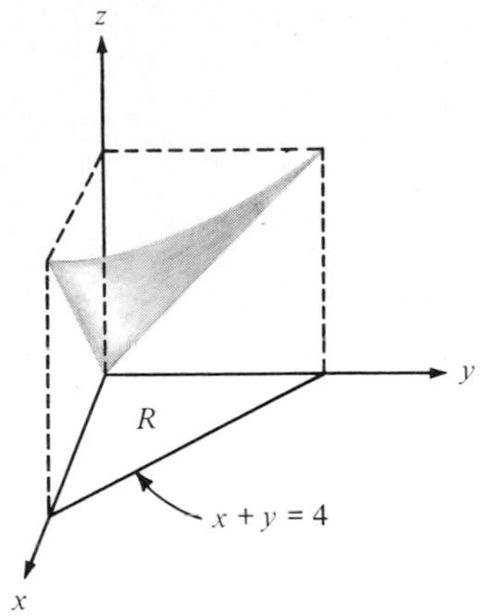


Figure 21.5.4

SOLUTION: Fig. 21.5.4 shows a sketch of the surface which lies over the triangle $R$ in the $xy$ plane bounded by the $x$-axis, the $y$-axis, and the line $x + y = 4$. We take

$$f(x, y) = \sqrt{x^2 + y^2}$$

Thus,

$$f_x(x, y) = \frac{x}{\sqrt{x^2 + y^2}} \quad \text{and} \quad f_y(x, y) = \frac{y}{\sqrt{x^2 + y^2}}$$

$$\sqrt{f_x^2(x, y) + f_y^2(x, y) + 1} = \sqrt{\frac{x^2}{x^2 + y^2} + \frac{y^2}{x^2 + y^2} + 1} = \sqrt{2}$$

Applying Theorem 21.5.1, the measure of the area of the surface is given by

$$A = \iint_R \sqrt{2}\, dA$$

$$= \sqrt{2} \int_0^4 \int_0^{4-x} dy\, dx$$

$$= \sqrt{2} \int_0^4 (4 - x) dx$$

$$= \sqrt{2} \left[ 4x - \frac{1}{2}x^2 \right]_0^4$$

$$= 8\sqrt{2}$$

Thus, the area of the surface is $8\sqrt{2}$ square units.

**8.** Find the area of the portion of the surface of the paraboloid $x^2 + y^2 = 3z$, which lies within the sphere $x^2 + y^2 + z^2 = 4z$.

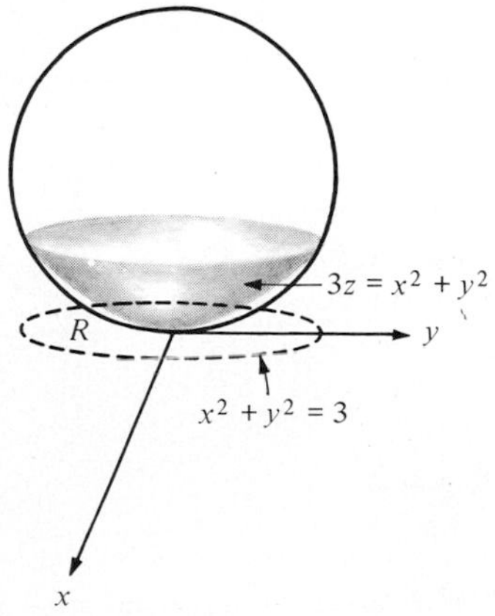


Figure 21.5.8

SOLUTION: We find the intersection of the paraboloid and the sphere. Eliminating $x^2$ and $y^2$ from the equations, we obtain $z^2 = z$. Therefore, the surfaces intersect in the planes $z = 0$ and $z = 1$. The curve of intersection in the surface $z = 1$ is the circle $x^2 + y^2 = 3, z = 1$. Completing the square on terms involving $z$ in the equation of the sphere, we have

$$x^2 + y^2 + (z - 2)^2 = 4$$

Thus, the sphere has center at $(0, 0, 2)$ and radius 2. Fig. 21.5.8 shows a sketch of the surface, which lies over the circle $x^2 + y^2 = 3$ in the $xy$ plane. We take

$$f(x, y) = \frac{1}{3}(x^2 + y^2)$$

Thus,

$$f_x(x, y) = \frac{2}{3}x \quad \text{and} \quad f_y(x, y) = \frac{2}{3}y$$

and

$$\sqrt{f_x^2(x, y) + f_y^2(x, y) + 1} = \sqrt{\frac{4}{9}x^2 + \frac{4}{9}y^2 + 1}$$

Therefore, the measure of the area of the surface is given by

$$A = \iint_R \sqrt{\frac{4}{9}(x^2 + y^2) + 1}\, dA$$

Using polar coordinates, we have

$$A = \int_0^{\sqrt{3}} \int_0^{2\pi} \sqrt{\frac{4}{9}r^2 + 1}\; r\, d\theta\, dr$$

$$= 2\pi \int_0^{\sqrt{3}} \sqrt{\frac{4}{9}r^2 + 1}\; r\, dr$$

$$= 2\pi\left(\frac{9}{8}\right)\frac{2}{3}\left[\left(\frac{4}{9}r^2 + 1\right)^{3/2}\right]_0^{\sqrt{3}}$$

$$= \frac{3}{2}\pi\left(\frac{7}{9}\sqrt{21} - 1\right)$$

Therefore, the area of the surface is $\frac{3}{2}\pi(\frac{7}{9}\sqrt{21} - 1)$ square units.

**12.** Find the area of the surface of revolution obtained by revolving the catenary $y = a\cosh(x/a)$ from $x = 0$ to $x = a$ about the $x$-axis.

SOLUTION: We apply Definition 21.5.2. Thus,

$$f(x) = a\cosh\left(\frac{x}{a}\right)$$

$$f'(x) = \sinh\left(\frac{x}{a}\right)$$

$$\sqrt{[f'(x)]^2 + 1} = \sqrt{\sinh^2 \frac{x}{a} + 1}$$

$$= \sqrt{\cosh^2 \frac{x}{a}}$$

$$= \cosh \frac{x}{a}$$

Then the measure of the surface area is

$$A = 2\pi \int_0^a a\cosh^2 \frac{x}{a}\, dx$$

$$= \pi a \int_0^a \left[1 + \cosh\left(\frac{2x}{a}\right)\right] dx$$

$$= \pi a \left[x + \frac{1}{2}a\sinh\left(\frac{2x}{a}\right)\right]_0^a$$

$$= \pi a\left(a + \frac{1}{2}a\sinh 2\right)$$

Hence, the surface area is $\frac{1}{2}\pi a^2(2 + \sinh 2)$ square units.

**16.** Find the area of the surface cut from the hyperbolic paraboloid $y^2 - x^2 = 6z$ by the cylinder $x^2 + y^2 = 36$.

SOLUTION: We take

$$f(x, y) = \frac{1}{6}(y^2 - x^2)$$

Then

$$f_x(x, y) = -\frac{1}{3}x \quad \text{and} \quad f_y(x, y) = \frac{1}{3}y$$

and

$$\sqrt{f_x^2(x, y) + f_y^2(x, y) + 1} = \sqrt{\frac{1}{9}(x^2 + y^2) + 1}$$

The surface lies over (and under) the circle $R$ in the $xy$ plane which is the graph of $x^2 + y^2 = 36$. Therefore, the measure of the surface area is

$$A = \iint_R \sqrt{\frac{1}{9}(x^2 + y^2) + 1}\, dA$$

Using polar coordinates, we have

$$\begin{aligned} A &= \int_0^6 \int_0^{2\pi} \sqrt{\frac{1}{9}r^2 + 1}\, r\, d\theta\, dr \\ &= 2\pi \int_0^6 \sqrt{\frac{1}{9}r^2 + 1}\, r\, dr \\ &= 2\pi\left(\frac{9}{2}\right)\left(\frac{2}{3}\right)\left(\frac{1}{9}r^2 + 1\right)^{3/2}\Big]_0^6 \\ &= 6\pi(5\sqrt{5} - 1) \end{aligned}$$

Thus, the surface area is $6\pi(5\sqrt{5} - 1)$ square units.

## 21.6 THE TRIPLE INTEGRAL

If $f$ is a continuous function of $x, y$, and $z$ on some region $S$ in $R^3$, then the *triple integral* of $f$ on $S$ is given by

$$\lim_{\|\Delta\| \to 0} \sum_{i=1}^{n} f(\bar{x}_i, \bar{y}_i, \bar{z}_i)\, \Delta_i V = \iiint_S f(x, y, z)\, dV$$

We use the triple integral to find the total mass of a solid with variable density. If $\rho(x, y, z)$ is the measure of the volume density of a solid that occupies the region $S$, then the measure of the mass of the solid is given by

$$M = \iiint_S \rho(x, y, z)\, dV$$

A special case occurs when $\rho(x, y, z) = 1$. In this case, the triple integral for mass gives the measure of the volume of the solid.

We may sometimes use a thrice-iterated integral to calculate the value of a triple integral. If $S$ is bounded by the planes $x = x_1$ and $x = x_2$, bounded by the cylinders $y = g_1(x)$ and $y = g_2(x)$ that are perpendicular to the $xy$ plane, and bounded by the surfaces $z = h_1(x, y)$ and $z = h_2(x, y)$, then

$$\iiint_S f(x, y, z)\,dV = \int_{x_1}^{x_2} \int_{g_1(x)}^{g_2(x)} \int_{h_1(x, y)}^{h_2(x, y)} f(x, y, z)\,dz\,dy\,dx$$

provided the functions $g_1, g_2, h_1$, and $h_2$ are smooth (i.e., they have continuous derivatives or partial derivatives).

The roles of $x, y$, and $z$ may sometimes be interchanged in the iterated integral used to evaluate a triple integral. Thus, there are six possible orders of integration that may be used to evaluate a triple integral.

## *Exercises 21.6*

**2.** Evaluate the iterated integral

$$\int_1^2 \int_y^{y^2} \int_0^{\ln x} ye^z\,dz\,dx\,dy$$

SOLUTION:

$$\int_1^2 \int_y^{y^2} \int_0^{\ln x} ye^z\,dz\,dx\,dy = \int_1^2 \int_y^{y^2} ye^z\Big]_{z=0}^{z=\ln x} dx\,dy$$

$$= \int_1^2 \int_y^{y^2} (yx - y)dx\,dy$$

$$= \int_1^2 \left[\frac{1}{2}yx^2 - xy\right]_{x=y}^{x=y^2} dy$$

$$= \int_1^2 \left[\left(\frac{1}{2}y^5 - y^3\right) - \left(\frac{1}{2}y^3 - y^2\right)\right] dy$$

$$= \left[\frac{1}{12}y^6 - \frac{1}{4}y^4 - \frac{1}{8}y^4 + \frac{1}{3}y^3\right]_1^2$$

$$= \frac{1}{12}(2^6 - 1) - \frac{3}{8}(2^4 - 1) + \frac{1}{3}(2^3 - 1)$$

$$= \frac{47}{24}$$

**8.** Evaluate the triple integral:

$$\iiint_S yz\,dV$$

if $S$ is the region bounded by the tetrahedron having vertices $(0, 0, 0)$, $(1, 1, 0)$, $(1, 0, 0)$, and $(1, 0, 1)$.

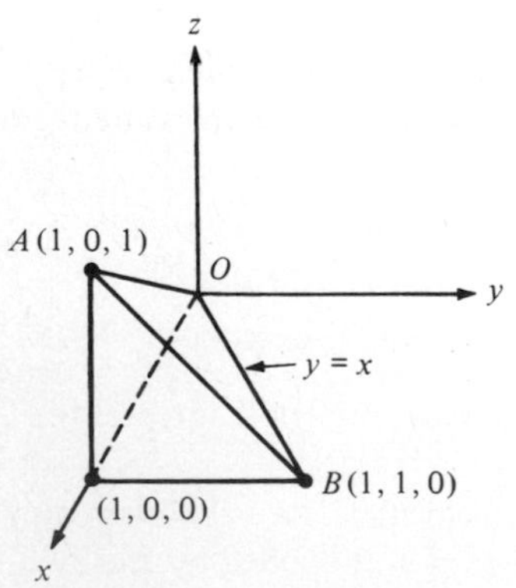

Figure 21.6.8

SOLUTION: In Fig. 21.6.8 we show a sketch of the region $S$, which is bounded by the planes $x = 0$ and $x = 1$, the cylinders $y = 0$ and $y = x$, and the surfaces $z = 0$ and the plane that contains points $O, A, B$, where $A = (1, 0, 1)$ and $B = (1, 1, 0)$. We find an equation of the plane. Let $\mathbf{A} = \mathbf{V}(\overrightarrow{OA})$ and $\mathbf{B} = \mathbf{V}(\overrightarrow{OB})$. Then

$$\mathbf{A} = \mathbf{i} + \mathbf{k} \quad \text{and} \quad \mathbf{B} = \mathbf{i} + \mathbf{j}$$

Thus, a normal vector to the plane is

$$\begin{aligned}\mathbf{A} \times \mathbf{B} &= (\mathbf{i} + \mathbf{k}) \times (\mathbf{i} + \mathbf{j}) \\ &= -\mathbf{i} + \mathbf{j} + \mathbf{k}\end{aligned}$$

Hence, the vector $-\mathbf{i} + \mathbf{j} + \mathbf{k}$ is a normal vector to the plane that contains the origin and the points $A$ and $B$. Thus, an equation of the plane is

$$\begin{aligned}-x + y + z &= 0 \\ z &= x - y\end{aligned}$$

Therefore, the given triple integral may be expressed as an iterated integral as follows

$$\begin{aligned}\iiint_S yz\, dV &= \int_0^1 \int_0^x \int_0^{x-y} yz\, dz\, dy\, dx \\ &= \frac{1}{2}\int_0^1 \int_0^x y(x-y)^2\, dy\, dx \\ &= \frac{1}{2}\int_0^1 \int_0^x (x^2y - 2xy^2 + y^3) dy\, dx \\ &= \frac{1}{2}\int_0^1 \left[\frac{1}{2}x^2y^2 - \frac{2}{3}xy^3 + \frac{1}{4}y^4\right]_{y=0}^{y=x} dx \\ &= \frac{1}{2}\int_0^1 \left[\frac{1}{2}x^4 - \frac{2}{3}x^4 + \frac{1}{4}x^4\right] dx \\ &= \frac{1}{24}\int_0^1 x^4\, dx \\ &= \frac{1}{120}\end{aligned}$$

12. Find the volume of the solid in the first octant bounded by the clinder $x^2 + z^2 = 16$, the plane $x + y = 2$, and the three coordinate planes.

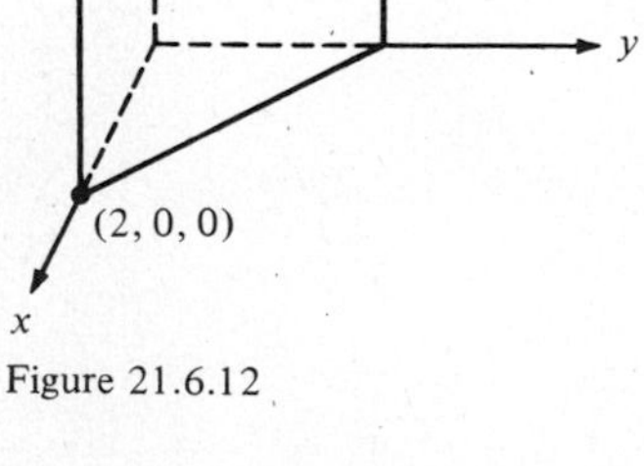

Figure 21.6.12

SOLUTION: Fig. 21.6.12 shows a sketch of the solid $S$, which is bounded by the planes $x = 0$ and $x = 2$, the cylinders $y = 0$ and $y = 2 - x$, and the surfaces $z = 0$ and $z = \sqrt{16 - x^2}$. Therefore, the measure of the volume of $S$ is

$$\begin{aligned}V &= \iiint_S dV \\ &= \int_0^2 \int_0^{2-x} \int_0^{\sqrt{16-x^2}} dz\, dy\, dx \\ &= \int_0^2 \int_0^{2-x} \sqrt{16 - x^2}\, dy\, dx \\ &= \int_0^2 (2 - x)\sqrt{16 - x^2}\, dx\end{aligned}$$

$$= 2\int_0^2 \sqrt{16 - x^2}\, dx - \int_0^2 x\sqrt{16 - x^2}\, dx \tag{1}$$

For the first integral in (1), let $x = 4 \sin \theta$. Then

$$\int_0^2 \sqrt{16 - x^2}\, dx = \int_0^{\pi/6} 16 \cos^2 \theta \, d\theta$$

$$= 8 \int_0^{\pi/6} (1 + \cos 2\theta) d\theta$$

$$= 8 \left[\theta + \frac{1}{2} \sin 2\theta\right]_0^{\pi/6}$$

$$= 8\left[\frac{1}{6}\pi + \frac{1}{4}\sqrt{3}\right]$$

$$= \frac{4}{3}\pi + 2\sqrt{3} \tag{2}$$

For the second integral in (1) we have

$$\int_0^2 x\sqrt{16 - x^2}\, dx = -\frac{1}{3}(16 - x^2)^{3/2}\Big]_0^2$$

$$= -\frac{1}{3}(24\sqrt{3} - 64)$$

$$= \frac{64}{3} - 8\sqrt{3} \tag{3}$$

Substituting from (2) and (3) into (1), we have

$$V = 2\left(\frac{4}{3}\pi + 2\sqrt{3}\right) - \left(\frac{64}{3} - 8\sqrt{3}\right)$$

$$= \frac{8}{3}\pi + 12\sqrt{3} - \frac{64}{3}$$

Therefore, the volume is $\frac{4}{3}(2\pi + 9\sqrt{3} - 16)$ cubic units.

**18.** Find the mass of the solid enclosed by the tetrahedron formed by the plane $100x + 25y + 16z = 400$ and the coordinate planes if the volume density varies as the distance from the $yz$ plane. The volume density is measured in slugs/ft$^3$.

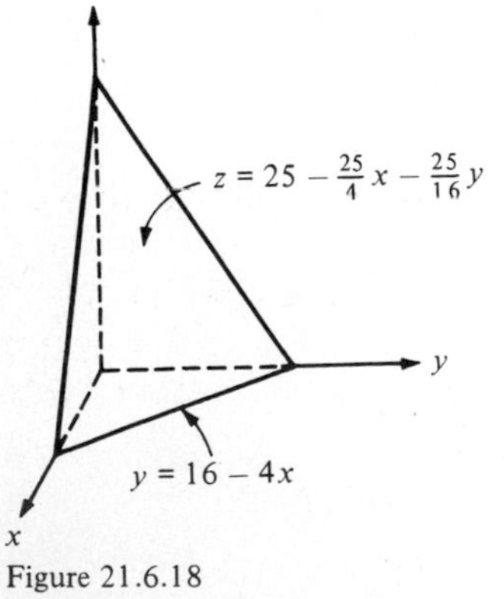

Figure 21.6.18

SOLUTION: Fig. 21.6.18 shows a sketch of the solid $S$. Let $\rho(x, y, z) = kx$ be the measure of the volume density at the point $(x, y, z)$. Then the measure of the total mass of the solid is given by

$$M = \lim_{\|\Delta\| \to 0} \sum_{i=1}^{n} \rho(\bar{x}_i, \bar{y}_i, \bar{z}_i)\, \Delta_i V$$

$$= \iiint_S kx \, dV$$

$$= \int_0^4 \int_0^{16-4x} \int_0^{25 - \frac{25}{4}x - \frac{25}{16}y} kx \, dz \, dy \, dx$$

$$= k \int_0^4 \int_0^{16-4x} x\left(25 - \frac{25}{4}x - \frac{25}{16}y\right) dy\, dx$$

$$= k \int_0^4 \left[25xy - \frac{25}{4}x^2 y - \frac{25}{32}xy^2\right]_0^{16-4x} dx$$

$$= k \int_0^4 \left[25x(16-4x) - \frac{25}{4}x^2(16-4x) - \frac{25}{32}x(16-4x)^2\right] dx$$

$$= k \int_0^4 \left[400x - 100x^2 - 100x^2 + 25x^3 - \frac{25}{2}x(16 - 8x + x^2)\right] dx$$

$$= k \int_0^4 \left[400x - 200x^2 + 25x^3 - 200x + 100x^2 - \frac{25}{2}x^3\right] dx$$

$$= k \int_0^4 \left[200x - 100x^2 + \frac{25}{2}x^3\right] dx$$

$$= k \left[100x^2 - \frac{100}{3}x^3 + \frac{25}{8}x^4\right]_0^4$$

$$= \frac{800k}{3}$$

Thus, the mass is $\frac{800}{3}k$ slugs.

## 21.7 THE TRIPLE INTEGRAL IN CYLINDRICAL AND SPHERICAL COORDINATES

Let $f$ be a continuous function of $r, \theta$, and $z$. Let $S$ be a region in $R^3$ bounded by the planes $\theta = \theta_1$ and $\theta = \theta_2$, bounded by the cylinders $r = g_1(\theta)$ and $r = g_2(\theta)$ that are perpendicular to the polar plane, and bounded by the surfaces $z = h_1(r, \theta)$ and $z = h_2(r, \theta)$. Then the triple integral of $f$ on $S$ is equivalent to an iterated integral with

$$\iiint_S f(r, \theta, z)\, dV = \iiint_S f(r, \theta, z)\, r\, dz\, dr\, d\theta$$

$$= \int_{\theta_1}^{\theta_2} \int_{g_1(\theta)}^{g_2(\theta)} \int_{h_1(r,\theta)}^{h_2(r,\theta)} f(r, \theta, z)\, r\, dz\, dr\, d\theta$$

provided the functions $g_1, g_2, h_1$, and $h_2$ are smooth.

Sometimes we may be able to interchange the roles of $r, \theta$, and $z$ in the iterated integral used to evaluate a triple integral.

Let $f$ be a continuous function of $\rho, \theta$, and $\phi$. Let $S$ be a region in $R^3$ bounded by the cones $\phi = \phi_1$ and $\phi = \phi_2$, bounded by the cylinders $\theta = \theta_1$ and $\theta = \theta_2$, and bounded by the surfaces $\rho = g_1(\theta, \phi)$ and $\rho = g_2(\theta, \phi)$. Then the triple integral of $f$ on $S$ is equivalent to an iterated integral with

$$\iiint_S f(\rho, \theta, \phi)\, dV = \iiint_S f(\rho, \theta, \phi)\, \rho^2 \sin\phi\, d\rho\, d\theta\, d\phi$$

$$= \int_{\phi_1}^{\phi_2} \int_{\theta_1}^{\theta_2} \int_{g_1(\theta,\phi)}^{g_2(\theta,\phi)} f(\rho, \theta, \phi)\, \rho^2 \sin\phi\, d\rho\, d\theta\, d\phi$$

provided the functions $g_1$ and $g_2$ are smooth. Sometimes we may be able to interchange the roles of $\theta, \phi$, and $\rho$ in the iterated integral.

*Exercises 21.7*

**4.** Evaluate the iterated integral

$$\int_{\pi/4}^{\pi/2}\int_{\pi/4}^{\phi}\int_{0}^{a\csc\theta} \rho^3 \sin^2\theta \sin\phi \, d\rho \, d\theta \, d\phi$$

SOLUTION:

$$\int_{\pi/4}^{\pi/2}\int_{\pi/4}^{\phi}\int_{0}^{a\csc\theta} \rho^3 \sin^2\theta \sin\phi \, d\rho \, d\theta \, d\phi = \frac{1}{4}a^4 \int_{\pi/4}^{\pi/2}\int_{\pi/4}^{\phi} \sin^2\theta \sin\phi \csc^4\theta \, d\theta \, d\phi$$

$$= \frac{1}{4}a^4 \int_{\pi/4}^{\pi/2} \sin\phi[-\cot\theta]_{\pi/4}^{\phi} d\phi$$

$$= -\frac{1}{4}a^4 \int_{\pi/4}^{\pi/2} \sin\phi(\cot\phi - 1)\, d\phi$$

$$= -\frac{1}{4}a^4 \int_{\pi/4}^{\pi/2} (\cos\phi - \sin\phi)\, d\phi$$

$$= -\frac{1}{4}a^4 [\sin\phi + \cos\phi]_{\pi/4}^{\pi/2}$$

$$= -\frac{1}{4}a^4 (1 - \sqrt{2})$$

$$= \frac{1}{4}a^4 (\sqrt{2} - 1)$$

**6.** If $S$ is the solid in the first octant bounded by the sphere $x^2 + y^2 + z^2 = 16$ and the coordinate planes, evaluate the triple integral $\iiint_S xyz \, dV$ by using three methods: **(a)** spherical coordinates, **(b)** rectangular coordinates, **(c)** cylindrical coordinates.

SOLUTION:

**(a)** In spherical coordinates we have

$$x = \rho \sin\phi \cos\theta \qquad y = \rho \sin\phi \sin\theta \qquad z = \rho \cos\phi$$

and

$$dV = \rho^2 \sin\phi \, d\rho \, d\theta \, d\phi$$

Thus,

$$\iiint_S xyz \, dV = \iiint_S \rho^5 \sin^3\phi \cos\phi \sin\theta \cos\theta \, d\rho \, d\theta \, d\phi \qquad \textbf{(1)}$$

Because $S$ is bounded by the point sphere $\rho = 0$ and the sphere $\rho = 4$, bounded by the cylinders $\theta = 0$ and $\theta = \frac{1}{2}\pi$, and bounded by the cones $\phi = 0$ and $\phi = \frac{1}{2}\pi$, the triple integral in (1) may be replaced by an iterated integral. Thus,

$$\iiint_S xyz\, dV = \int_0^{\pi/2} \int_0^{\pi/2} \int_0^4 \rho^5 \sin^3\phi \cos\phi \sin\theta \cos\theta \, d\rho \, d\theta \, d\phi$$

$$= \frac{1}{6} \cdot 4^6 \int_0^{\pi/2} \int_0^{\pi/2} \sin^3\phi \cos\phi \sin\theta \cos \, d\theta \, d\phi$$

$$= \frac{1}{3} \cdot 2^{11} \int_0^{\pi/2} \sin^3\phi \cos\phi \left[\frac{1}{2} \sin^2\theta\right]_0^{\pi/2} d\phi$$

$$= \frac{1}{3} \cdot 2^{10} \int_0^{\pi/2} \sin^3\phi \cos\phi \, d\phi$$

$$= \frac{1}{3} \cdot 2^{10} \left[\frac{1}{4} \sin^4\phi\right]_0^{\pi/2}$$

$$= \frac{1}{3} \cdot 2^8$$

$$= \frac{256}{3}$$

**(b)** In rectangular coordinates, $S$ is bounded by the surface $z = 0$ and $z = \sqrt{16 - x^2 - y^2}$, bounded by the cylinders $y = 0$ and $y = \sqrt{16 - x^2}$, and bounded by the planes $x = 0$ and $x = 4$. Thus,

$$\iiint_S xyz\, dV = \int_0^4 \int_0^{\sqrt{16-x^2}} \int_0^{\sqrt{16-x^2-y^2}} xyz \, dz \, dy \, dx$$

$$= \frac{1}{2} \int_0^4 \int_0^{\sqrt{16-x^2}} xy(16 - x^2 - y^2) dy \, dx$$

$$= \frac{1}{2} \int_0^4 x \left[8y^2 - \frac{1}{2}x^2y^2 - \frac{1}{4}y^4\right]_{y=0}^{y=\sqrt{16-x^2}} dx$$

$$= \frac{1}{2} \int_0^4 x \left[8(16 - x^2) - \frac{1}{2}x^2(16 - x^2) - \frac{1}{4}(16 - x^2)^2\right] dx$$

$$= \frac{1}{8} \int_0^4 x(16 - x^2)[32 - 2x^2 - (16 - x^2)] \, dx$$

$$= \frac{1}{8} \int_0^4 x(x^2 - 16)^2 \, dx$$

$$= \frac{1}{8} \cdot \frac{1}{2} \cdot \frac{1}{3} (x^2 - 16)^3 \Big]_0^4$$

$$= \frac{1}{3} \cdot 2^{-4} \cdot 2^{12}$$

$$= \frac{1}{3} \cdot 2^8$$

$$= \frac{256}{3}$$

(c) In cylindrical coordinates, we have $x = r\cos\theta$, $y = r\sin\theta$, and $dV = r\,dz\,dr\,d\theta$. Thus,

$$\iiint_S xyz\,dV = \iiint_S r^3 \sin\theta\cos\theta\, z\,dz\,dr\,d\theta \tag{2}$$

Because $S$ is bounded by the surface $z = 0$ and $z = \sqrt{16 - r^2}$, bounded by the cylinders $r = 0$ and $r = 4$, and bounded by the planes $\theta = 0$ and $\theta = \frac{1}{2}\pi$, the triple integral in (2) may be replaced by an iterated integral. Thus,

$$\begin{aligned}
\iiint_S xyz\,dV &= \int_0^{\pi/2}\int_0^4\int_0^{\sqrt{16-r^2}} r^3\sin\theta\cos\theta\, z\,dz\,dr\,d\theta \\
&= \frac{1}{2}\int_0^{\pi/2}\int_0^4 \sin\theta\cos\theta\, r^3(16 - r^2)dr\,d\theta \\
&= \frac{1}{2}\int_0^{\pi/2} \sin\theta\cos\theta \left[4r^4 - \frac{1}{6}r^6\right]_0^4 d\theta \\
&= \frac{1}{2}\left[2^2\cdot 2^8 - \frac{1}{6}\cdot 2^{12}\right]\int_0^{\pi/2}\sin\theta\cos\theta\,d\theta \\
&= \frac{1}{2}\cdot 2^{10}\left[1 - \frac{2}{3}\right]\int_0^{\pi/2}\sin\theta\cos\theta\,d\theta \\
&= \frac{1}{3}\cdot 2^9\left[\frac{1}{2}\sin^2\theta\right]_0^{\pi/2} \\
&= \frac{1}{3}\cdot 2^8 \\
&= \frac{256}{3}
\end{aligned}$$

**10.** Use cylindrical coordinates to find the moment of inertia of the solid bounded by a right circular cylinder of altitude $h$ feet and radius $a$ feet, with respect to the axis of the cylinder. The volume density varies as the distance from the axis of the cylinder, and it is measured in slugs/ft$^3$.

SOLUTION: Take the $z$-axis as the axis of the cylinder with the base of the cylinder in the polar plane. We are given that the measure of the volume density is $\rho(r, \theta, z) = kr$. If $S$ is the region occupied by the solid, then the measure of the moment of inertia with respect to the axis of the cylinder is given by

$$\begin{aligned}
I_z &= \lim_{\|\Delta\|\to 0}\sum_{i=1}^{n} \bar{r}_i^{\,2}(k\bar{r}_i)\,\Delta_i V \\
&= k\iiint_S r^3\cdot r\,dr\,d\theta\,dz
\end{aligned}$$

$$= k \int_0^h \int_0^{2\pi} \int_0^a r^4 \, dr \, d\theta \, dz$$

$$= \frac{1}{5} ka^5 \int_0^h \int_0^{2\pi} d\theta \, dz$$

$$= \frac{2}{5} \pi ka^5 \int_0^h dz$$

$$= \frac{2}{5} \pi kha^5$$

Thus, the moment of inertia is $\frac{2}{5}\pi kha^5$ slug-ft$^2$.

**14.** Use spherical coordinates to find the mass of a spherical solid of radius $a$ feet if the volume density at each point is proportional to the distance of the point from the center of the sphere. The volume density is measured in slugs/ft$^3$.

SOLUTION: Take the center of the sphere at the origin. We are given that the measure of the volume density is $k\rho$. Thus, the measure of the mass is given by

$$M = \lim_{\|\Delta\| \to 0} \sum_{i=1}^{n} k\bar{\rho}_i \, \Delta_i V$$

$$= k \iiint_S \rho \, dV$$

$$= k \int_0^{\pi} \int_0^{2\pi} \int_0^a \rho(\rho^2 \sin \phi \, d\rho \, d\theta \, d\phi)$$

$$= \frac{1}{4} ka^4 \int_0^{\pi} \int_0^{2\pi} \sin \phi \, d\theta \, d\phi$$

$$= \frac{1}{4} ka^4 \cdot 2\pi \int_0^{\pi} \sin \phi \, d\phi$$

$$= \frac{1}{2} \pi ka^4 \cdot 2$$

$$= \pi ka^4$$

Thus the mass is $\pi ka^4$ slugs.

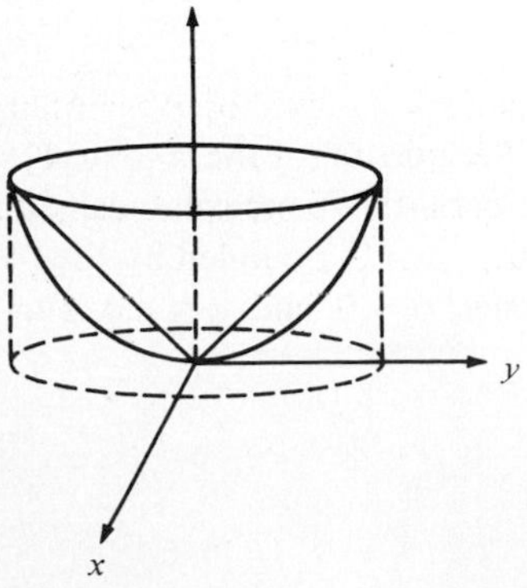

Figure 21.7.16

**16.** Use the coordinate system that you decide is best to find the moment of inertia with respect to the $z$-axis of the homogeneous solid inside the paraboloid $x^2 + y^2 = z$ and outside the cone $x^2 + y^2 = z^2$. The constant volume density is $k$ slugs/ft$^3$.

SOLUTION: Eliminating $x$ and $y$ from the given equations, we have $z^2 = z$. Thus, the given surfaces intersect in the planes $z = 0$ and $z = 1$. Fig. 21.7.16 shows a sketch of the solid $S$ and its projection on the $xy$ plane. We take a cylindrical coordinate system. Thus, the solid $S$ is bounded below by the paraboloid $z = r^2$ and bounded above by the cone $z = r$, bounded by the cylinders $r = 0$ and $r = 1$, and bounded by the planes $\theta = 0$ and $\theta = 2\pi$. Therefore, the measure of the moment of inertia with respect to the $z$-axis is given by

$$I_z = \lim_{\|\Delta\| \to 0} \sum_{i=1}^{n} k\bar{r}_i^2 \, \Delta_i V$$

$$= k \iiint_S r^2 \, dV$$

$$= k \int_0^{2\pi} \int_0^1 \int_{r^2}^{r} r^2 (r \, dz \, dr \, d\theta)$$

$$= k \int_0^{2\pi} \int_0^1 r^3 (r - r^2) dr \, d\theta$$

$$= k \int_0^{2\pi} \left[ \frac{1}{5} r^5 - \frac{1}{6} r^6 \right]_0^1 d\theta$$

$$= \frac{1}{30} k \int_0^{2\pi} d\theta$$

$$= \frac{1}{15} \pi k$$

The moment of inertia is $\frac{1}{15}\pi k$ slug-ft$^2$.

**20.** Evaluate the iterated integral by using either cylindrical or spherical coordinates.

$$\int_0^2 \int_0^{\sqrt{4-y^2}} \int_0^{\sqrt{4-x^2-y^2}} \frac{1}{x^2 + y^2 + z^2} \, dz \, dx \, dy$$

SOLUTION: The given iterated integral is equivalent to a triple integral over a region $S$ which is bounded by the surfaces

$$z = 0 \quad \text{and} \quad z = \sqrt{4 - x^2 - y^2}$$

bounded by the cylinders

$$x = 0 \quad \text{and} \quad x = \sqrt{4 - y^2}$$

and bounded by the planes

$$y = 0 \text{ and } y = 2$$

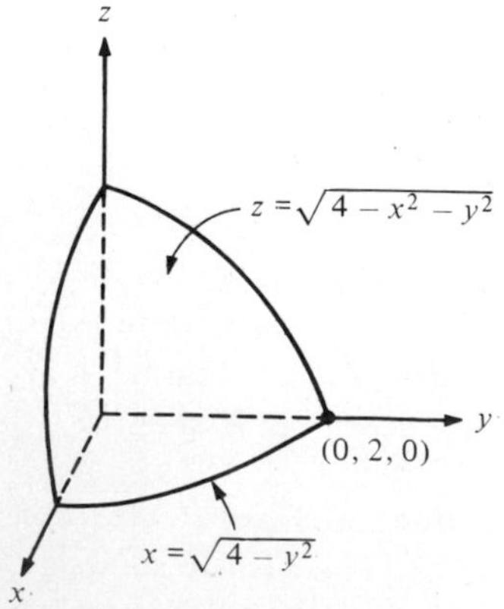

Figure 21.7.20

Fig. 21.7.20 shows a sketch of the region $S$, which is the portion of the spherical solid enclosed by $x^2 + y^2 + z^2 = 4$ that is in the first octant. We use spherical coordinates. Thus, $S$ is bounded by the spheres $\rho = 0$ and $\rho = 2$, bounded by the cylinders $\theta = 0$ and $\theta = \frac{1}{2}\pi$, and bounded by the cones $\phi = 0$ and $\phi = \frac{1}{2}\pi$. Furthermore, in spherical coordinates we have

$$x^2 + y^2 + z^2 = \rho^2$$

and

$$dV = \rho^2 \sin \phi \, d\rho \, d\theta \, d\phi$$

Therefore,

$$\int_0^2 \int_0^{\sqrt{4-y^2}} \int_0^{\sqrt{4-x^2-y^2}} \frac{1}{x^2+y^2+z^2}\, dz\, dx\, dy = \int_0^{\pi/2} \int_0^{\pi/2} \int_0^2 \frac{1}{\rho^2}\rho^2 \sin\phi\, d\rho\, d\theta\, d\phi$$

$$= 2\int_0^{\pi/2} \int_0^{\pi/2} \sin\phi\, d\theta\, d\phi$$

$$= \pi \int_0^{\pi/2} \sin\phi\, d\phi$$

$$= \pi$$

## *Review Exercises*

In Exercises 1-8, evaluate the given iterated integral.

4. $\int_0^{\pi} \int_0^{3(1+\cos\theta)} r^2 \sin\theta\, dr\, d\theta$

SOLUTION:

$$\int_0^{\pi} \int_0^{3(1+\cos\theta)} r^2 \sin\theta\, dr\, d\theta = \frac{1}{3}\int_0^{\pi} \sin\theta\, [r^3]_0^{3(1+\cos\theta)}\, d\theta$$

$$= 9\int_0^{\pi} (1+\cos\theta)^3 \sin\theta\, d\theta$$

$$= 9\left(-\frac{1}{4}\right)[(1+\cos\theta)^4]_0^{\pi}$$

$$= 36$$

8. $\int_0^{a} \int_0^{\pi/2} \int_0^{\sqrt{a^2-z^2}} zre^{-r^2}\, dr\, d\theta\, dz$

SOLUTION:

$$\int_0^{a} \int_0^{\pi/2} \int_0^{\sqrt{a^2-z^2}} zre^{-r^2}\, dr\, d\theta\, dz = -\frac{1}{2}\int_0^{a} \int_0^{\pi/2} z[e^{-r^2}]_0^{\sqrt{a^2-z^2}}\, d\theta\, dz$$

$$= -\frac{1}{2}\int_0^{a} \int_0^{\pi/2} (ze^{z^2-a^2} - z)d\theta\, dz$$

$$= -\frac{1}{4}\pi \int_0^{a} (ze^{z^2-a^2} - z)\, dz$$

$$= -\frac{1}{8}\pi[e^{z^2-a^2} - z^2]_0^a$$

$$= -\frac{1}{8}\pi(1 - a^2 - e^{-a^2})$$

12. Evaluate the multiple integral

$$\iiint_S y \cos(x+z)\,dV$$

$S$ is the region bounded by the cylinder $x = y^2$ and the planes $x + z = \frac{1}{2}\pi$, $y = 0$, and $z = 0$.

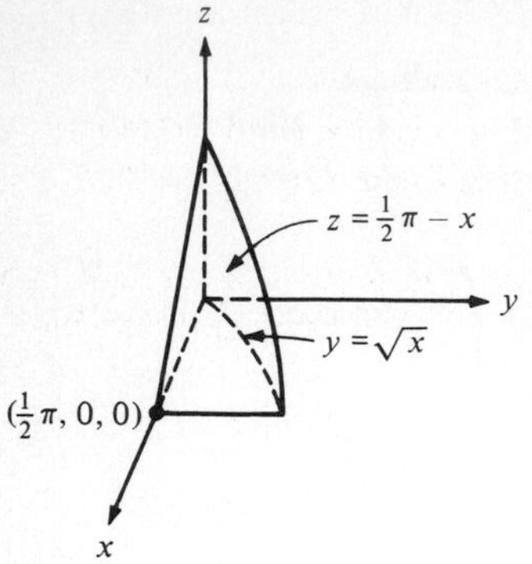

Figure 21.12R

SOLUTION: Fig. 21.12R shows a sketch of the solid $S$, which is bounded below by the plane $z = 0$, bounded above by the plane $z = \frac{1}{2}\pi - x$, and bounded on the sides by the cylinders $y = 0$ and $y = \sqrt{x}$ and by the planes $x = 0$ and $x = \frac{1}{2}\pi$. Therefore,

$$\iiint_S y\cos(x+z)\,dV = \int_0^{\pi/2}\int_0^{\sqrt{x}}\int_0^{\pi/2-x} y\cos(x+z)\,dz\,dy\,dx$$

$$= \int_0^{\pi/2}\int_0^{\sqrt{x}} y[\sin(x+z)]_{z=0}^{z=\pi/2-x}\,dy\,dx$$

$$= \int_0^{\pi/2}\int_0^{\sqrt{x}} y(1-\sin x)\,dy\,dx$$

$$= \frac{1}{2}\int_0^{\pi/2} x(1-\sin x)\,dx$$

$$= \frac{1}{2}\int_0^{\pi/2} x\,dx - \frac{1}{2}\int_0^{\pi/2} x\sin x\,dx$$

$$= \frac{1}{4}x^2\Big]_0^{\pi/2} - \frac{1}{2}[-x\cos x + \sin x]_0^{\pi/2}$$

$$= \frac{1}{16}\pi^2 - \frac{1}{2}[1]$$

$$= \frac{1}{16}(\pi^2 - 8)$$

16. Evaluate the iterated integral by reversing the order of integration.

$$\int_0^1\int_0^{\cos^{-1}y} e^{\sin x}\,dx\,dy$$

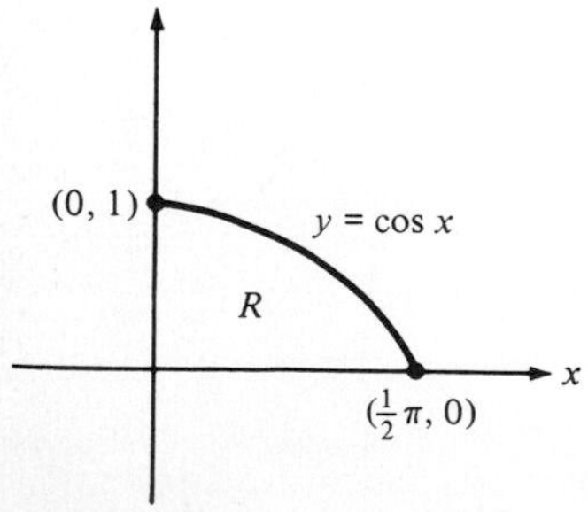

Figure 21.16R

SOLUTION: The region $R$ in the $xy$ plane on which the double integral is taken is bounded by the curves $x = 0$ and $x = \cos^{-1}y$ and the lines $y = 0$ and $y = 1$. A sketch of the region is shown in Fig. 21.16R. We may regard $R$ as being bounded by the curves $y = 0$ and $y = \cos x$ and the lines $x = 0$ and $x = \frac{1}{2}\pi$. Thus,

$$\int_0^1\int_0^{\cos^{-1}y} e^{\sin x}\,dx\,dy = \int_0^{\pi/2}\int_0^{\cos x} e^{\sin x}\,dy\,dx$$

$$= \int_0^{\pi/2} \cos x e^{\sin x}\,dx$$

$$= e^{\sin x}\Big]_0^{\pi/2}$$

$$= e - 1$$

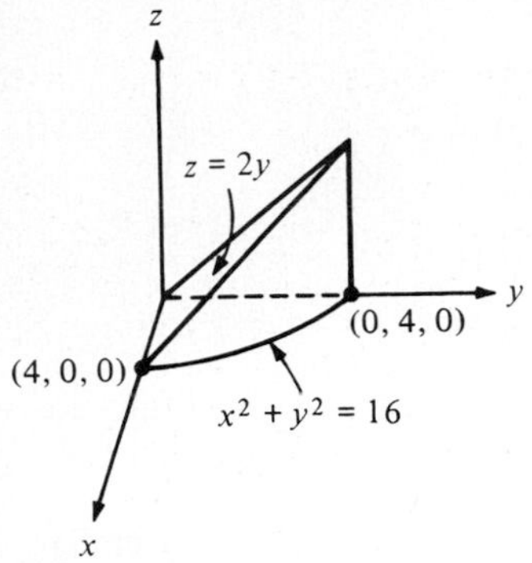

Figure 21.20 R

**20.** Use double integration to find the volume of the solid above the $xy$ plane bounded by the cylinder $x^2 + y^2 = 16$ and the plane $z = 2y$ by two methods: **(a)** integrating first with respect to $x$, and **(b)** integrating first with respect to $y$.

SOLUTION: Fig. 21.20R shows the half of the solid that is in the first octant. We take $R$ to be the first quadrant part of the circle $x^2 + y^2 = 16$, and the measure of the volume is given by

$$V = 2 \lim_{\|\Delta\| \to 0} \sum_{i=1}^{n} \bar{z}_i \, \Delta_i A$$

$$= 2 \iint_R 2y \, dA$$

**(a)** We regard $R$ as being bounded by the lines $y = 0$ and $y = 4$, and bounded by the curves $x = 0$ and $x = \sqrt{16 - y^2}$. Thus,

$$V = 2 \int_0^4 \int_0^{\sqrt{16-y^2}} 2y \, dx \, dy$$

$$= 4 \int_0^4 y\sqrt{16 - y^2} \, dy$$

$$= -\frac{4}{3}(16 - y^2)^{3/2}\Big]_0^4$$

$$= \frac{256}{3}$$

Thus, the volume is $\frac{256}{3}$ cubic units.

**(b)** We regard $R$ as being bounded by the lines $x = 0$ and $x = 4$ and the curves $y = 0$ and $y = \sqrt{16 - x^2}$. Thus,

$$V = 2 \int_0^4 \int_0^{\sqrt{16-x^2}} 2y \, dy \, dx$$

$$= 2 \int_0^4 (16 - x^2) dx$$

$$= 2\left[16x - \frac{1}{3}x^3\right]_0^4$$

$$= \frac{256}{3}$$

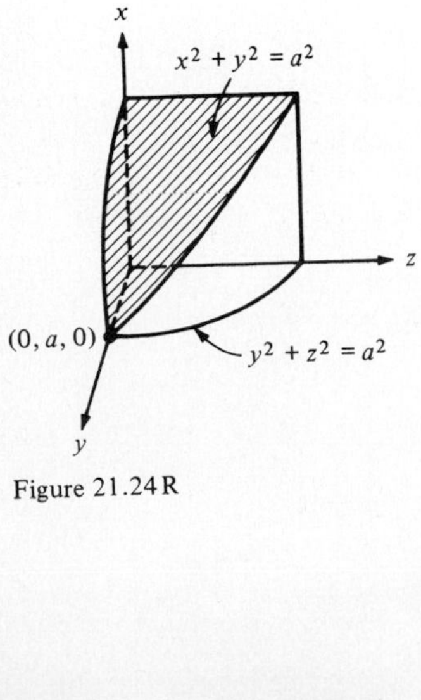

Figure 21.24 R

**24.** Find the area of the surface of the part of the cylinder $x^2 + y^2 = a^2$ that lies inside the cylinder $y^2 + z^2 = a^2$.

SOLUTION: Fig. 21.24R shows a sketch of that part of the surface that lies in the first octant. This is one-eighth of the entire surface. We take the $yz$ plane as the horizontal plane and apply Theorem 21.5.1 with $x$ replaced by $z$. Solving the equation of the cylinder $x^2 + y^2 = a^2$ for $x$, we obtain $x = \pm\sqrt{a^2 - y^2}$. We take

$$f(y, z) = \sqrt{a^2 - y^2}$$

Thus,

$$f_y(y, z) = \frac{-y}{\sqrt{a^2 - y^2}} \quad \text{and} \quad f_z(y, z) = 0$$

Hence,

$$\sqrt{f_y^2(y, z) + f_z^2(y, z) + 1} = \sqrt{\frac{y^2}{a^2 - y^2} + 1}$$
$$= \frac{a}{\sqrt{a^2 - y^2}}$$

Let $R$ be the region in the first quadrant of the $yz$ plane bounded by the circle $y^2 + z^2 = a^2$. The measure of the surface area is given by

$$A = 8 \iint_R \frac{a}{\sqrt{a^2 - y^2}}\, dA$$
$$= 8a \int_0^a \int_0^{\sqrt{a^2 - y^2}} \frac{1}{\sqrt{a^2 - y^2}}\, dz\, dy$$
$$= 8a \int_0^a dy$$
$$= 8a^2$$

Thus, the surface area is $8a^2$ square units.

28. Find the center of mass of the lamina in the shape of the region bounded by the parabolas $x^2 = 4 + 4y$ and $x^2 = 4 - 8y$ is the area density at any point is $kx^2$ slugs/ft$^2$.

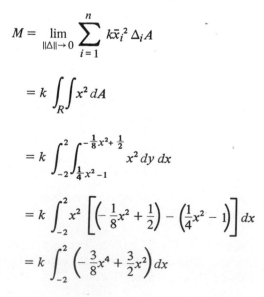

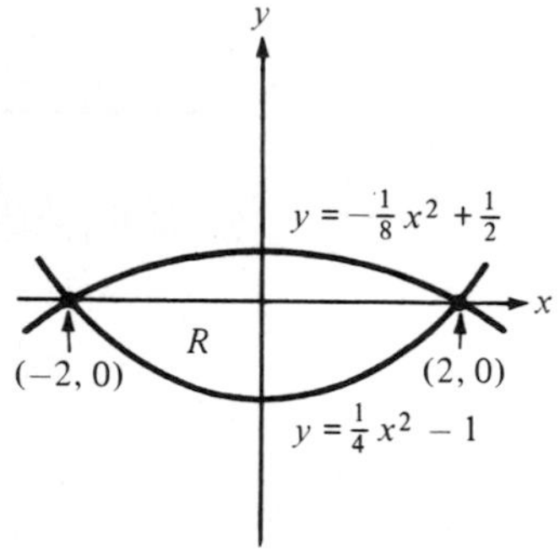

Figure 21.28R

SOLUTION: Fig. 21.28R shows a sketch of the region $R$ which is bounded by the lines $x = -2$ and $x = 2$, bounded below by the curve $y = \frac{1}{4}x^2 - 1$, and bounded above by the curve $y = -\frac{1}{8}x^2 + \frac{1}{2}$. Thus, the measure of the total mass is given by

$$M = \lim_{\|\Delta\| \to 0} \sum_{i=1}^{n} k\bar{x}_i^2\, \Delta_i A$$
$$= k \iint_R x^2\, dA$$
$$= k \int_{-2}^{2} \int_{\frac{1}{4}x^2 - 1}^{-\frac{1}{8}x^2 + \frac{1}{2}} x^2\, dy\, dx$$
$$= k \int_{-2}^{2} x^2 \left[\left(-\frac{1}{8}x^2 + \frac{1}{2}\right) - \left(\frac{1}{4}x^2 - 1\right)\right] dx$$
$$= k \int_{-2}^{2} \left(-\frac{3}{8}x^4 + \frac{3}{2}x^2\right) dx$$

$$= k\left[-\frac{3}{40}x^5 + \frac{1}{2}x^3\right]_{-2}^{2}$$

$$= \frac{16}{5}k \qquad (1)$$

Furthermore,

$$M \cdot \bar{x} = \lim_{\|\Delta\| \to 0} \sum_{i=1}^{n} k\bar{x}_i^2 \bar{x}_i \, \Delta_i A$$

$$= k \iint_R x^3 \, dA$$

$$= k \int_{-2}^{2} \int_{\frac{1}{4}x^2 - 1}^{-\frac{1}{8}x^2 + \frac{1}{2}} x^3 \, dy \, dx$$

$$= k \int_{-2}^{2} \left(-\frac{3}{8}x^5 + \frac{3}{2}x^3\right) dx$$

$$= k\left[-\frac{1}{16}x^6 + \frac{3}{8}x^4\right]_{-2}^{2}$$

$$= 0 \qquad (2)$$

and

$$M \cdot \bar{y} = \lim_{\|\Delta\| \to 0} \sum_{i=1}^{n} k\bar{x}_i^2 \bar{y}_i \, \Delta_i A$$

$$= k \iint_R x^2 y \, dA$$

$$= k \int_{-2}^{2} \int_{\frac{1}{4}x^2 - 1}^{-\frac{1}{8}x^2 + \frac{1}{2}} x^2 y \, dy \, dx$$

$$= \frac{1}{2}k \int_{-2}^{2} x^2 \left[\left(-\frac{1}{8}x^2 + \frac{1}{2}\right)^2 - \left(\frac{1}{4}x^2 - 1\right)^2\right] dx$$

$$= \frac{1}{2}k \int_{-2}^{2} x^2 \left[\left(-\frac{1}{8}\right)^2 (x^2 - 4)^2 - \left(\frac{1}{4}\right)^2 (x^2 - 4)^2\right] dx$$

$$= -\frac{3}{128}k \int_{-2}^{2} x^2 (x^2 - 4)^2 \, dx$$

$$= -\frac{3}{128}k \int_{-2}^{2} (x^6 - 8x^4 + 16x^2) \, dx$$

$$= -\frac{3}{128}k\left[\frac{1}{7}x^7 - \frac{8}{5}x^5 + \frac{16}{3}x^3\right]_{-2}^{2}$$

$$= -\frac{16}{35}k \qquad (3)$$

Substituting from (1) into (2) and (3), we obtain

$$\frac{16}{5}k\bar{x} = 0 \quad \text{and} \quad \frac{16}{5}k\bar{y} = -\frac{16}{35}k$$

$$\bar{x} = 0 \qquad\qquad \bar{y} = -\frac{1}{7}$$

Therefore, the center of mass is $(0, -\frac{1}{7})$.

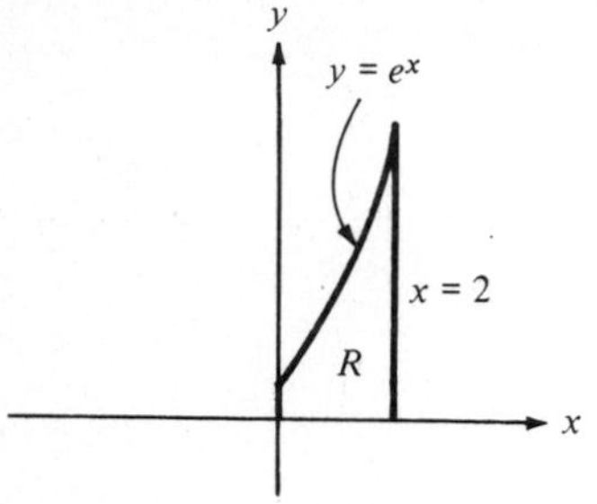

Figure 21.32R

**32.** Find the moment of inertia about the $y$-axis of the lamina in the shape of the region bounded by the curve $y = e^x$, the line $x = 2$, and the coordinate axes, if the area density at any point is $xy$ slugs/ft$^2$.

SOLUTION: A sketch of the region $R$ is shown in Fig. 21.32R. The measure of the moment of inertia about the $y$-axis is given by

$$I_y = \lim_{\|\Delta\| \to 0} \sum_{i=1}^{n} \bar{x}_i^{\,2}(\bar{x}_i \bar{y}_i)\, \Delta_i A$$

$$= \iint_R x^3 y\, dA$$

$$= \int_0^2 \int_0^{e^x} x^3 y\, dy\, dx$$

$$= \frac{1}{2}\int_0^2 x^3 e^{2x}\, dx \qquad (1)$$

We use integration by parts with

$$u = x^3 \qquad dv = e^{2x} dx$$

$$du = 3x^2 dx \qquad v = \frac{1}{2}e^{2x}$$

Thus,

$$\int x^3 e^{2x} dx = \frac{1}{2}x^3 e^{2x} - \frac{3}{2}\int x^2 e^{2x}\, dx \qquad (2)$$

We integrate by parts two more times and obtain from (2)

$$\int x^3 e^{2x} dx = \frac{1}{2}x^3 e^{2x} - \frac{3}{4}x^2 e^{2x} + \frac{3}{4}xe^{2x} - \frac{3}{8}e^{2x} \qquad (3)$$

Substituting from (3) into (1), we obtain

$$I_y = \frac{1}{2}\left[\frac{1}{2}x^3 e^{2x} - \frac{3}{4}x^2 e^{2x} + \frac{3}{4}xe^{2x} - \frac{3}{8}e^{2x}\right]_0^2$$

$$= \frac{1}{16}[e^{2x}(4x^3 - 6x^2 + 6x - 3)]_0^2$$

$$= \frac{1}{16}(17e^4 + 3)$$

Thus, the moment of inertia is $\frac{1}{16}(17e^4 + 3)$ slugs-ft$^2$.

**40.** Find the center of mass of the solid bounded by the sphere $x^2 + y^2 + z^2 - 6z = 0$ and the cone $x^2 + y^2 = z^2$, and above the cone, if the volume density at any point is $kz$ slugs/ft$^3$.

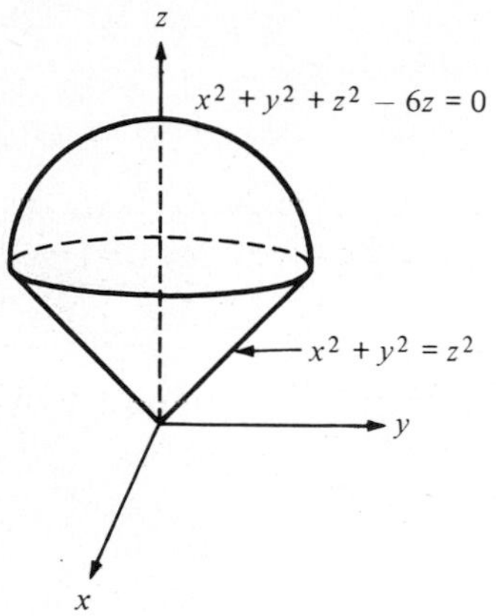

Figure 21.40 R

SOLUTION: Eliminating $x^2 + y^2$ from the given equations, we have $z(z - 3) = 0$. Thus, the surfaces intersect in the planes $z = 0$ and $z = 3$. Furthermore, by completing the square we have for the equation of the sphere

$$x^2 + y^2 + (z - 3)^2 = 9$$

Thus, the sphere has center $(0, 0, 3)$ and radius 3. Fig. 21.40R shows a sketch of the solid $S$. We use spherical coordinates. Because

$$x^2 + y^2 + z^2 = \rho^2 \quad \text{and} \quad z = \rho \cos \phi$$

the given equation of the sphere becomes

$$\rho^2 = 6\rho \cos \phi$$
$$\rho = 6 \cos \phi$$

in spherical coordinates. For the equation of the cone we have by adding $z^2$ to both sides

$$x^2 + y^2 + z^2 = 2z^2$$
$$\rho^2 = 2\rho^2 \cos^2 \phi$$
$$\cos^2 \phi = \frac{1}{2}$$
$$\cos \phi = \frac{1}{2}\sqrt{2}$$
$$\phi = \frac{1}{4}\pi$$

Thus, the region $S$ is bounded by the surfaces $\rho = 0$ and $\rho = 6 \cos \phi$, bounded by the cones $\phi = 0$ and $\phi = \frac{1}{4}\pi$, and bounded by the planes $\theta = 0$ and $\theta = 2\pi$. Therefore, the measure of the mass of the solid is given by

$$M = \lim_{\|\Delta\| \to 0} \sum_{i=1}^{n} k\bar{z}_i \, \Delta_i V$$

$$= \iiint_S kz \, dV$$

We have $z = \rho \cos \phi$ and $dV = \rho^2 \sin \phi \, d\rho \, d\theta \, d\phi$. Thus,

$$M = k \int_0^{\pi/4} \int_0^{2\pi} \int_0^{6 \cos \phi} \rho^3 \sin \phi \cos \phi \, d\rho \, d\theta \, d\phi$$

$$= \frac{1}{4}k \int_0^{\pi/4} \int_0^{2\pi} \sin \phi \cos \phi (6 \cos \phi)^4 \, d\theta \, d\phi$$

$$= 2^2 \cdot 3^4\, k \int_0^{\pi/4} \int_0^{2\pi} \sin\phi \cos^5 \phi \, d\theta \, d\phi$$

$$= 2^3 \cdot 3^4 \pi k \int_0^{\pi/4} \sin\phi \cos^5 \phi \, d\phi$$

$$= -2^2 \cdot 3^3 \pi k [\cos^6 \phi]_0^{\pi/4}$$

$$= \frac{1}{2} \cdot 3^3 \cdot 7\pi k \qquad (1)$$

Furthermore,

$$M \cdot \bar{z} = \lim_{\|\Delta\| \to 0} \sum_{i=1}^{n} k \bar{z}_i \, \bar{z}_i \, \Delta_i V$$

$$= k \iiint_S z^2 \, dV$$

$$= k \int_0^{\pi/4} \int_0^{2\pi} \int_0^{6\cos\phi} \rho^4 \cos^2 \phi \sin\phi \, d\rho \, d\theta \, d\phi$$

$$= \frac{1}{5} \cdot 2^5 \cdot 3^5\, k \int_0^{\pi/4} \int_0^{2\pi} \cos^7 \phi \sin\phi \, d\theta \, d\phi$$

$$= \frac{1}{5} \cdot 2^6 \cdot 3^5 \, \pi k \int_0^{\pi/4} \cos^7 \phi \sin\phi \, d\phi$$

$$= -\frac{1}{5} \cdot 2^3 \cdot 3^5 \, \pi k [\cos^8 \phi]_0^{\pi/4}$$

$$= \frac{1}{2} \cdot 3^6 \pi k \qquad (2)$$

Substituting from (1) into (2), we obtain

$$\frac{1}{2} \cdot 3^3 \cdot 7\pi k \bar{z} = \frac{1}{2} \cdot 3^6 \pi k$$

$$\bar{z} = \frac{1}{7} \cdot 3^3 = \frac{27}{7}$$

Because $S$ is symmetric with respect to the $z$-axis and the density function is also symmetric with respect to the $z$-axis, the center of mass lies on the $z$-axis. Thus, the center of mass is $(0, 0, \frac{27}{7})$.

# Appendix

## CHAPTER TESTS

**TEST FOR CHAPTER 17** **(60 minutes)** **Solutions on page 883.**

1. Find a unit vector **u** which is in the same direction as the vector with representation $\overrightarrow{PQ}$ and express **u** as a linear combination of the unit vectors **i** and **j** if $P = (-3, 1)$ and $Q = (1, 3)$.

2. Find the vector projection of $3\mathbf{i} - 2\mathbf{j}$ onto $2\mathbf{i} + \mathbf{j}$.

3. Find $d^2y/dx^2$ without eliminating the parameter $t$ if $x = t^2$ and $y = t^3 + t$.

4. Find $|\mathbf{R}'(t)|$ and simplify your answer if $\mathbf{R}(t) = e^{2t}(\cos t\,\mathbf{i} + \sin t\,\mathbf{j})$.

5. Find the curvature of the ellipse $x^2 + 4y^2 = 40$ at the point $(2, 3)$.

6. Use a definite integral with polar coordinates to find the length of the curve $r = 1 + \cos\theta$ from the point where $\theta = 0$ to the point where $\theta = 3\pi/2$.

7. A particle is moving along the curve having the parametric equations $x = 2t^2$ and $y = t^3$; where $t$ represents time. Find the particular values of the vectors **V**, **A**, **T**, and **N** and the scalars $ds/dt, A_T, A_N$, and $K$ when $t = 1$.

**TEST FOR CHAPTER 18** **(60 minutes)** **Solutions on page 885.**

1. Let $\mathbf{A} = \langle 2, 1, -1\rangle$ and $\mathbf{B} = \langle -1, 2, 1\rangle$.
 (a) Find $|\mathbf{A} + \mathbf{B}|$.
 (b) Find $\mathbf{A} \cdot \mathbf{B}$.
 (c) Find $\mathbf{A} \times \mathbf{B}$.

2. Let $P = (1, -1, 2)$ and $Q = (-3, 1, 0)$.
 (a) Find symmetric equations of the line $PQ$.
 (b) Find an equation of the plane that is the perpendicular bisector of line segment $PQ$.

3. Use vectors to find each of the following distances.
 (a) The distance between the point $(2, -1, 1)$ and the plane $x + 2y - 2z + 6 = 0$.
 (b) The distance between the point $(2, -1, 1)$ and the line $x = 1 + 3t, y = 2t, z = 6t$.

4. Draw a sketch of the surface whose Cartesian equation is given; give the coordinates of the intercept points; and give the name of the surface.

$$z = 4 - 4x^2 - y^2$$

5. Let $C$ be the curve that is the intersection of the given surfaces. Find an equation of the cylinder perpendicular to the $xy$ plane that contains the curve $C$. Identify the curve $C$ by name and draw a sketch of the projection of $C$ on the $xy$ plane. Label the intercept points on the graph.

$$z^2 = x^2 + y^2 - 4 \qquad y + z = 2$$

6. Find the curvature of the following curve at the point where $t = 1$.

$$x = t^2 \qquad y = t + \frac{1}{3}t^3 \qquad z = t - \frac{1}{3}t^3$$

7. Find an equation of the surface of revolution generated by revolving the curve $y = x^3$ in the $xy$ plane about the $y$-axis.

8. Find a Cartesian equation for the surface whose equation is given in spherical coordinates. Describe the surface completely.

$$\rho = 4 \sin \phi \cos \theta + 2 \cos \phi$$

**TEST FOR CHAPTER 19** **(50 minutes)** **Solutions on page 886.**

1. Draw a sketch of the region in the $xy$ plane that is the *domain* of the function $f$. Give the coordinates of the intercepts of the boundary curve.

$$f(x, y) = \sqrt{16 - 4x^2 + y^2}$$

2. Determine whether or not $\lim_{(x, y) \to (0, 0)} f(x, y)$ exists, and prove your answer.

$$f(x, y) = \frac{xy}{x + y}$$

3. Show that $f$ is continuous at the origin

$$f(x, y) = \begin{cases} \dfrac{x^4}{x^2 + y^2} & \text{if } (x, y) \neq (0, 0) \\ 0 & \text{if } (x, y) = (0, 0) \end{cases}$$

4. Use any theorems to find $f_x(x, y)$ and simplify your answer.

$$f(x, y) = \tan^{-1}\left(\frac{y^2}{x^2}\right)$$

5. Use any theorems to find $f_2(x, y)$ and factor your answer completely.

$$f(x, y) = x^2 y^3 e^{x/y}$$

6. Find $\dfrac{\partial^2 z}{\partial x^2}$.

$$z = \ln(x^2 + y^2)$$

7. Let $u = x^2 + xy^2$, $x = r\cos\theta$, and $y = r\sin\theta$. Use the chain rule to find $\partial u/\partial\theta$ and $\partial^2 u/\partial\theta^2$ and express your answer in terms of $x$ and $y$.

8. The formula $V = \pi r^2 h$ is used to calculate the volume of a solid in the shape of a right circular cylinder. The radius is 5 cm with a possible error of 0.1 cm, and the altitude is 10 cm with a possible error of 0.2 cm. Use the total differential of $V$ to approximate the relative error in $V$.

## TEST FOR CHAPTER 20

**(75 minutes)** **Solutions on page 888**

1. Let $f(x, y, z) = x^2 + y^2 + 3xz$.
   (a) Find the rate of change of the function $f$ at point $P$ in the direction of $\overrightarrow{PQ}$ if $P = (1, 2, -1)$ and $Q = (2, 0, 1)$.
   (b) Find the unit vector $u$ such that $D_u f(1, 2, -1)$ has an absolute maximum value and find the maximum value of $D_u f(1, 2, -1)$.

2. Find an equation of the plane that is tangent to the surface $x^2 + 2y^2 - z^2 = 2z$ at the point $(1, 1, 1)$.

3. At every point on the plane $x + y - z = 4$, the temperature is $T$ degrees with $T = x^2 yz$.
   (a) Use the method of Lagrange multipliers to find the critical point of $T$ that is not on a coordinate plane.
   (b) Eliminate one of the variables and use the second derivative test to determine whether the above critical point yields a relative maximum value, a relative minimum value, or no relative extremum for $T$.

4. If $p$ dollars is the price per unit of one commodity and $q$ dollars is the price per unit of another commodity, then $100x$ units of the first commodity and $100y$ units of the second commodity are demanded, with $x = 8 - 2p + q$ and $y = 6 + p - q$. If it costs \$2 to produce each unit of the first commodity and \$1 to produce each unit of the second commodity, find the prices that a monopolist should set for each commodity in order to make the most profit.

5. Show that the given vector is a gradient and find a function having the given gradient.

$$\sin y\,\mathbf{i} + (x\cos y + ze^{yz})\mathbf{j} + ye^{yz}\mathbf{k}$$

6. A particle is moved along the curve $y = 2x^3$ from the origin to the point $(1, 2)$ by the force field $\mathbf{F}$, where $\mathbf{F}(x, y) = xy\,\mathbf{i} + x^3\mathbf{j}$. Find the measure of the total work done.

7. Show that the given line integral is independent of the path and evaluate the line integral in two ways: **(a)** by using the potential function, and **(b)** by using any sectionally smooth curve from point $A$ to point $B$.

$$\int_C (3x^2y + y^2)dx + (x^3 + 2xy)dy$$

$$A = (1, 0) \quad \text{and} \quad B = (-1, 2)$$

**TEST FOR CHAPTER 21** **(70 minutes)** **Solutions on page 891.**

1. A lamina is bounded by the lines $y = 2x$, $y = 2$, and $x = 0$. At every point the measure of the area density is given by $kxy^2$.
   **(a)** Set up an iterated integral of the form $\int\int f(x, y)dx\,dy$ that gives the measure of $M$, the mass, but do not evaluate the integral.
   **(b)** Assume that $M$ is known and set up an iterated integral of the form $\int\int f(x, y)dy\,dx$ that gives $\bar{y}$, the $y$-coordinate of the center of mass, but do not evaluate the integral.

2. Let $V$ be the measure of the volume of the solid that is bounded by the cone $z^2 = x^2 + y^2$ and the cylinder $x^2 + y^2 = 2x$.
   **(a)** Set up an iterated integral of the form $\int\int f(x, y)dy\,dx$ that gives $V$, but do not evaluate the integral.
   **(b)** Set up an iterated integral of the form $\int\int f(r, \theta)dr\,d\theta$ that gives $V$, but do not evaluate the integral.

3. Let $S$ be the part of the surface $x^2 + z^2 = 4$ that is in the first octant and bounded by the coordinate planes and the cylinder $y = z^2$. Set up an iterated integral of the form $\int\int f(x, y)dy\,dx$ that gives the measure of the area of $S$, but do not evaluate the integral.

4. Let $S$ be the solid in the first octant that is bounded by $x^2 + y = 4$, $9y = z^2$, and the coordinate planes. Let the volume density of $S$ be proportional to the distance from the $xy$ plane.
   **(a)** Set up an iterated integral of the form $\int\int\int f(x, y, z)dx\,dy\,dz$ that gives the measure of the mass of $S$, but do not evaluate the integral.
   **(b)** Set up an iterated integral of the form $\int\int\int f(x, y, z)dy\,dz\,dx$ that gives the measure of the moment of inertia of $S$ about the $z$-axis, but do not evaluate the integral.

5. Let $S$ be the solid that is above the plane $z = 1$ and inside the sphere $x^2 + y^2 + z^2 = 4$. Let the volume density of $S$ be given by $(x^2 + y^2 + z^2)^{-1/2}$.
   **(a)** Set up an iterated integral with cylindrical coordinates that gives the measure of the volume of $S$, but do not evaluate the integral.
   **(b)** Set up an iterated integral with spherical coordinates that gives the measure of the mass of $S$, but do not evaluate the integral.

6. Evaluate the integral.

   (a) $$\int_0^1 \int_{x^2}^{x} xy^2\,dy\,dx$$

   (b) $$\int_0^{\pi/2} \int_0^{\pi} \int_0^{\cos\phi} \rho^2 \sin^2\phi\,d\rho\,d\theta\,d\phi$$

## SOLUTIONS FOR CHAPTER TESTS

### SOLUTIONS FOR TEST 17

1. $\mathbf{V}(\overrightarrow{PQ}) = \langle 1, 3\rangle - \langle -3, 1\rangle$
$= \langle 4, 2\rangle$
$= 2\langle 2, 1\rangle$

$|\mathbf{V}(\overrightarrow{PQ})| = 2\sqrt{2^2 + 1^2}$
$= 2\sqrt{5}$

$$\mathbf{u} = \frac{\mathbf{V}(\overrightarrow{PQ})}{|\mathbf{V}(\overrightarrow{PQ})|} = \frac{2\langle 2, 1\rangle}{2\sqrt{5}} = \frac{2}{5}\sqrt{5}\,\mathbf{i} + \frac{1}{5}\sqrt{5}\,\mathbf{j}$$

2. $\mathbf{A} = \langle 3, -2\rangle$ and $\mathbf{B} = \langle 2, 1\rangle$

$$\mathbf{A_B} = \frac{\mathbf{A}\cdot\mathbf{B}}{\mathbf{B}\cdot\mathbf{B}}\mathbf{B} = \frac{\langle 3, -2\rangle\cdot\langle 2, 1\rangle}{\langle 2, 1\rangle\cdot\langle 2, 1\rangle}\langle 2, 1\rangle = \frac{4}{5}\langle 2, 1\rangle = \frac{8}{5}\mathbf{i} + \frac{4}{5}\mathbf{j}$$

3. $$y' = \frac{dy}{dx} = \frac{\frac{dy}{dt}}{\frac{dx}{dt}} = \frac{3t^2+1}{2t}$$

$$\frac{d^2y}{dx^2} = \frac{dy'}{dx} = \frac{\frac{dy'}{dt}}{\frac{dx}{dt}} = \frac{\frac{(2t)(6t) - (3t^2+1)(2)}{4t^2}}{2t} = \frac{3t^2 - 1}{4t^3}$$

4. $\mathbf{R}'(t) = e^{2t}\langle -\sin t, \cos t\rangle + 2e^{2t}\langle \cos t, \sin t\rangle$
$= e^{2t}\langle -\sin t + 2\cos t, 2\sin t + \cos t\rangle$

$|\mathbf{R}'(t)| = e^{2t}\sqrt{(-\sin t + 2\cos t)^2 + (2\sin t + \cos t)^2}$
$= e^{2t}\sqrt{\sin^2 t - 4\sin t\cos t + 4\cos^2 t + 4\sin^2 t + 4\sin t\cos t + \cos^2 t}$
$= e^{2t}\sqrt{5\sin^2 t + 5\cos^2 t}$
$= e^{2t}\sqrt{5}$

5. $2x + 8y\,D_xy = 0$

$$D_xy = \left(-\frac{1}{4}\right)\frac{x}{y}$$

$D_xy = -\frac{1}{6}$ when $x = 2$ and $y = 3$

$$D_x^2y = -\frac{1}{4}\left[\frac{y - x\,D_xy}{y^2}\right]$$

$$D_x{}^2y = -\frac{1}{4}\left[\frac{3 - 2\left(-\frac{1}{6}\right)}{3^2}\right] = -\frac{5}{54} \quad \text{when } x = 2 \text{ and } y = 3$$

$$\begin{aligned} K &= \frac{|D_x{}^2y|}{[1 + (D_xy)^2]^{3/2}} \\ &= \frac{\frac{5}{54}}{\left(1 + \frac{1}{36}\right)^{3/2}} \\ &= \frac{20}{37^{3/2}} \end{aligned}$$

6. $$\begin{aligned} L &= \int_0^{3\pi/2} \sqrt{r^2 + \left(\frac{dr}{d\theta}\right)^2}\, d\theta \\ &= \int_0^{3\pi/2} \sqrt{(1 + \cos\theta)^2 + (-\sin\theta)^2}\, d\theta \\ &= \int_0^{3\pi/2} \sqrt{1 + 2\cos\theta + \cos^2\theta + \sin^2\theta}\, d\theta \\ &= \int_0^{3\pi/2} \sqrt{2(1 + \cos\theta)}\, d\theta \\ &= \int_0^{3\pi/2} \sqrt{4\cos^2\left(\frac{1}{2}\theta\right)}\, d\theta \\ &= 2\int_0^{3\pi/2} \left|\cos\left(\frac{1}{2}\theta\right)\right| d\theta \\ &= 2\left[\int_0^{\pi} \cos\left(\frac{1}{2}\theta\right) d\theta - \int_{\pi}^{3\pi/2} \cos\left(\frac{1}{2}\theta\right) d\theta\right] \\ &= 2\left[2\sin\frac{1}{2}\theta\right]_0^{\pi} - 2\left[2\sin\frac{1}{2}\theta\right]_{\pi}^{3\pi/2} \\ &= 4\left[\sin\frac{1}{2}\pi - \sin 0\right] - 4\left[\sin\frac{3}{4}\pi - \sin\frac{1}{2}\pi\right] \\ &= 8 - 2\sqrt{2} \end{aligned}$$

7. $$\begin{aligned} \mathbf{R}(t) &= 2t^2\mathbf{i} + t^3\mathbf{j} \\ \mathbf{V}(t) &= 4t\mathbf{i} + 3t^2\mathbf{j} \\ \mathbf{A}(t) &= 4\mathbf{i} + 6t\mathbf{j} \end{aligned}$$

$$\begin{aligned} \mathbf{V}(1) &= 4\mathbf{i} + 3\mathbf{j} \\ \mathbf{A}(1) &= 4\mathbf{i} + 6\mathbf{j} \end{aligned}$$

$$\mathbf{T}(1) = \frac{\mathbf{V}(1)}{|\mathbf{V}(1)|} = \frac{1}{5}(4\mathbf{i} + 3\mathbf{j})$$

$$\mathbf{N}(1) = \frac{1}{5}(-3\mathbf{i} + 4\mathbf{j})$$

$$\frac{ds}{dt} = |\mathbf{V}(1)| = 5$$

$$A_T = \mathbf{A} \cdot \mathbf{T} = (4\mathbf{i} + 6\mathbf{j}) \cdot \left(\frac{4}{5}\mathbf{i} + \frac{3}{5}\mathbf{j}\right)$$

$$= \frac{16}{5} + \frac{18}{15}$$

$$= \frac{34}{5}$$

$$A_N = \mathbf{A} \cdot \mathbf{N} = (4\mathbf{i} + 6\mathbf{j}) \cdot \left(\frac{-3}{5}\mathbf{i} + \frac{4}{5}\mathbf{j}\right)$$

$$= \frac{-12}{5} + \frac{24}{5}$$

$$= \frac{12}{5}$$

$$K = \frac{A_N}{\left(\frac{ds}{dt}\right)^2}$$

$$= \frac{\frac{12}{5}}{5^2}$$

$$= \frac{12}{125}$$

## SOLUTIONS FOR TEST 18

1. (a) $|\mathbf{A} + \mathbf{B}| = |\langle 1, 3, 0\rangle| = \sqrt{10}$

   (b) $\mathbf{A} \cdot \mathbf{B} = 2(-1) + 1(2) + (-1)(1) = -1$

   (c) $$\mathbf{A} \times \mathbf{B} = \begin{vmatrix} \mathbf{i} & \mathbf{j} & \mathbf{k} \\ 2 & 1 & -1 \\ -1 & 2 & 1 \end{vmatrix} = \begin{vmatrix} 1 & -1 \\ 2 & 1 \end{vmatrix}\mathbf{i} - \begin{vmatrix} 2 & -1 \\ -1 & 1 \end{vmatrix}\mathbf{j} + \begin{vmatrix} 2 & 1 \\ -1 & 2 \end{vmatrix}\mathbf{k}$$

   $$= 3\mathbf{i} - \mathbf{j} + 5\mathbf{k}$$

2. (a) $\mathbf{V}(\overrightarrow{PQ}) = \langle -4, 2, -2\rangle = -2\langle 2, -1, 1\rangle$

   $$\frac{x-1}{2} = \frac{y+1}{-1} = \frac{z-2}{1}$$

   (b) Midpoint $= (-1, 0, 1)$
   $\mathbf{N} = \langle 2, -1, 1\rangle$

   $$2(x+1) - (y-0) + (z-1) = 0$$
   $$2x - y + z + 1 = 0$$

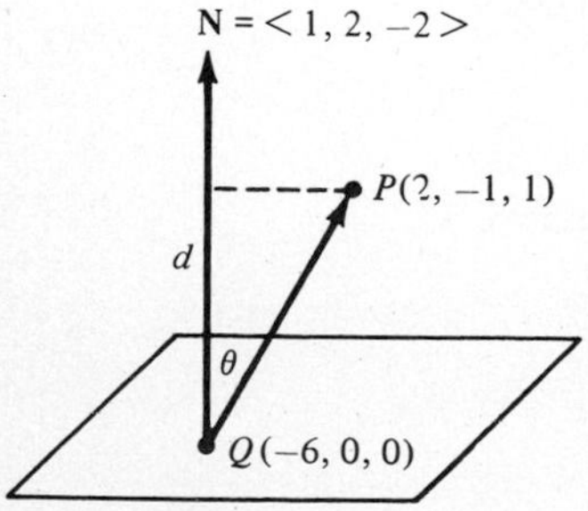

Figure 18.3(a)T

3. (a) See Fig. 18.3(a)T. To find point $Q$, let $y = 0, z = 0$. Thus, $Q = (-6, 0, 0)$.

   $$\mathbf{V}(\overrightarrow{QP}) = \langle 8, -1, 1\rangle$$
   $$\mathbf{N} = \langle 1, 2, -2\rangle$$
   $$d = |\mathbf{V}(\overrightarrow{QP})||\cos\theta|$$
   $$= \frac{|\mathbf{N} \cdot \mathbf{V}(\overrightarrow{QP})|}{|\mathbf{N}|}$$
   $$= \frac{4}{3}$$

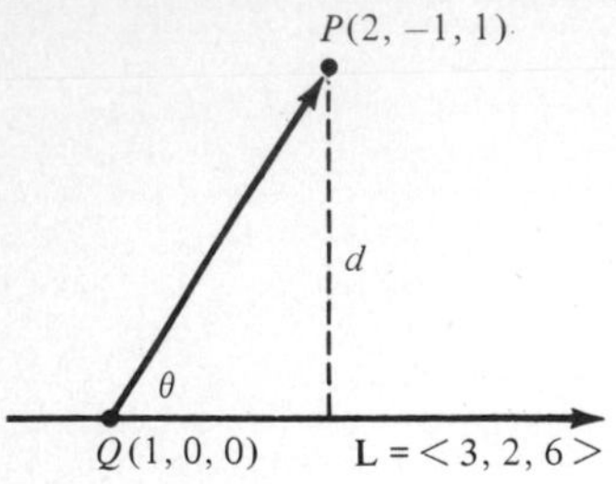

Figure 18.3(b)T

**(b)** See Fig. 18.3(b)T. To find $Q$, let $t = 0$. Thus, $Q = (1, 0, 0)$.

$$\mathbf{V}(\overrightarrow{QP}) = \langle 1, -1, 1\rangle$$
$$\mathbf{L} = \langle 3, 2, 6\rangle$$
$$d = |\mathbf{V}(QP)| \sin\theta$$
$$= \frac{|\mathbf{L} \times \mathbf{V}(\overrightarrow{QP})|}{|\mathbf{L}|}$$
$$= \frac{|\langle -8, -3, 5\rangle|}{|\langle 3, 2, 6\rangle|}$$
$$= \sqrt{2}$$

4. See Fig. 18.4T. The surface is an elliptic paraboloid.

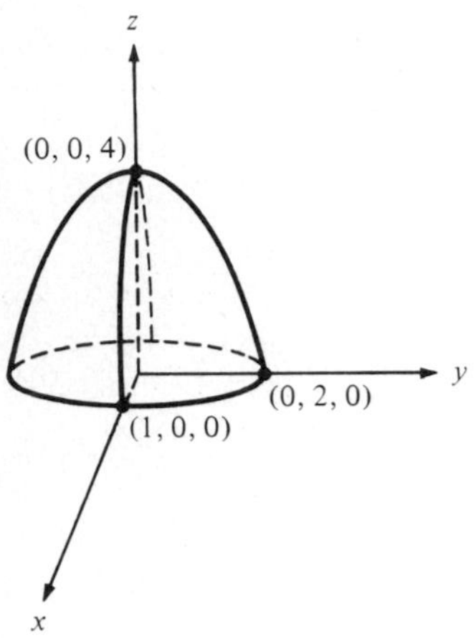

Figure 18.4T

5. Eliminate $z$ from the given equations. Thus,

$$(2 - y)^2 = x^2 + y^2 - 4$$
$$x^2 = -4(y - 2)$$

The curve is a parabola. See Fig. 18.5T.

6. $K(1) = \dfrac{|\mathbf{R}'(1) \times \mathbf{R}''(1)|}{|\mathbf{R}'(1)|^3}$

$$\mathbf{R}'(t) = \langle 2t, 1 + t^2, 1 - t^2\rangle$$
$$\mathbf{R}''(t) = \langle 2, 2t, -2t\rangle$$
$$\mathbf{R}'(1) = \langle 2, 2, 0\rangle$$
$$\mathbf{R}''(1) = \langle 2, 2, -2\rangle$$
$$\mathbf{R}'(1) \times \mathbf{R}''(1) = \langle -4, 4, 0\rangle$$
$$K(1) = \frac{|\langle -4, 4, 0\rangle|}{|\langle 2, 2, 0\rangle|^3}$$
$$= \frac{4\sqrt{2}}{(2\sqrt{2})^3}$$
$$= \frac{1}{4}$$

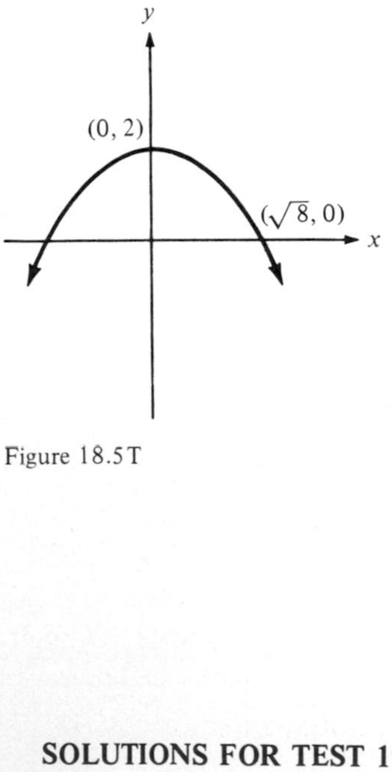

Figure 18.5T

7. $x = y^{1/3} = f(y)$
$$x^2 + z^2 = [f(y)]^2$$
$$x^2 + z^2 = y^{2/3}$$

8. Multiply by $\rho$. Thus,

$$\rho^2 = 4\rho \sin\phi \cos\theta + 2\rho\cos\phi$$
$$x^2 + y^2 + z^2 = 4x + 2z$$
$$(x - 2)^2 + y^2 + (z - 1)^2 = 5$$

The graph is a sphere with center at $(2, 0, 1)$ and radius $\sqrt{5}$.

## SOLUTIONS FOR TEST 19

1. domain $= \{(x, y) \mid 16 - 4x^2 + y^2 \geqslant 0\}$
$$= \{(x, y) \mid 4x^2 - y^2 \leqslant 16\}$$

See Fig. 19.1.T.

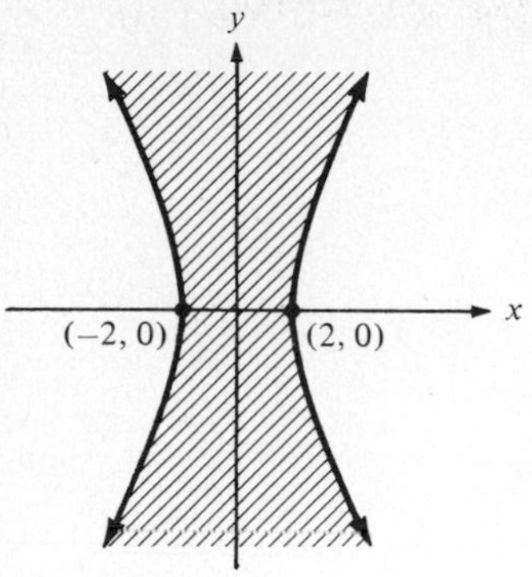

Figure 19.1 T

2. Let $S$ be the line $y = -x$.

$$\lim_{\substack{(x,y)\to(0,0) \\ (P \text{ in } S)}} f(x,y) = \lim_{x\to 0} \frac{-x^2}{0}$$

does not exist. Thus,

$$\lim_{(x,y)\to(0,0)} f(x,y)$$

does not exist.

3. Because $f(0,0) = 0$, we must show that $\lim_{(x,y)\to(0,0)} f(x,y) = 0$. We show that for any $\epsilon > 0$ there is some $\delta > 0$ such that

$$\left|\frac{x^4}{x^2+y^2}\right| < \epsilon \quad \text{if } 0 < \sqrt{x^2+y^2} < \delta$$

Because

$$\left|\frac{x^4}{x^2+y^2}\right| \leqslant x^2 \left|\frac{x^2+y^2}{x^2+y^2}\right| = x^2 \leqslant x^2+y^2 < \delta^2 \qquad \text{whenever } 0 < \sqrt{x^2+y^2} < \delta$$

we take $\delta = \sqrt{\epsilon}$.

4. $$f_x(x,y) = \frac{1}{1+\left(\frac{y^2}{x^2}\right)^2} \cdot \frac{-2y^2}{x^3}$$

$$= \frac{-2xy^2}{x^4+y^4}$$

5. $$f_2(x,y) = x^2\left[y^3 e^{x/y}\left(-\frac{x}{y^2}\right) + e^{x/y}(3y^2)\right]$$

$$= x^2 e^{x/y}[-xy + 3y^2]$$

$$= x^2 y e^{x/y}(3y - x)$$

6. $$\frac{\partial z}{\partial x} = \frac{2x}{x^2+y^2}$$

$$\frac{\partial^2 z}{\partial x^2} = \frac{(x^2+y^2)(2) - 2x(2x)}{(x^2+y^2)^2}$$

$$= \frac{2(y^2-x^2)}{(x^2+y^2)^2}$$

7. $$\frac{\partial u}{\partial \theta} = \frac{\partial u}{\partial x}\frac{\partial x}{\partial \theta} + \frac{\partial u}{\partial y}\frac{\partial y}{\partial \theta}$$

$$= (2x+y^2)(-r\sin\theta) + (2xy)(r\cos\theta)$$

$$= (2x+y^2)(-y) + (2xy)(x)$$

$$= -2xy - y^3 + 2x^2y$$

$$\frac{\partial^2 u}{\partial \theta^2} = (-2y+4xy)(-r\sin\theta) + (-2x-3y^2+2x^2)(r\cos\theta)$$

$$= (-2y+4xy)(-y) + (-2x-3y^2+2x^2)(x)$$

$$= 2x^3 - 2x^2 - 7xy^2 + 2y^2$$

8. We must find $dv/v$ when $r = 5$, $h = 10$, $dr = 0.1$, and $dh = 0.2$.

$$\frac{dv}{v} = \frac{\pi(2rh\,dr + r^2\,dh)}{\pi r^2 h}$$
$$= 2\frac{dr}{r} + \frac{dh}{h}$$
$$= 2\left(\frac{0.1}{5}\right) + \frac{0.2}{10}$$
$$= 0.06$$

**SOLUTIONS FOR TEST 20**

1. (a) $\nabla f(x, y, z) = (2x + 3z)\mathbf{i} + (2y)\mathbf{j} + (3x)\mathbf{k}$

$\nabla f(1, 2, -1) = -\mathbf{i} + 4\mathbf{j} + 3\mathbf{k}$

$\mathbf{V}(\overrightarrow{PQ}) = \mathbf{i} - 2\mathbf{j} + 2\mathbf{k}$

$$\mathbf{u} = \frac{\mathbf{V}(\overrightarrow{PQ})}{|\mathbf{V}(\overrightarrow{PQ})|} = \frac{1}{3}(\mathbf{i} - 2\mathbf{j} + 2\mathbf{k})$$

$$D_{\mathbf{u}} f(1, 2, -1) = \mathbf{u} \cdot \nabla f(1, 2, -1)$$
$$= \frac{1}{3}(\mathbf{i} - 2\mathbf{j} + 2\mathbf{k}) \cdot (-\mathbf{i} + 4\mathbf{j} + 3\mathbf{k})$$
$$= -1$$

(b) $$\mathbf{u} = \frac{\nabla f(1, 2, -1)}{|\nabla f(1, 2, -1)|}$$
$$= \frac{1}{\sqrt{26}}(-\mathbf{i} + 4\mathbf{j} + 3\mathbf{k})$$

The maximum value of $D_{\mathbf{u}} f(1, 2, -1)$ is given by

$|\nabla f(1, 2, -1)| = \sqrt{26}$

2. $F(x, y, z) = x^2 + 2y^2 - z^2 - 2z = 0$

$\nabla F(x, y, z) = \langle 2x, 4y, -2z - 2\rangle$

$\nabla F(1, 1, 1) = \langle 2, 4, -4\rangle$

The plane is

$$\langle 2, 4, -4\rangle \cdot \langle x - 1, y - 1, z - 1\rangle = 0$$
$$2(x - 1) + 4(y - 1) - 4(z - 1) = 0$$
$$x + 2y - 2z - 1 = 0$$

3. (a) $F(x, y, z, \lambda) = x^2yz + \lambda(x + y - z - 4)$

$F_x(x, y, z, \lambda) = 2xyz + \lambda = 0$

$F_y(x, y, z, \lambda) = x^2z + \lambda = 0$

$F_z(x, y, z, \lambda) = x^2y - \lambda = 0$

$F_\lambda(x, y, z, \lambda) = x + y - z - 4 = 0$ **(1)**

Thus, eliminating $\lambda$ we have

$2xyz = x^2z$ and $x^2y = -x^2z$

$2y = x$ $\qquad -y = z$

Substituting into (1), we get

$$2y + y + y - 4 = 0$$
$$y = 1$$

The critical point is $(2, 1, -1)$.

**(b)** Eliminating $z$, we have

$$T = x^2y(x + y - 4)$$
$$= x^3y + x^2y^2 - 4x^2y$$

Thus,

$$T_x = 3x^2y + 2xy^2 - 8xy$$
$$T_y = x^3 + 2x^2y - 4x^2$$
$$T_{xx} = 6xy + 2y^2 - 8y$$
$$T_{yy} = 2x^2$$
$$T_{xy} = 3x^2 + 4xy - 8x$$

$$T_{xx}(2, 1) \cdot T_{yy}(2, 1) - T_{xy}{}^2(2, 1) = 6 \cdot 8 - 4^2 = 32$$

Because $T_{xx}(2, 1) > 0$ and $T_{xx}(2, 1) \cdot T_{yy}(2, 1) - T_{xy}{}^2(2, 1) > 0$, $T$ has a relative minimum value at the point $(2, 1)$.

**4.** Let $100S$ dollars be the profit.

$$S = xp + yq - 2x - y$$
$$= (8 - 2p + q)p + (6 + p - q)q - 2(8 - 2p + q) - (6 + p - q)$$
$$= -2p^2 + 2pq - q^2 + 11p + 5q - 22$$

$$\frac{\partial S}{\partial p} = -4p + 2q + 11$$

$$\frac{\partial S}{\partial q} = 2p - 2q + 5$$

If $\partial S/\partial p = 0$ and $\partial S/\partial q = 0$, then adding the equations gives

$$-2p + 16 = 0$$
$$p = 8$$
$$q = 10.5$$

At the critical point, we have

$$\frac{\partial^2 S}{\partial p^2} = -4 \qquad \frac{\partial^2 S}{\partial q^2} = -2 \qquad \frac{\partial^2 S}{\partial p \partial q} = 2$$

Because

$$\frac{\partial^2 S}{\partial p^2} < 0 \text{ and } \left(\frac{\partial^2 S}{\partial p^2}\right)\left(\frac{\partial^2 S}{\partial q^2}\right) - \left(\frac{\partial^2 S}{\partial p \partial q}\right)^2 = (-4)(-2) - 2^2 = 4 > 0$$

we conclude that the maximum profit results if the prices are \$8 for the first commodity and \$10.50 for the second.

**5.** $M(x, y, z) = \sin y \qquad N(x, y, z) = x \cos y + ze^{yz} \qquad R(x, y, z) = ye^{yz}$

Then

$$M_y(x, y, z) = \cos y \qquad N_x(x, y, z) = \cos y$$
$$M_z(x, y, z) = 0 \qquad R_x(x, y, z) = 0$$
$$N_z(x, y, z) = yze^{yz} + e^{yz} \qquad R_y(x, y, z) = yze^{yz} + e^{yz}$$

Hence, the given vector is a gradient of $\phi$, where

$$\phi_x(x, y, z) = \sin y \quad (1)$$
$$\phi_y(x, y, z) = x \cos y + ze^{yz} \quad (2)$$
$$\phi_z(x, y, z) = ye^{yz} \quad (3)$$

Integrating (2) with respect to $y$ gives

$$\phi(x, y, z) = x \sin y + e^{yz} + C(x, z) \quad (4)$$
$$\phi_x(x, y, z) = \sin y + C_x(x, z) \quad (5)$$

From (5) and (1), we have

$$C_x(x, z) = 0$$
$$C(x, z) = K(z)$$

Substituting in (4), we get

$$\phi(x, y, z) = x \sin y + e^{yz} + K(z) \quad (6)$$
$$\phi_z(x, y, z) = ye^{yz} + K'(z) \quad (7)$$

From (7) and (3), we have

$$K'(z) = 0$$
$$K(z) = C$$

Substituting in (6) gives

$$\phi(x, y, z) = x \sin y + e^{yz} + C$$

6. A vector equation of $C$ is

$$\mathbf{R}(t) = t\mathbf{i} + 2t^3\mathbf{j} \qquad 0 \leqslant t \leqslant 1$$

Thus,

$$\mathbf{R}'(t) = \mathbf{i} + 6t^2\mathbf{j}$$
$$\mathbf{F}(t, 2t^3) = 2t^4\mathbf{i} + t^3\mathbf{j}$$
$$\mathbf{F}(t, 2t^3) \cdot \mathbf{R}'(t) = 2t^4 + 6t^5$$

Thus,

$$W = \int_0^1 (2t^4 + 6t^5)\,dt$$
$$= \frac{2}{5}t^5 + t^6\Big]_0^1$$
$$= \frac{7}{5}$$

7. $M(x, y) = 3x^2y + y^2 \qquad N(x, y) = x^3 + 2xy$

Then

$$M_y(x, y) = 3x^2 + 2y \quad \text{and} \quad N_x(x, y) = 3x^2 + 2y$$

Because $M_y(x, y) = N_x(x, y)$, the given integral is independent of the path.

(a) $\phi_x(x, y) = 3x^2y + y^2 \quad (1)$

$\phi_y(x, y) = x^3 + 2xy \quad (2)$

Thus, integrating (1) with respect to $x$, we get

$$\phi(x, y) = x^3y + xy^2 + C(y)$$
$$\phi_y(x, y) = x^3 + 2xy + C'(y) \quad (3)$$

From (2) and (3), we take $C'(y) = 0$. Thus, a potential function is

$$\phi(x, y) = x^3 + 2xy$$

and the given line integral is

$$\phi(-1, 2) - \phi(1, 0) = (-1 - 4) - 1 = -6$$

**(b)** Let $C_1$ be the $x$-axis from $(1, 0)$ to $(-1, 0)$ and let $C_2$ be the line $x = -1$ from $(-1, 0)$ to $(-1, 2)$. Because $y = 0$ over $C_1$, we have

$$\int_{C_1} (3x^2y + y^2)dx + (x^3 + 2xy)dy = 0$$

Because $dx = 0$ over $C_2$, we have

$$\int_{C_2} (3x^2y + y^2)dx + (x^3 + 2xy)dy = \int_0^2 (-1 - 2y)dy$$

$$= -y - y^2]_0^2$$

$$= -2 - 4$$

$$= -6$$

## SOLUTIONS FOR TEST 21

**1.** See Fig. 21.1T.

**(a)** $M = \int_0^2 \int_0^{y/2} kxy^2 \, dx \, dy$

**(b)** $\bar{y} = \dfrac{1}{M} \int_0^1 \int_{2x}^2 kxy^3 dy \, dx$

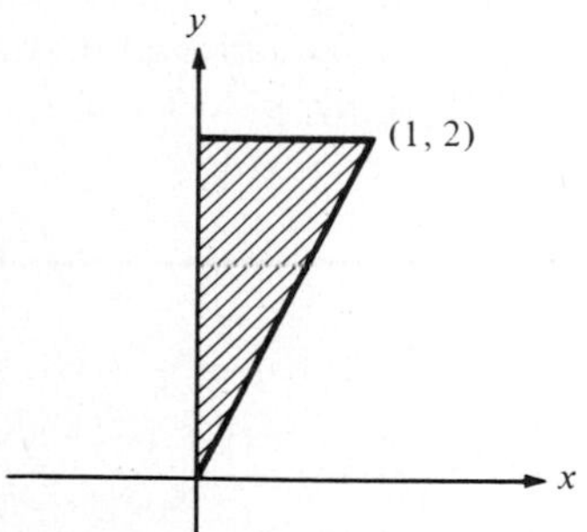

Figure 21.1T

**2.** See Fig. 21.2T which shows the upper half of the solid

**(a)** $V = 2 \int_0^2 \int_{-\sqrt{2x - x^2}}^{\sqrt{2x - x^2}} \sqrt{x^2 + y^2} \, dy \, dx$

**(b)** $V = 2 \int_{-\pi/2}^{\pi/2} \int_0^{2\cos\theta} r^2 \, dr \, d\theta$

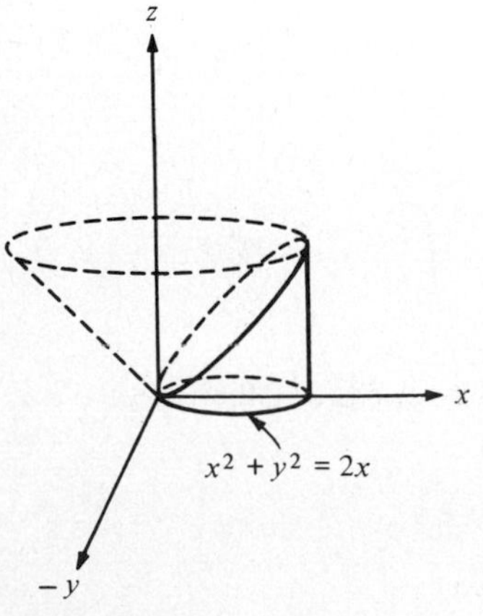

Figure 21.2T

**3.** See Fig. 21.3T. We eliminate $z$ from the given equations to find the projection of the surface onto the $xy$ plane. Thus, we have

$$x^2 + y = 4$$

is a boundary of the region in the $xy$ plane over which the surface lies. Solving the equation $x^2 + z^2 = 4$ for $z$, we obtain

$$z = \pm\sqrt{4 - x^2}$$

We take $f(x, y) = \sqrt{4 - x^2}$. Thus,

$$f_x(x, y) = \frac{-x}{\sqrt{4 - x^2}} \quad \text{and} \quad f_y(x, y) = 0$$

Hence,

$$\sqrt{f_x^2(x,y)+f_y^2(x,y)+1} = \sqrt{\frac{x^2}{4-x^2}+1} = \frac{2}{\sqrt{4-x^2}}$$

Thus, the area is given by

$$A = \int_0^2 \int_0^{4-x^2} \frac{2}{\sqrt{4-x^2}}\, dy\, dx$$

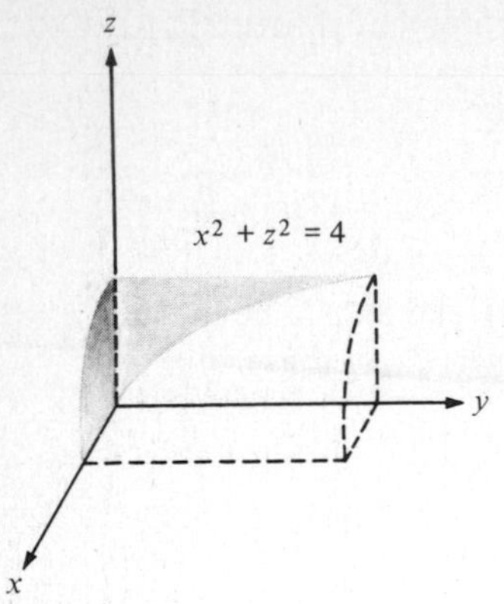


Figure 21.3T

4. See Fig. 21.4T.

(a) $M = \int_0^6 \int_{z^2/9}^4 \int_0^{\sqrt{4-y}} kz\, dx\, dy\, dz$

(b) $I_z = \int_0^2 \int_0^{3\sqrt{4-x^2}} \int_{z^2/9}^{4-x^2} kz(x^2+y^2)dy\, dz\, dx$

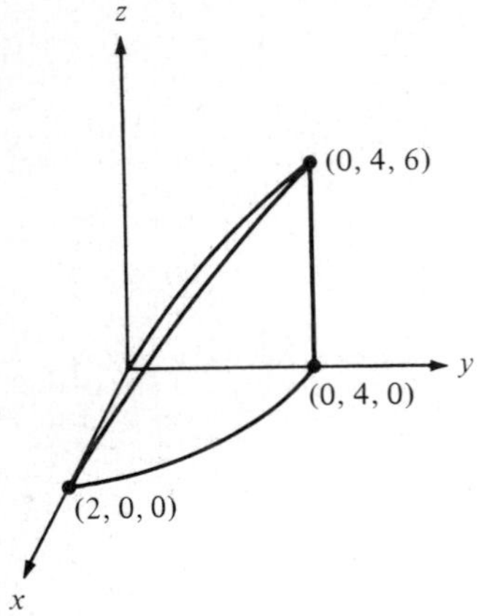


Figure 21.4T

5. See Fig. 21.5T.

(a) Eliminating $z$, we have $x^2+y^2=3$ or $r=\sqrt{3}$, which is the equation of the boundary of the projection of the solid onto the polar plane.

$$V = \int_0^{2\pi} \int_0^{\sqrt{3}} \int_1^{\sqrt{4-r^2}} r\, dz\, dr\, d\theta$$

(b) Because $z = \rho \cos\phi$, the plane $z = 1$ is given by $\rho\cos\phi = 1$. An equation of the sphere is $\rho = 2$. The intersection of these two boundaries must satisfy $2\cos\phi = 1$ or $\phi = \pi/3$. The density function is given by $\rho^{-1}$. Thus,

$$M = \int_0^{\pi/3} \int_0^{2\pi} \int_{1/\cos\phi}^2 \rho \sin\phi\, d\rho\, d\theta\, d\phi$$

6. (a)

$$\int_0^1 \int_{x^2}^x xy^2\, dy\, dx = \frac{1}{3}\int_0^1 \left[ xy^3 \right]_{x^2}^x dx$$
$$= \frac{1}{3}\int_0^1 x(x^3 - x^6)dx$$
$$= \frac{1}{3}\left[\frac{1}{5}x^5 - \frac{1}{8}x^8\right]_0^1$$
$$= \frac{1}{40}$$

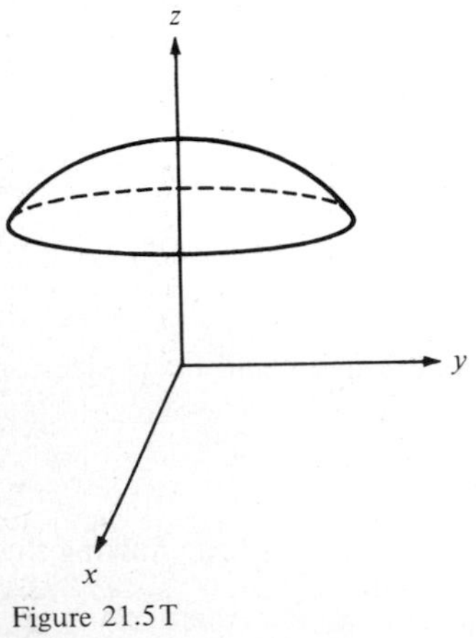


Figure 21.5T

(b)

$$\int_0^{\pi/2} \int_0^{\pi} \int_0^{\cos\phi} \rho^2 \sin^2\phi\, d\rho\, d\theta\, d\phi = \frac{1}{3}\int_0^{\pi/2}\int_0^{\pi} \sin^2\phi \cos^3\phi\, d\theta\, d\phi$$
$$= \frac{1}{3}\pi \int_0^{\pi/2} \sin^2\phi(1-\sin^2\phi)\cos\phi\, d\phi$$
$$= \frac{1}{3}\pi\left[\frac{1}{3}\sin^3\phi - \frac{1}{5}\sin^5\phi\right]_0^{\pi/2}$$
$$= \frac{1}{3}\pi\left[\frac{1}{3}-\frac{1}{5}\right]$$
$$= \frac{2}{45}\pi$$

76 77 78 79 9 8 7 6 5 4 3 2 1